Engineering Materials Technology

Engineering Materials Technology

Second Edition

W. Bolton

Newnes
An imprint of Butterworth-Heinemann Ltd
Linacre House, Jordan Hill, Oxford OX2 8DP

A member of the Reed Elsevier group

OXFORD LONDON BOSTON
MUNICH NEW DELHI SINGAPORE SYDNEY
TOKYO TORONTO WELLINGTON

First published 1989
Reprinted 1991
Second edition 1993

British Library Cataloguing in Publication Data
Bolton, W. (William), *1933–*
Engineering materials technology – 2Rev. ed.
1. Materials
I. Title
620.11

ISBN 0 7506 1740 3

Typeset by Vision Typesetting, Manchester
Printed and bound in Great Britain

Contents

Preface

This book has been written for students of mechanical and production engineering who need a firm grasp of the principles and practical consideration that underlie the proper selection of engineering materials. It aims to provide:

- A basic knowledge of the structure of materials.
- An understanding of their properties, including fracture, fatigue and creep behaviour.
- An appreciation of the characteristics of ferrous and non-ferrous metals and of polymeric, ceramic and composite materials in relation to their structure that allows an informed choice to be made.
- A guide to the criteria that merit consideration when materials and processes are selected for particular types of product.

This body of information is covered in five main sections: The structure of materials (Chapters 1 to 4); Properties of materials (Chapters 5 to 9); Metals (Chapters 10 to 13); Non-metallic materials; (Chapters 14 to 16); and Selection (Chapters 18 to 20).

The book is designed for use on degree courses and courses leading to BTEC HNC/HND. It covers a range of BTEC units, notably:

Engineering materials III (U84/266) – Chapters 3, 4, 5, 11, 12, 13
Bearing materials III (U84/267) – Chapter and Sections 12.7, 13.8 and 20.5
Cast irons III (U84/268) – Section 11.4
Ceramics III (U84/269) – Sections 4.4 and 15.5
Corrosion prevention III (U84/270) – Chapter 9
Non-ferrous alloys III (U84/272) – Chapter 13
Polymeric materials III (U84/273) – Chapters 4 and 15 and Section 17.1
Welded and brazed metallic materials III (U84/276) – Chapter 17
Heat resisting and maraging steels IV (U84/271) – Chapter 11
Stainless steels IV (U84/274) – Chapter 11
Tool materials IV (U84/275) – Sections 16.3 and 20.5
Composites NIII (U86/327) – Chapter 16

Engineering materials H (U86/326) – Chapters 2, 6, 7, 8, 18, 19, 20
Applied materials technology H (U86/323) – Chapters 18, 19, 20

Acknowledgements are due to the large number of companies that have supplied me with information and also to those publishers who have given me permission to reproduce from their publications. Every effort has been made to acknowledge sources of material used – if in any case I have not made full acknowledgement I hope that my apologies will be accepted.

Second edition

For the second edition, the book has been extended by a more detailed consideration of materials selection and additions to give a better coverage of materials testing, composites and joining materials. The more detailed consideration of materials selection has involved a complete restructuring and expansion of what was just a single chapter to give three chapters concerning the selection of materials for specific properties, the selection of processes and then the selection of materials, and processes, for components.

This book more than covers the new BTEC units Materials for engineering 171J and 172J and the General National Vocational Qualifications (GNVQs) in Manufacturing, Level 2 Select and test materials and Level 3 Select materials.

PART ONE

The structure of materials

1

Basic structure of materials

1.1 Atoms and molecules

All matter is made up of atoms. A material that is made up of just one type of atom is called an *element*. Hydrogen, carbon, copper and iron are examples of elements. An *atom* is the smallest particle of an element that has the characteristics of that element. We can thus talk of a copper atom or an iron atom. Atoms themselves are made up of other particles which are not characteristics of the element but are basic building blocks out of which all atoms are constructed. Each atom is composed of a nucleus, which is positively charged, and electrons, which are negatively charged. We can think of an atom as being a very small nucleus which contains virtually all the mass of the atom, surrounded by a cloud of electrons which occupies most of the space of the atom.

The term *molecule* is used to describe groups of atoms which tend to exist together in a stable form. Thus, for instance, hydrogen tends to exist in a stable form as a combination of two hydrogen atoms rather than as just individual atoms. Some molecules may exist as combinations of atoms from a number of different elements. Water, for example, consists of molecules each of which is made up of two hydrogen atoms and an oxygen atom.

Atomic structure

The atom consists of a positively-charged nucleus surrounded by negatively-charged electrons. Attractive forces occur between the electrons and the nucleus (opposite charges attract) and repulsive forces occur between the electrons (like charges repel). For such forces to result in stable atoms, Bohr proposed a model for atomic structure in which electrons move in fixed orbits round the nucleus, like planets orbiting the sun. Only certain orbits are possible and only in these orbits is there stability. With this model, electrons in orbits close to the nucleus have stronger forces of attraction holding them in orbit than electrons which are in orbits further out. The force of attraction between oppositely-charged particles is inversely proportional to the square of the distance between their centres; in other words, if the radius of the orbit is doubled then the force is reduced to a quarter of its value. Thus, electrons in close orbits are much more difficult to remove from an atom than those in further out orbits.

Elements differ from each other in the charge carried by the nucleus. Since the charge is carried by protons this is the same as saying they differ according to the number of protons. Atoms of the same element all having the same number of protons and this number is specific to that element. The *atomic number* of an element is the number of protons in its atoms. An atom will normally carry no net charge and thus the number of electrons will be characteristic of the element concerned. Hydrogen has just one electron, carbon has twelve. The Bohr Model proposed that the electrons existed in orbits, these sometimes being referred to as shells. The innermost orbit, sometimes called the K shell, can hold just two electrons before it is full, the second orbit out, called the L shell, requires eight electrons to be full. Thus hydrogen with just one electron has this electron in the innermost orbit – the K shell (Figure 1.1). This K shell for hydrogen is only partially occupied. Helium has two electrons. These both occupy the K shell and fill it. Lithium has three electrons. Two of the electrons occupy and fill the K shell and the remaining electron is in the L shell which is thus only partially filled. Carbon with six electrons has two filling the K shell.

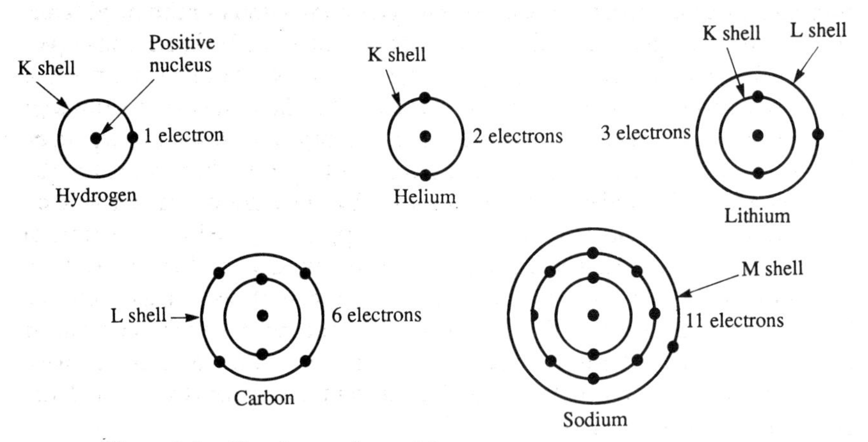

Figure 1.1 Simple atomic model

Table 1.1 shows how the shells are occupied, according to this Bohr Model of the atom for elements with atomic numbers up to eighteen.

Elements with full outer shells, e.g. helium with its full K shell and neon with its full K and L shells, are ones which are generally referred to as inert. They do not combine with other elements to form compounds. In comparison,

Table 1.1 *Bohr Model shells*

Atomic number	*Element*	*Electronic configuration* K	L	M
1	Hydrogen	1		
2	Helium	2		
3	Lithium	2	1	
4	Beryllium	2	2	
5	Boron	2	3	
6	Carbon	2	4	
7	Nitrogen	2	5	
8	Oxygen	2	6	
9	Fluorine	2	7	
10	Neon	2	8	
11	Sodium	2	8	1
12	Magnesium	2	8	2
13	Aluminium	2	8	3
14	Silicon	2	8	4
15	Phosphorus	2	8	5
16	Sulphur	2	8	6
17	Chlorine	2	8	7
18	Argon	2	8	8

elements with just one electron in their outer shell are particularly active, combining readily with other elements to form compounds – lithium with one electron in the L shell and sodium with one electron in the M shell are examples of such elements. In forming compounds, i.e. bonds with other elements, the aim appears to be to achieve full outer shells. Thus helium is inert because it already has a full outer shell and does not form compounds because it would be impossible for it to do this and still achieve a full outer shell. Sodium, on the other hand, could achieve a full outer shell by donating its one outer shell electron to another element. Chlorine just happens to need one electron to complete a shell and thus sodium forms a compound with chlorine, sodium chloride (NaCl) or common salt, with the result that both sodium and chlorine then achieve the desired full outer shell state. Another way this state can be achieved is by two atoms sharing electrons. Hydrogen atoms combine with themselves to give a molecule, H_2, with each atom sharing an electron with the other so that each then has a full shell.

Therefore, elements are characterized by different numbers of electrons, with these electrons occupying a shell structure around the nucleus. Only electrons outside full shells take part in interactions between atoms, the aim of such interactions being to achieve full shells.

The *valency* of an element is equal to either the number of electrons in the outermost shell or the number of electrons needed to fill the shell. Thus, sodium has a valency of one since it has just one electron in its outermost shell. Chlorine is just one electron short of a complete shell and so also has a valency of one. Oxygen has two electrons in the K shell and six in the L shell. As it needs two further electrons to complete its outermost shell it has a valency of two. Carbon has two electrons in the K shell and four electrons in the L shell. This means it has a valency of four.

Quantum numbers

The above discussion has been in terms of the Bohr Model of the atom. This model has long since been supplanted but the concept of shells still applies to modern atomic models. Each electron has four quantum numbers, denoted by n, l, m_l and m_s, with each electron in an atom having its own characteristic set of quantum numbers (*Pauli exclusion principle*). n is called the principal quantum number and can have an integer value of 1, 2, 3, 4, etc; l is called the angular momentum quantum number and can have an integer value in the range 0 to $(n-1)$; m_l is the magnetic quantum number and can have an integer value in the range 0 to $\pm l$; m_s is the spin quantum number and can have the value $+\frac{1}{2}$ or $-\frac{1}{2}$.

Consider the situation with n equal to 1. A consequence of this value is, since $(n-1)$ is zero, that l is 0. Because l is zero then m_l is zero. Since m_s can be either $+\frac{1}{2}$ or $-\frac{1}{2}$, then there are just two possible combinations of quantum numbers that can occur with n equal to one. This means that for the $n=1$, or K shell, there are just two possible electrons.

Consider the situation with n equal to 2, the condition for the L shell. l can have the values 0 or 1, m_l the value 0, -1 or $+1$, and m_s $+\frac{1}{2}$ or $-\frac{1}{2}$.

The possible combinations of quantum numbers for $n=2$ are thus:

l	m_l	m_s	
0	0	$+\frac{1}{2}$	2s subshell
0	0	$-\frac{1}{2}$	
1	0	$+\frac{1}{2}$	2p subshell
1	0	$-\frac{1}{2}$	
1	1	$+\frac{1}{2}$	
1	1	$-\frac{1}{2}$	
1	-1	$+\frac{1}{2}$	
1	-1	$-\frac{1}{2}$	

There are therefore, a total of eight possible combinations of quantum numbers for the $n=2$ shell and this shell can accommodate eight electrons. The shell is considered to be made up of two subshells, the s shell where $l=0$ and the *p* shell where $l=1$.

If n equals 3, the condition for the M shell, then the quantum numbers give eighteen possible combinations for the shell. These can be considered to be in three subshells: the 3s with two possible configurations, the 3p with six possibilities and the 3d with ten. The s shell is where $l=0$, the p shell where $l=1$ and the d shell where $l=2$.

Table 1.2 *Electronic configuration of elements*

		Electronic configuration				
		K	*L*		*M*	
		n=1	*n=2*		*n=3*	
Atomic number	*Element*	*1s*	*2s*	*2p*	*3s*	*3p*
1	Hydrogen	1				
2	Helium	2				
3	Lithium	2	1			
4	Beryllium	2	2			
5	Boron	2	2	1		
6	Carbon	2	2	2		
7	Nitrogen	2	2	3		
8	Oxygen	2	2	4		
9	Fluorine	2	2	5		
10	Neon	2	2	6		
11	Sodium	2	2	6	1	
12	Magnesium	2	2	6	2	
13	Aluminium	2	2	6	2	1
14	Silicon	2	2	6	2	2
15	Phosphorus	2	2	6	2	3
16	Sulphur	2	2	6	2	4
17	Chlorine	2	2	6	2	5
18	Argon	2	2	6	2	6

Table 1.2 shows the electronic configurations of elements with atomic numbers up to eighteen.

Size of atoms

An atom is not something with a firm boundary. If you think of the Bohr Atom model as being similar to the solar system then the edge of the atom might be compared with the orbit of the outer planet, i.e. Pluto. However, modern views of the atom lead to a less-defined picture with an electron only having a certain probability that it can be found at a certain distance from the nucleus, there no longer being the certainty that it will be in some particular well-defined orbit. The size of the atom might then be defined by the boundary, within which there is a 90 per cent or perhaps 99 per cent chance of finding the outer electron. Another way is by specifying the boundary in terms of how close atoms are packed in solids. Thus, the radius of an atom would be half the distance between centres of atoms when they are packed together in some crystal structure. With such a method of specifying size, account must be taken of whether the structure has the atom as an ion rather than as a neutral atom. Removing an electron from an atom, to give a positive ion, will reduce the effective size of the atom while adding an electron, to give a negative ion, will increase the size.

Table 1.3 *Sizes of atoms*

Element	*Atomic number*	*Atom radius (nm)*
Aluminium	13	0.143
Cadmium	48	0.150
Carbon	6	0.071
Chlorine	17	0.107
Chromium	24	0.125
Cobalt	27	0.125
Copper	29	0.128
Hydrogen	1	0.046
Iron	26	0.124
Lead	82	0.175
Magnesium	12	0.160
Manganese	25	0.112
Molybdenum	42	0.136
Nickel	28	0.125
Nitrogen	7	0.071
Oxygen	8	0.060
Phosphorus	15	0.109
Sodium	11	0.186
Tin	50	0.158
Titanium	22	0.147
Tungsten	74	0.137
Vanadium	23	0.132
Zinc	30	0.133

Table 1.3 shows the sizes generally specified for some common atoms.

Bonds

The way in which atoms join together, or molecules join together, is called bonding. The following are the different types of bonding (Figure 1.2) that can occur; the type of bonding existing within a material determines many of its properties.

Ionic bonding

An individual atom is electrically neutral, having as much positive charge as negative charge. However, if an atom loses an electron it must then have a net positive charge. It is then referred to as an *ion* – in this case, a positive ion. If an atom gains an extra electron it ends up with a net negative charge, becoming a negative ion. In an ionically bonded material an atom of one element gives an electron to an atom of another element. One element then consists of positive ions and the other of negative ions. Unlike-charged bodies attract each other thus, there is a force of attraction between the atoms of the two constituent elements in the material.

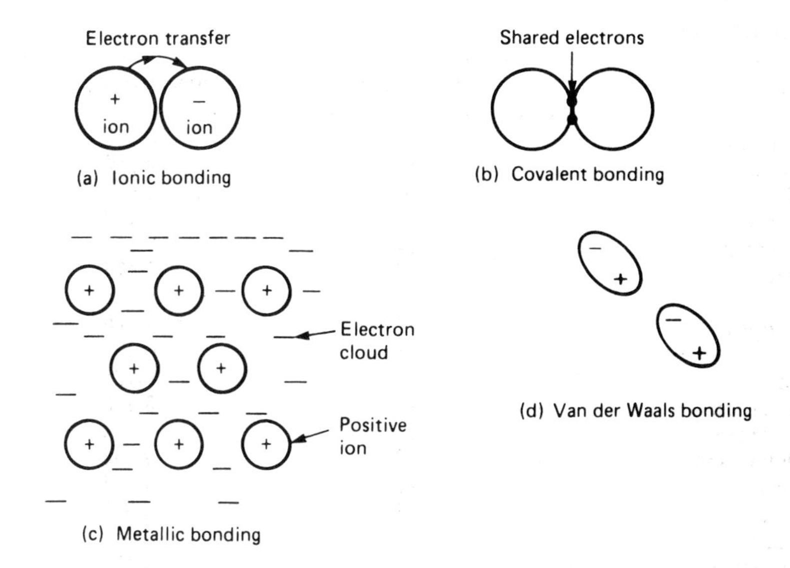

Figure 1.2 Types of bonds

Sodium chloride, common salt, is an example of an ionic bonded material. In order for the sodium and the chlorine atoms to assume the 'full shell' configuration for their atoms the sodium has to lose an electron and the chlorine gain one. The transfer of an electron from a sodium atom to a chlorine atom enables this to happen and provides a bond between the atoms (Figure 1.2a). Sodium chloride exists as a vast structure of sodium and chlorine ions rather than just as a single pair of ions. In such an array like-charged ions repel each other and unlike-charged ions attract each other. Thus sodium ions repel sodium ions, chlorine ions repel chlorine ions, but sodium and chlorine ions attract each other. A stable structure results because sodium and chlorine ions alternate and the attractive forces are just enough to overcome the repulsive forces.

With ionic bonding there is no directionality of forces and hence no directionality of bonding. Materials with this form of bonding have high melting and boiling points since the bond is a very strong one.

Covalent bonding

Ionic bonding is not possible with some elements because the energy required to remove electrons in outer shells is too great. An alternative way by which atoms of such elements can realize the 'full shell' configuration is by sharing electrons with neighbours. Thus a hydrogen molecule involves two hydrogen atoms, each atom having just one electron, sharing a pair of electrons (Figure 1.2b). This type of bond, called a covalent bond, can be thought of as two positive ions held together by the pair of electrons located between them, each ion being attracted to the electrons.

An atom may share electrons with more than one other atom, if by doing so it can achieve the 'full shell' configuration. Thus, for example, carbon needs four electrons to obtain a 'full shell', whereas chlorine requires only one. Thus the compound carbon tetrachloride, CCl_4, consists of a carbon atom bonded by covalent bonds to four chlorine atoms. Each chlorine atom shares one of the carbon electrons and the carbon atom obtains a share in one electron from each of the four chlorine atoms (Figure 1.3). A convenient way of representing such bonds is by a line linking atoms, each line representing a pair of electrons being shared.

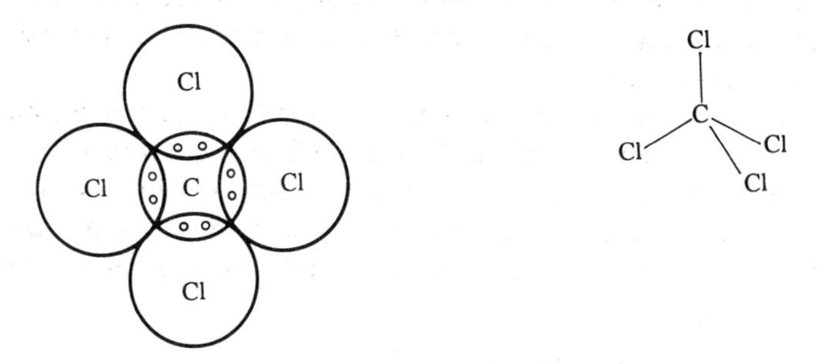

Figure 1.3 The carbon tetrachloride molecule

The atoms in some molecules can have bonds involving more than one pair of shared electrons. Where two pairs are shared the bond is said to be a double bond, where three pairs a triple bond. Oxygen molecules are composed of two oxygen atoms with two pairs of shared electrons (Figure 1.4a). Carbon dioxide molecules consist of a carbon atom with two oxygen atoms, the bonds between each oxygen atom and the carbon being double bonds.

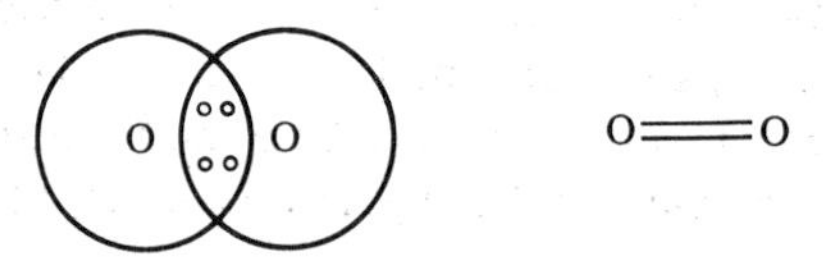

Figure 1.4 The oxygen molecule

Covalent bonds are directional bonds, the bond only occurring in the direction of the shared pair, or pairs, of electrons. These directions are determined by the repulsive forces that occur between the pair involved in the bond and other electrons in the outer shell of the atoms. Covalent bonds are very strong bonds. Diamond is an example of a material formed as a result of covalent bonds, the bonds being between carbon atoms, with each atom forming bonds with four other atoms.

Metallic bonding
Metals have atoms from which electrons are readily released. Thus, for example, copper with an atomic number of 29 has full K, L and M shells and just one electron in the N shell. This odd electron is only very loosely attached to the copper atom. In solid copper these electrons become detached from the atoms leaving a positive copper ion. The electrons that have been detached do not combine with any ion but remain as a cloud of negative charge floating between the ions. The result is bonding, the positive ions being held together by their attraction to the cloud or negative charge in which they are embedded (see Figure 1.2c). Such bonds are not directional and are, in general, weaker than ionic or covalent bonds.

Because the metal ions are not bonded directly to each other but to the electron cloud, a metallic solid can be formed from a mixture of two or more metallic elements, the result being called an alloy. Because of this form of bonding it is not necessary for the constituents of an alloy to be present in fixed proportions, as is necessary with compounds and structures formed by ionic or covalent bonding.

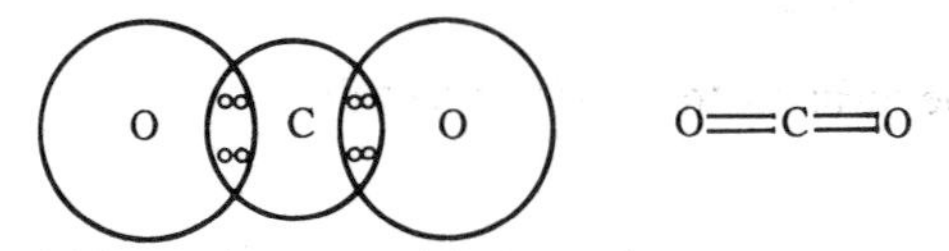

Figure 1.5 The carbon dioxide molecule

Van der Waals bonding
Carbon dioxide has a molecule formed by a carbon atom bonding with covalent bonds to two oxygen atoms (Figure 1.5). Since the molecule has electrons already participating in a molecular bond there are no electrons available to participate in further bonds. One might imagine from this that solid carbon dioxide would not be possible since it would require bonds

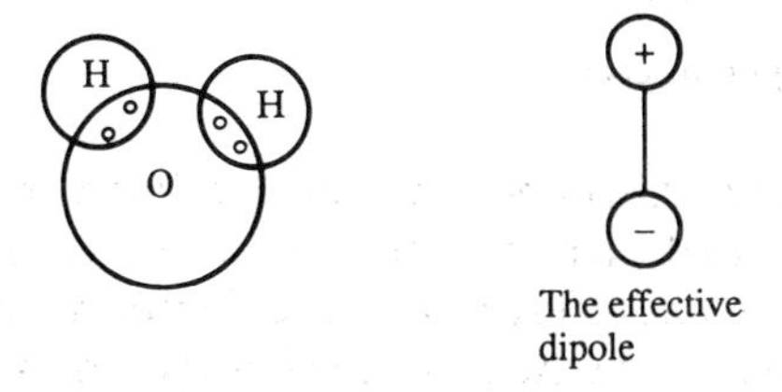

Figure 1.6 The water dipole

between carbon dioxide molecules. However, solid carbon dioxide is possible, as are many other solids formed from similar molecules. Molecular solids are held together by forces which depend on the way the charge is distributed within a molecule.

The water molecule consists of two hydrogen atoms with bonds to an oxygen atom, as shown in Figure 1.6. The electrons tend to be more concentrated at the oxygen end of the molecule than the hydrogen ends, which can be conceived as essentially bare protons. The result is what is termed an electric *dipole*. Forces can occur between such dipoles such that bonding can occur when the dipoles align themselves to result in unlike-charged ends in close proximity (Figure 1.7). Such bonding forces are much weaker than ionic, covalent or metallic bonds.

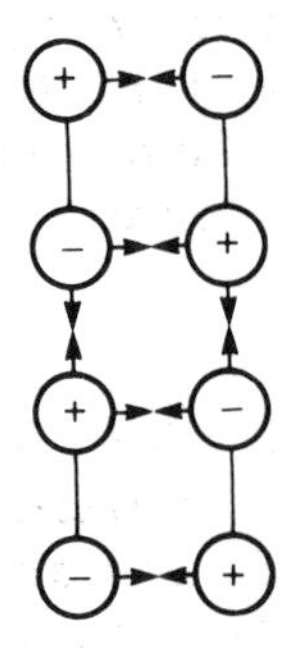

Figure 1.7 Bonding between dipoles

It is also possible to have dipole forces used for bonding where molecules are not permanent dipoles like the water molecule. The hydrogen atom can be considered to consist of a single electron orbiting a proton. At some point we can conceive the electron being at a particular position and the hydrogen atom being a dipole orientated in a particular direction. This instantaneous dipole can exert forces on neighbouring hydrogen atoms. It will attract the unlike charge and repel the like charge with the result that a neighbouring hydrogen atom becomes a dipole orientated in such a way that it is attracted to the initial dipole. A bond then occurs. This form of bond is called a van der Waals bond. Such bonds are much weaker than ionic, covalent or metallic bonds.

A comparable effect occurs when a charged piece of plastic is brought close to a small piece of (uncharged) paper – the paper becomes attracted to the plastic. This is because the charge on the paper becomes redistributed as a result of the presence of the plastic; the charge of the same sign as that on the plastic being repelled to the remote parts of the paper while the opposite sign is attracted to the nearer ends of the paper. The result is that there is a smaller distance between the unlike charges and a greater distance between like charges. Hence a net attractive force occurs.

Ionic, covalent and metallic bonds are termed *primary bonds*, while the bonds resulting form either permanent or induced dipoles are called *secondary bonds*. Primary bonds are much stronger than secondary bonds.

States of matter

Materials can be classified as having three possible states: gas, liquid or solid. Solids have definite shapes and volumes. Liquids have definite volumes but can alter their shape to take up the shape of a containing vessel. Gases have no definite volume or shape but expand until they fill any container in which they are placed. In general, solids have higher densities than their liquids and these in turn have higher densities than their gases.

A simple model which can be used to describe solids, liquids and gases is to consider a solid as being a well-packed arrangement of atoms or molecules, a liquid as being a jumbled heap of such particles and a gas as being the particles widely separated from each other. Such a model can explain the changes in density occurring when there is a change of state.

1.2 Crystals

The form of crystals and the way in which they grow can be explained if matter is considered to be made up of small particles which are packed together in a regular manner, as in Figure 1.8. The dotted lines in Figure 1.8 enclose what is called the *unit cell.* In this case the unit cell is a cube. The unit cell is the geometric figure which illustrates the grouping of the particles in the solid. This group is repeated many times in space within a crystal, which can be considered to be made up, in the case of a *simple cubic crystal*, of a large number of these unit cells stacked together. Figure 1.9 shows that portion of the stacked spheres that is within the unit cell. The crystal is considered to consist of large numbers of particles arranged in a regular, repetitive pattern, known as the *space lattice.*

It is this regular, repetitive pattern of particles that characterizes crystalline material. A solid having no such order in the arrangement of its constituent particles is said to be *amorphous*.

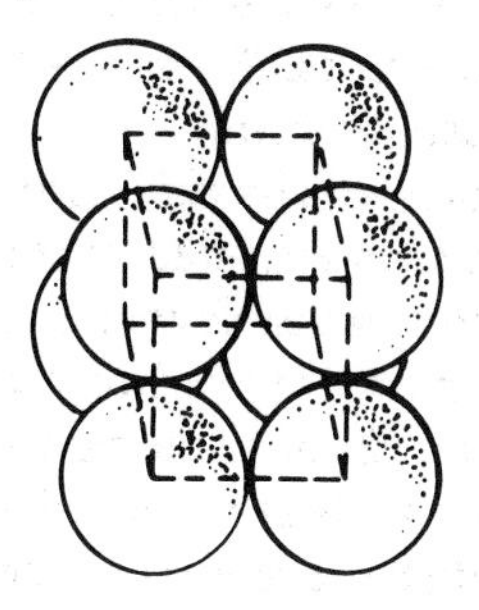

Figure 1.8 A simple cubic structure

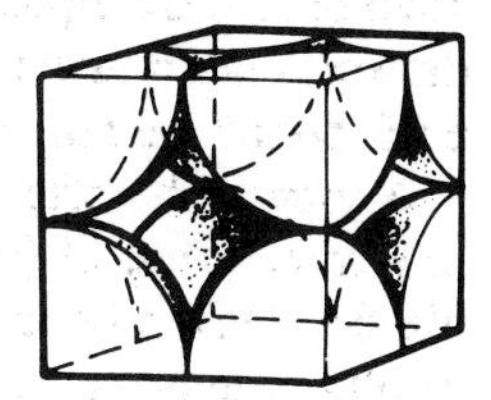

Figure 1.9 The simple cubic structure unit cell

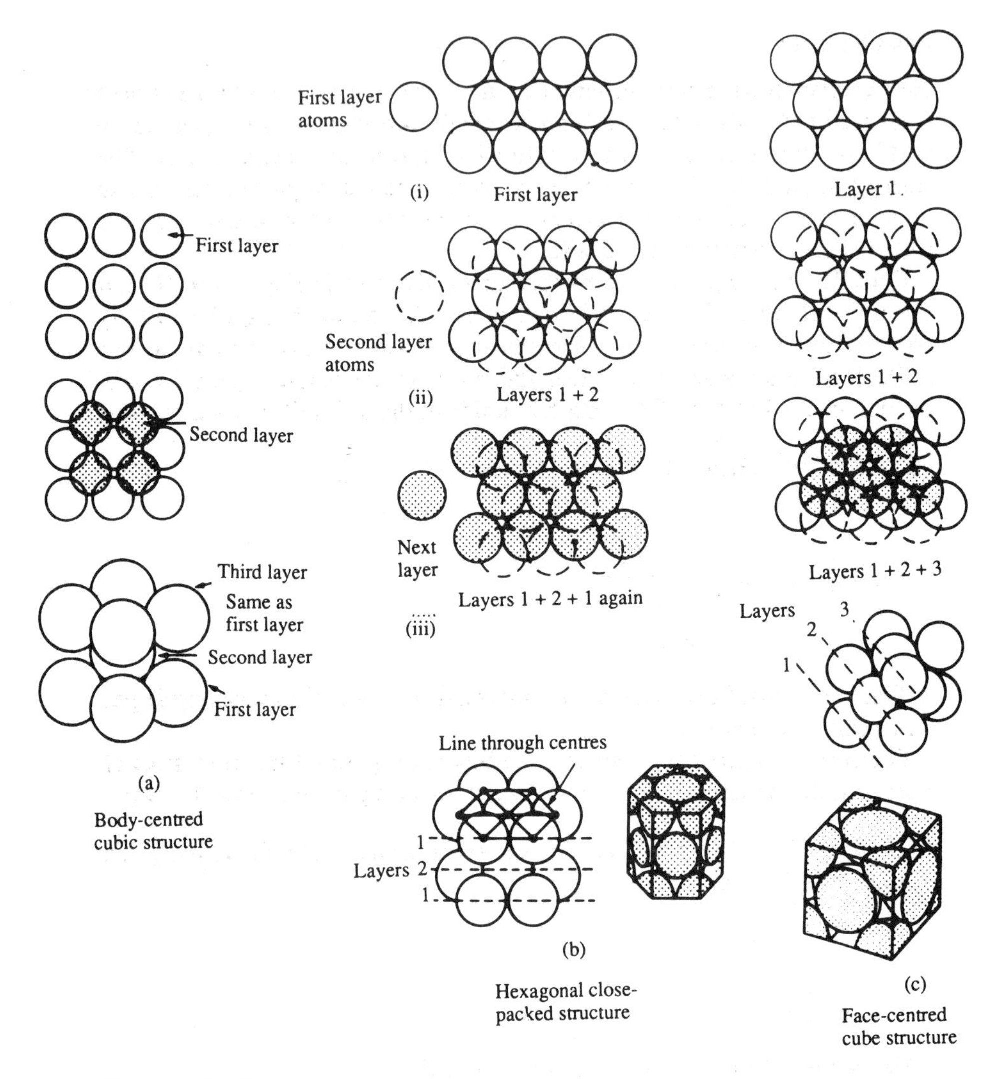

Figure 1.10 Common crystal structures

The simple cubic crystal shape is arrived at by stacking spheres in one particular way (Figure 1.8). By stacking spheres in different ways, other crystal shapes can be produced (Figure 1.10). With the simple cubic unit cell the centres of the spheres lie at the corners of a cube. With the *body-centred* cubic unit cell the cell is slightly more complex than the simple cubic cell in that it has an extra sphere in the centre of the cell. The *face-centred* cubic cell is another modification of the simple cubic cell, having spheres at the centre of each face of the cube. Another common arrangement is the *hexagonal close-packed* structure.

Packing factor

An important feature of the different forms of crystal structure is the amount of free space within the structure. This affects the movement of foreign atoms to within the structure and the ease with which the structure can deform. The fraction of a unit cell that is occupied by atoms is called the *packing factor*. The larger the packing factor the greater the fraction of the unit cell occupied and so the smaller the amount of free space.

Consider the simple cubic structure unit cell shown in Figure 1.9. Within the unit cell there is effectively one atom (this figure being obtained by counting up the volumes of the pieces of atom considered to lie within the unit cell). If the radius of an atom is r then the total volume occupied by the atoms in the unit cell is $1 \times \frac{4\pi r^3}{3}$. The length of a side of the unit cell is 2r and its total volume is $(2r)^3$. Hence the packing fraction is:

$$\text{Packing fraction} = \frac{1 \times \frac{4\pi r^3}{3}}{(2r)^3}$$

$$= 0.52$$

This means that 52 per cent of the unit cell is occupied, the remaining 48 per cent being free space.

For the face-centred cubic structure, shown in Figure 1.10c, the number of atoms within the unit cell is four and hence the occupied volume is $4 \times \frac{4\pi r^3}{3}$. The unit cell has a face diagonal of length $4r$ and hence a side of length L given by (using Pythagoras):

$$(4r)^2 = L^2 + L^2$$

$$L = \frac{4r}{\sqrt{2}}$$

Hence the volume of the unit cell is $(4r/\sqrt{2})^3$.

$$\text{Packing fraction} = \frac{4 \times \frac{4\pi r^3}{3}}{(4r/\sqrt{2})}$$

$$= 0.74$$

Thus compared with the simple cubic structure the face-centred cubic structure is more closely packed.

Table 1.4 shows the packing factors for the different unit cells.

The hexagonal close-packed and the face-centred cubic structures are therefore the most close-packed structures. Metallic solids tend to form the most densely packed structures although some will give the other structures because of the existence of some directionality in the way the atoms bond together.

Table 1.4 *Packing fractions*

Unit cell structure	*Packing fraction*
Simple cubic	0.52
Face-centred cubic	0.74
Body-centred cubic	0.68
Hexagonal close-packed	0.74

Structure and properties

The way in which atoms, or molecules, are packed together in a material is called the structure of that material. Sodium chloride – common salt – is composed of sodium ions, positive, and chlorine ions, negative. These are arranged in the simple cubic structure illustrated in Figure 1.9. A single crystal of sodium chloride, however, consists of large numbers of sodium and chlorine ions all bonded together by ionic bonds, in this cubic form into an enormous structure. Figure 1.11 illustrates part of this structure. Such a structure is three-dimensional in the sense that ions are held in place by bonds in three dimensions.

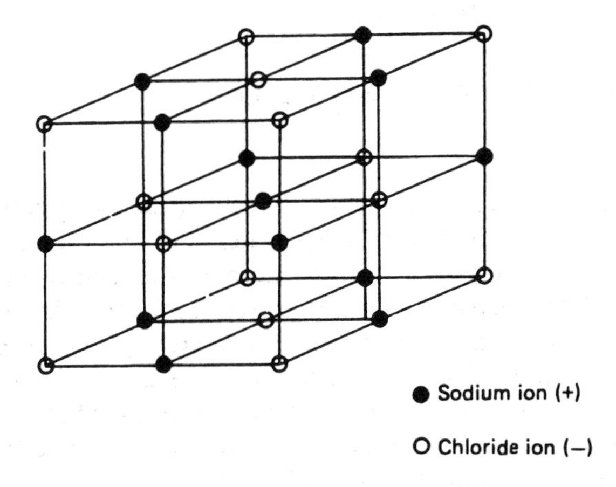

Figure 1.11 Part of a giant structure – sodium chloride crystal – showing the position of the ions

Diamond is an example of a crystal structure based solely on carbon atoms. Each atom is bound, by covalent bonding, to four other carbon atoms in a tetrahedral arrangement. Figure 1.12 illustrates this arrangement and shows how an enormous three-dimensional structure can be built up. The strength of these bonds between the carbon atoms and the uniform arrangement in which each atom is held in its place in the structure makes diamond a very hard material.

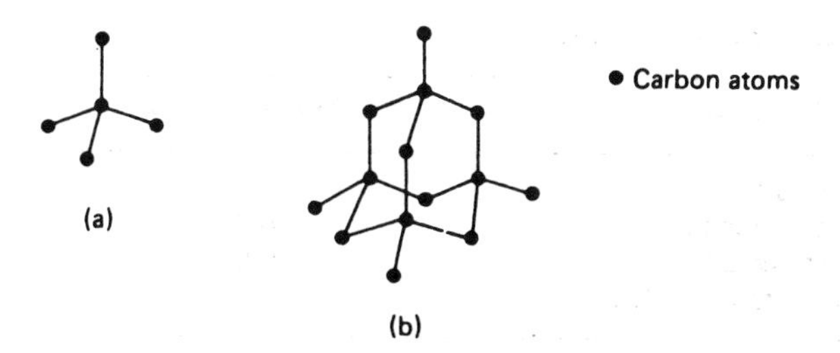

Figure 1.12 (a) The bonding arrangement for the carbon atoms in diamond, (b) Part of the three-dimensional structure of diamond

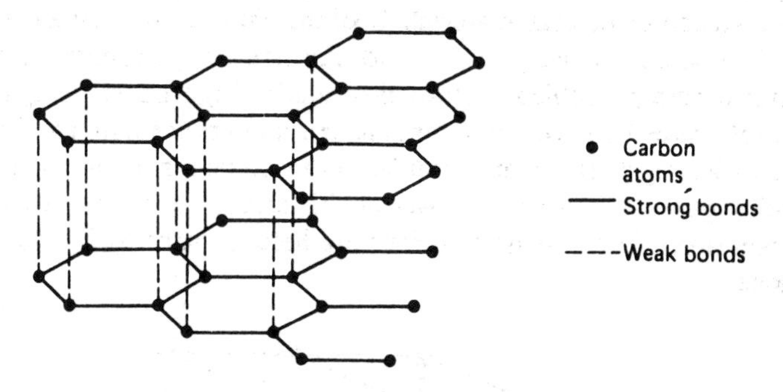

Figure 1.13 The layered structure of graphite

Graphite – the lead in pencils – is, by contrast, a very soft material. It is because layers of graphite can be easily removed that it finds a use as pencil lead. Graphite also, like diamond, consists only of carbon atoms. However, the way in which the carbon atoms are arranged in the solid is different and it is because of this difference that carbon has different properties in graphite and in diamond. Figure 1.13 shows the structure of graphite. It can be considered to be a 'layered' structure since the atoms are strongly bonded together with covalent bonds, in two-dimensional layers, with only very weak bonds, van der Waal bonds, between the atoms in different layers.

The way in which atoms, or molecules, are packed together thus markedly affects the properties.

Interatomic forces in crystals

A simple model for a crystal, whatever the form of the crystal, is that of an orderly array of spheres to represent the atoms with the spheres linked one-with-another by springs to represent the interatomic bonds. Figure 1.14 shows such a model. To move one atom aside from its equilibrium position means stretching or compressing springs. For simplicity, consider the stretching and

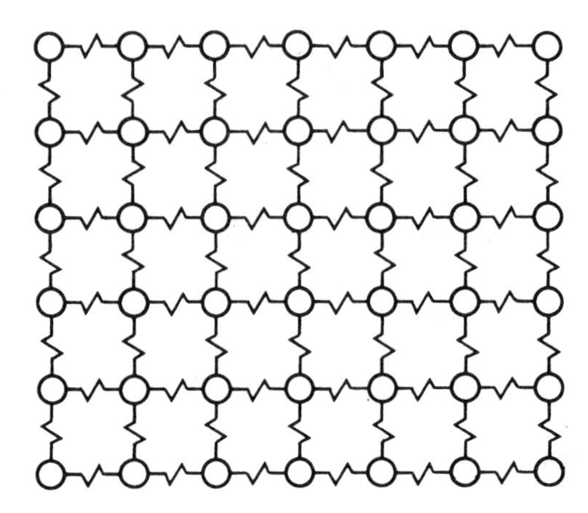

Figure 1.14 Simple model of a crystal

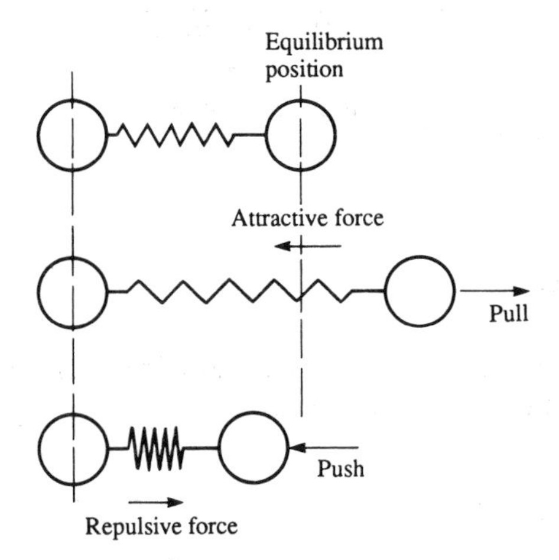

Figure 1.15 Interatomic forces in model

compressing of just one spring (Figure 1.15). Assuming that the spring obeys Hooke's law then the forces required to extend or compress it are proportional to the extension or compression. If one atom is pulled to increase the distance between the atoms then the spring will exert an attractive force on the atom and endeavour to attract it back to its equilibrium position. If the atoms are pushed together then the spring will exert a repulsive force in an endeavour to restore the atom to its equilibrium position. A graph of these forces plotted against the amount of stretching or compressing of the spring will look like Figure 1.16a. An alternative way of drawing this graph is shown in Figure 1.16b. At the equilibrium position the force acting on an atom is zero. If the separation of atoms is increased there is an attractive force which increases as the separation

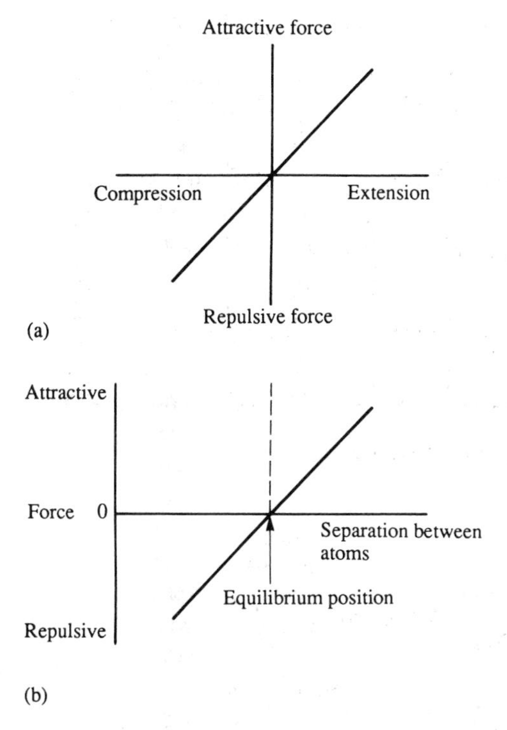

Figure 1.16 Interatomic forces in model

increases. If the separation of the atoms is decreased there is a repulsive force which increases as the separation is decreased.

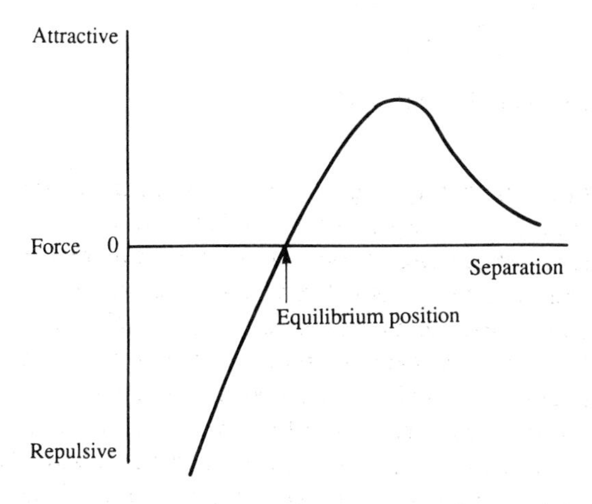

Figure 1.17 Interatomic forces in a crystal

In general the force-separation graph between two atoms in a crystal is considered to have the form shown in Figure 1.17. This, at small displacements from the equilibrium position, is rather like the result derived from a consideration of the simple balls and springs model.

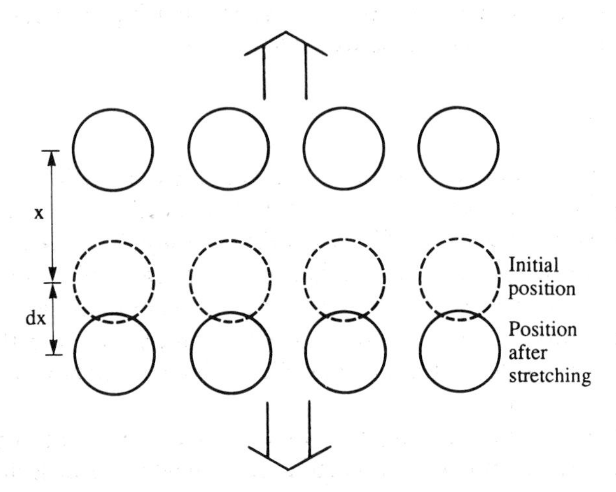

Figure 1.18 Stretching the crystal model

Consider stretching the simple model of the crystal shown in Figure 1.14. Figure 1.18 shows the edges of two planes of atoms which are at right angles to the stretching forces. If the cross-sectional area of the material is A and the atomic separation is x, then the number of atoms in such an area is A/x^2. The action of stretching causes each pair of atoms across the plane to resist being stretched. If the attractive force between each pair of atoms is dF when the material is stretched by an amount dx, then:

$$\text{total resisting force} = \frac{A\,dF}{x^2}$$

and so

$$\text{stress} = \frac{\text{force}}{\text{area}} = \frac{dF}{x^2}$$

$$\text{strain} = \text{extension per unit length} = \frac{dx}{x}$$

The tensile modulus (Young's modulus) E is stress/strain, hence:

$$E = \frac{dF/x^2}{dx/x}$$

$$E = \frac{1}{x}\frac{dF}{dx}$$

The tensile modulus is thus inversely proportional to the separation of the atoms and directly proportional to the gradient of the force-separation graph for the interatomic forces. The modulus is a characteristic of the interactions between the atoms in a material. This model of a crystalline solid must, however, be taken as only representing a very simple view of a solid and so the above result is only an indication of the relationship between the tensile modulus and the interactions between the atoms.

1.3 Liquids

A characteristic of liquids is that they can be made to flow. Some liquids flow more easily than others and the property used to describe the ease with which a liquid flows is called *viscosity*. The lower the viscosity the more easily a liquid will flow. For example, a heavy machine oil has a higher viscosity than water and does not, therefore, pour out of a can as quickly as water. Viscosity depends on temperature; the higher the temperature the lower the viscosity and so the more easily the liquid flows. Warm machine oil flows more readily than cold machine oil.

Viscosity is an important property when it comes to pouring liquid metals, glasses or polymers into moulds. Liquid metals tend to have, at their melting points, viscosities similar to that of water at room temperature and they flow fairly easily. Glasses and polymers, however, can have much higher viscosities and so less readily flow into all parts of a mould.

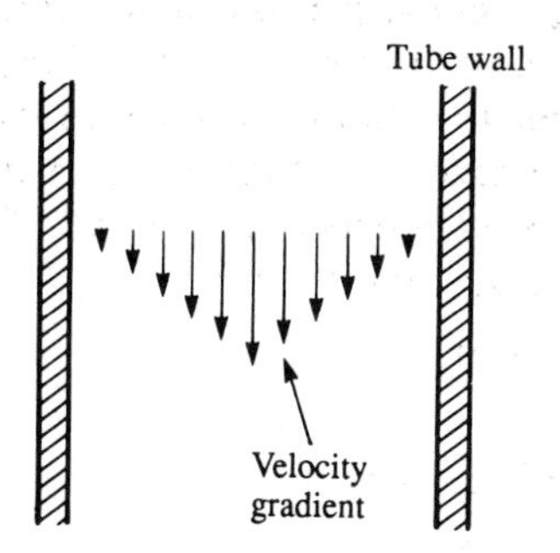

Figure 1.19 Viscous flow

The simplest form of flow is called *streamline* or *laminar* flow in that the flow can be considered in terms of layers of liquid sliding past each other. If liquid flow through a pipe is considered, the layer of liquid in contact with the wall of the tube is stationary and successive layers have increasing velocities so that a velocity gradient exists from the wall of the tube to the centre (Figure 1.19). The size of this velocity gradient (dv/dx) depends on the shearing force per unit area (F/A) used to drive the liquid through the pipe, i.e. slide one layer of liquid over another, and the viscosity η of the liquid.

$$\frac{F}{A} = \eta \frac{dv}{dx}$$

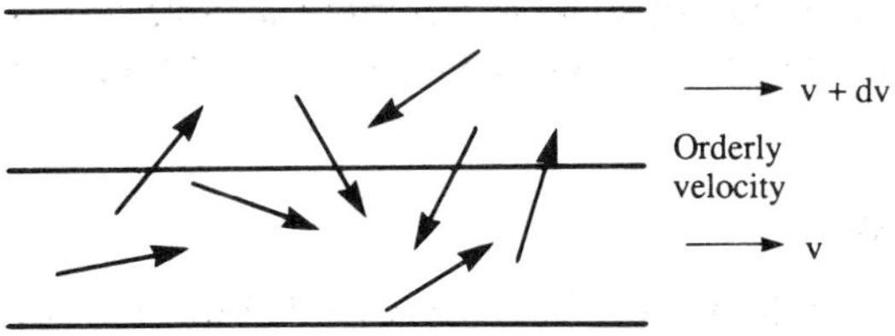

v + dv
Orderly
velocity
v

Figure 1.20 Molecular motion with laminar flow

A liquid consists of large numbers of molecules in random motion; the higher the temperature the greater the average molecular speed. When there is laminar flow there is an orderly velocity in a particular direction, superimposed on top of the random molecular velocity, such that molecules in one layer have an orderly velocity slightly greater than that in the adjacent layer (Figure 1.20). The random motion of the molecules carries them across the layers in both directions. This means that the layer with the faster orderly velocity will lose some of its molecules and gain some of the molecules from the slower moving layer. The net result is that its molecules will have a lower average orderly velocity. We can think of this faster moving layer of liquid being dragged back by the existence of the adjacent slower moving layer. It is this drag which we call viscosity.

We can explain the decrease in viscosity resulting from an increase in temperature by considering that a higher temperature means a higher average random molecular speed – more molecules move per second from one layer to another. Not only do molecules in a slower moving layer move into a faster moving layer but molecules in a faster moving layer move into a slower moving layer. The effect of this movement between layers being increased is that a greater number of molecules are accelerated by the pressure used to move the liquid. An increased rate of flow means a lower viscosity.

The viscosity of many liquids is independent of the velocity gradient in the liquid. This means that the velocity gradient is proportional to the applied pressure. If the pressure is doubled the velocity gradient is doubled. The liquid flows no more easily with a high velocity gradient than with a low velocity gradient. Such liquids are called *Newtonian liquids*. There are however liquids, notably polymers, which decrease in viscosity when the velocity gradient increases. Non-drip paints are such liquids. In the tin the paint has a high viscosity. When it is being brushed onto a surface, i.e. a velocity gradient is occurring, the viscosity decreases and the paint flows quite easily. When the brushing ceases, i.e. the velocity gradient is reduced, the paint becomes more viscous and so does not so readily flow and drip.

Polymers are long chain molecules. These long chains become easily tangled with each other. The paint in the tin consists of tangled molecules and because of this does not flow easily. When the paint is brushed the act of brushing aligns many molecules so that their chains point in the same direction. This allows them to slide more easily over one another and so the viscosity decreases.

1.4 Surfaces

Molecules in the bulk of a liquid are surrounded by other molecules and are subjected to attractive forces which are roughly the same in all directions. At the surface of the liquid, however, the molecules have no liquid molecules above them, only below resulting in a net attractive force downwards into the liquid on surface molecules. If the surface of a liquid is to be increased more molecules have to be moved into the surface against the attractive force. Energy is thus required to increase surface area. This is referred to as the *free surface energy*. The free surface energy is defined as the energy required to produce unit area of surface. It has the symbol γ and units of J/m^2.

Water, at room temperature, has a free surface energy of about 0.070 J/m^2. The free surface energies of liquid metals, however, are much higher, e.g. molten aluminium 0.50 J/m^2 and molten iron 1.50 J/m^2.

In the same way that liquids have free surface energies, so also do solids. Atoms in the bulk of a solid are subject to attractive forces in all directions while those in the surface are subject to only inward-directed forces. Thus to increase the surface area of a solid requires energy. This is discussed later in connection with the propagation of cracks through solids. When a crack propagates, new solid surfaces are produced and this requires energy.

Friction

When one object moves or tends to move, either by sliding or rolling, against another the force opposing this is called *friction*. In the case of sliding, the tangential force necessary to start relative movement of the objects is called static friction and the force necessary to maintain relative motion is called kinetic or dynamic friction. Rolling friction occurs when an object rolls over the surface of another object.

There are two basic laws of friction:

1 The frictional force is proportional to the normal force or load.
2 The frictional force is independent of the apparent area of contact between the sliding surfaces.

A consequence of the first law is that for any particular pair of surfaces the ratio of frictional force to normal force is a constant. This constant is called the *coefficient of friction* (μ).

$$\mu = \frac{\text{frictional force } F}{\text{normal force } N}$$

The second law uses the term 'apparent area of contact' in relation to the sliding surfaces. This is because when two surfaces are in contact the real contact between them occurs at only a limited number of discrete points. No matter how smooth a surface, on a molecular scale it is very irregular. Hence contact tends to occur at only the peaks of the surface irregularities (Figure 1.21). The real area of contact is only a very small fraction of the apparent area of contact. It is these small, real, contact areas that have to carry the load between the surfaces.

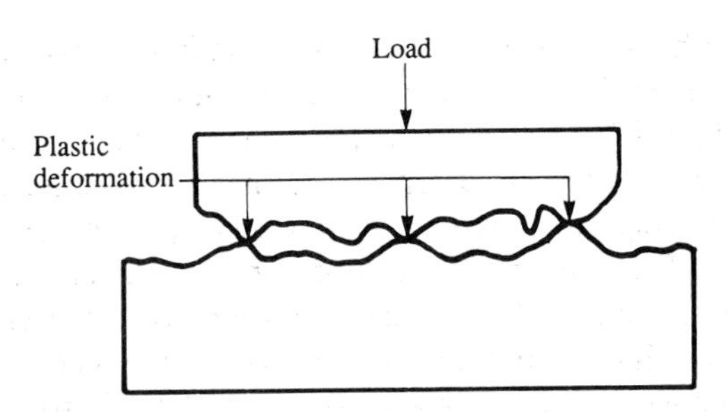

Figure 1.21 Real contact points between surfaces

Because the real area of contact between surfaces is so small the pressures at the contact points will be very high, even under light loading. In the case of metals, this pressure will generally be high enough to cause appreciable plastic deformation. The greater the load the more the material will deform and so the greater will the real area of contact become. The points of contact crush down plastically until the area of contact is sufficiently large to support the load at the yield pressure.

The true area of contact A is given by

$$\mathrm{A} = \frac{\text{normal force } N}{\text{yield pressure p}}$$

The yield pressure is equal to the indentation hardness value obtained for the metal as a result of a hardness test such as the Brinell test. A consequence of the above equation is that the true area of contact is proportional to the normal force and hence the second law of friction could be written as: the frictional force is proportional to the real area of contact between the sliding surfaces.

The pressures at the real points of contact between two surfaces are very high and as a consequence strong adhesion takes place between the two surfaces at these points. With metals the process is cold welding. When the surfaces slide over each other these junctions between the surfaces must be sheared. The frictional force thus arises from the force to shear junctions and the force required to plough the asperities of one surface through those of the other surface. In general this 'ploughing' term is relatively small and most of the frictional force is due to the shearing of junctions. Thus, to a reasonable approximation:

$$\text{Shear strength} = \frac{\text{frictional force } (F)}{\text{true area } (A)}$$

$$= \frac{\mu N}{N/\text{p}}$$

$$\text{Hence} \qquad \mu = \frac{\text{shear strength}}{\text{yield pressure}}$$

The shear strength concerned in the above equation is that of the softer material of a sliding pair. Thus for a low coefficient of friction, and hence low frictional forces, a low yield strength and high yield pressure, i.e. hardness, is required. Such a combination would appear to be impossible since metals with a high hardness have a high shear strength. If, however, a hard metal is coated with a thin layer of soft metal, the load can be borne by the hard metal, since the asperities would penetrate the soft layer, but the greatest area of contact would be with the soft layer. Consequently, an increase in load may produce only a very small change in the real area of contact with the hard surface. The result is a low coefficient of friction. Thus surface coatings and surface treatments can be used to give low frictional forces.

To keep frictional forces low between plain bearing surfaces the materials used are generally compounded to give either small particles of a hard phase embedded in a soft matrix or softer phase material dispersed throughout a hard matrix. With the hard particles within the soft matrix it is considered that the hard particles support the load while the sliding takes place within a thin smeared film of the softer material. There is some doubt as to whether this is actually the mechanism that occurs. Materials based on this principle are the white metals. In the case of the softer materials being dispersed throughout the hard matrix, when sliding occurs the soft material becomes smeared over the surface and the result is similar to that produced by a soft metal layer being used to coat a hard surface. Copper–lead alloys are examples of this type of material.

Frictional forces involve only the surface layers of materials. However, most metals exposed to air become coated with an oxide film or adsorbed gas. In addition the surfaces may be coated with a layer of dirt. All such layers affect the frictional forces. Lubricants are used to interpose films between moving surfaces so that friction is reduced. The lubricant film reduces the number and area of the metallic junctions between the sliding surfaces.

The *wear* of surfaces can be explained in terms of the adhesion occurring between the points of real contact between surfaces. The term 'wear' is used for the unintentional removal of material from two rubbing surfaces. Four different forms of wear situation can be considered to exist.

1 When the junctions between the surfaces are weaker than the sliding materials, shearing occurs at the interface and there is little transfer of metal from one surface to the other, i.e. little wear occurs. An example of this is a tin-base alloy sliding on steel.
2 When the junctions are stronger than one of the metals but weaker than the other, shearing takes place a small distance within the softer material. Wear of the softer material thus occurs and eventually a film of softer material builds up on the harder surface. An example of this is a lead-base alloy sliding on steel.
3 When the junctions are stronger than both metals, shearing will take place mainly in the softer of the two metals but some fragments of the harder metal may be ploughed out. An example of this is copper sliding on steel.
4 When both the sliding surfaces are the same, the process of deformation and sliding causes the junction material to work harder. As a consequence of this shearing occurs within both the metals and considerable wear can occur.

Rolling friction is considered to mainly occur as a result of the energy lost by the elastic deformation and recovery of the surface over which the rolling occurs. It is an elastic hysteresis effect. The surface becomes elastically deformed as the ball begins to roll over it and after the ball has passed the surface recovers but the deformation of the surface is greater than the energy released when it recovers. The result is an energy loss, this being the rolling friction loss. In the above discussion it has been assumed that plastic deformation of the surface does not occur as the ball rolls over it, e.g. a permanent groove being produced. In such a case energy would be needed for the plastic deformation. When a ball bearing rolls in its race, only elastic deformation occurs and since for the steels involved the hysteresis losses are very small the rolling resistance is very small. Typically the rolling coefficient of friction is about 0.001. This is not the only source of friction with a ball and roller bearing since the balls are surrounded by a cage to keep them apart and prevent them rubbing on each other. Friction between the balls and the cage can be greater than that between the balls and the race, hence the use of a lubricant to reduce it. The material used for the cage must also be one that has a low coefficient of friction with respect to the steel used for the ball bearings.

With metals sliding on metals the true area of contact between them is proportional to the normal force and so the frictional force is proportional to the normal force. Hence the coefficient of friction, the ratio of frictional force to normal force, is a constant and independent of the value of the normal force. With polymers, sliding on metals or polymers this is not generally true. Polymers deform visco-elastically (see section 8.4) and the true area of contact depends on time as well as the normal force. A consequence of this is that the frictional force is proportional to (normal force)^x, where x has a value of about three-quarters, and hence a coefficient of friction that tends to decrease with increasing load.

Table 1.5 *Kinetic coefficients of friction*

Materials involved	*Coefficient of kinetic friction*
Mild steel on mild steel	0.5
White metal on steel	0.5
Copper–lead alloy on steel	0.2
Phosphor bronze on steel	0.4
Nylon on nylon	0.3
Nylon on steel	0.2
PTFE on PTFE	0.05
Bronze impregnated with PTFE on steel	0.05

Table 1.5 gives typical values of the kinetic coefficient of sliding friction.

Problems

1 Describe the basic characteristics of the three states of matter.
2 Describe the basic types of bonding between atoms or molecules.
3 Explain the terms *unit cell* and *space lattice*.
4 How do differences of structure explain the different properties of diamond and graphite.
5 How does a crystalline material differ in structure from an amorphous material?
6 Explain how differences in the way ions are packed in a solid can lead to the different structure of simple cubic, face-centred cubic, body-centred cubic and close-packed hexagonal.
7 Derive the packing factor for the body-centred cubic structure.
8 Explain how the tensile modulus for a material can be deduced from a graph of its interatomic forces plotted against atomic separation.
9 Explain viscosity, and its dependence on temperature, in terms of molecular motion within a liquid.
10 Describe and explain the viscous behaviour of non-drip paints.
11 Explain what is meant by the term *free surface* energy.
12 Explain the relationship of frictional forces to the real areas of contact between sliding surfaces.
13 Explain how frictional forces can be reduced by
(a) A blend of hard and soft phase materials.
(b) surface coatings.
14 Explain how wear arises.

2

Structure of metals

2.1 Metals as crystalline

Metals are crystalline substances. Figure 2.1 shows a section of a metal. The term *grain* is used to describe the crystals within the metal. A grain is merely a crystal without its geometrical shape and flat faces because its growth was impeded by contact with other crystals. Within a grain the arrangement of particles is just as regular and repetitive as within a crystal with smooth faces. A simple model of a metal with its grains is given by the raft of bubbles on the surface of a liquid (Figure 2.2). The bubbles pack together in an orderly and repetitive manner but if 'growth' is started at a number of centres then 'grains' are produced. At the boundaries between the 'grains' the regular pattern

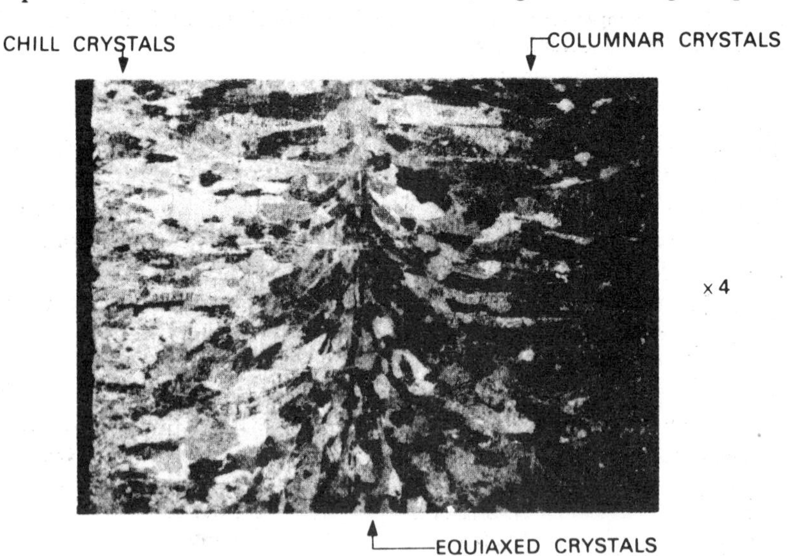

Figure 2.1 Cross-section of a small aluminium ingot (From Monks, H. A. and Rochester, D. C., *Technician Structure and Properties of Metals*, Cassell)

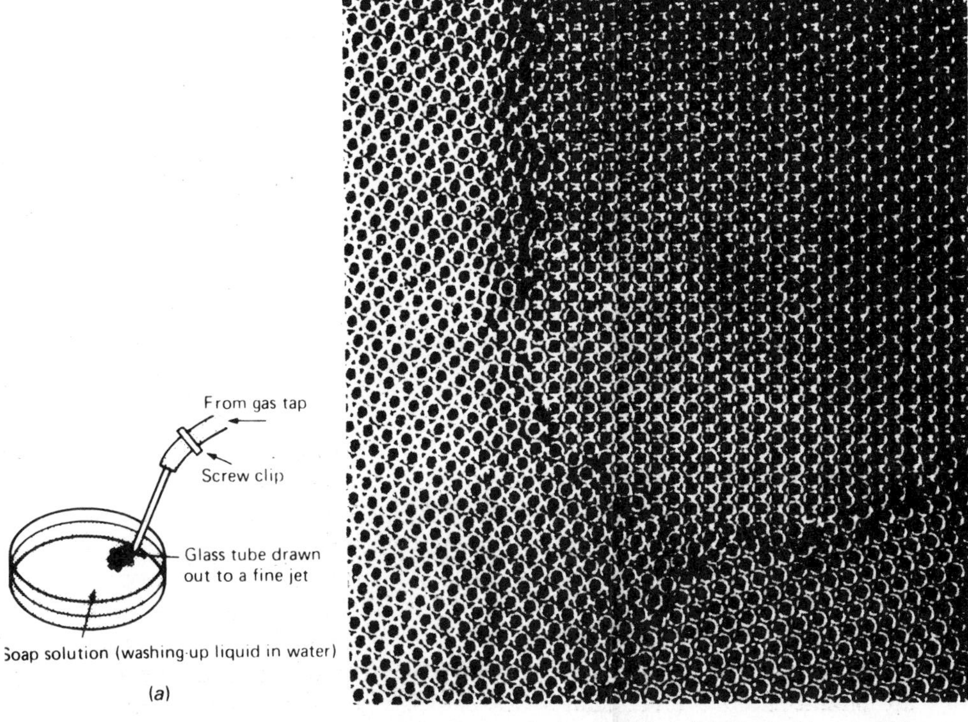

Figure 2.2 (a) Simple arrangement for producing bubbles, (b) 'Grains' in a bubble raft (Courtesy of the Royal Society)

breaks down as the pattern changes from the orderly pattern of one 'grain' to that of the next 'grain'.

The grains in the surface of a metal are not generally visible. They can be made visible by careful etching of the surface with a suitable chemical. The chemical preferentially attacks the grain boundaries.

Examples of the different forms of crystal structure adopted by metallic elements are shown in Table 2.1.

Table 2.1

Body-centred cubic	*Face-centred cubic*	*Hexagonal close-packed*
Chromium	Aluminium	Beryllium
Molybdenum	Copper	Cadmium
Niobium	Lead	Magnesium
Tungsten	Nickel	Zinc

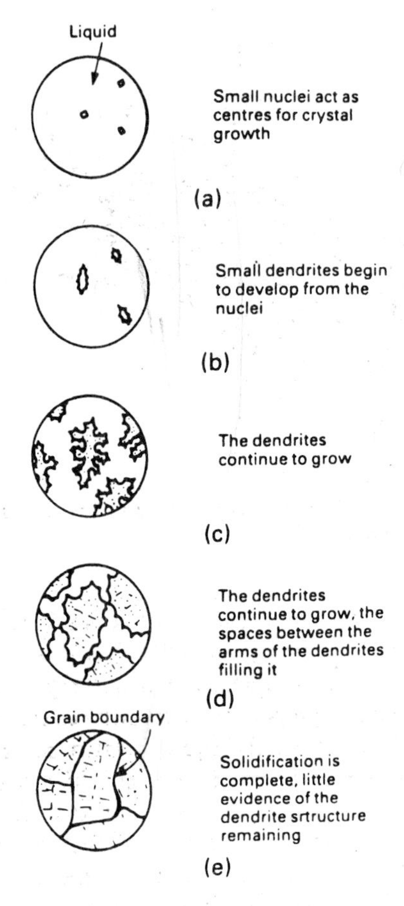

Figure 2.3 Solidification of a metal

Growth of metal crystals

Figure 2.3 shows the various stages that can occur when a metal solidifies. Crystallization, whether with metals or any other substances, occurs round small nuclei, which may be impurity particles. The initial crystals that form have the shape of the crystal pattern into which the metal normally solidifies, e.g. face-centred cubic in the case of copper. However, as the crystal grows it tends to develop spikes. The shape of the growing crystal thus changes into a 'tree-like' growth called a *dendrite* (Figure 2.4). As the dendrite grows so the spaces between the arms of the dendrite fill up. Outward growth of the dendrites ceases when the growing arms meet other dendrite arms. Eventually the entire liquid solidifies. When this happens there is little trace of the dendrite structure, only the grains into which the dendrites have grown.

Why do metals tend to grow from the melt as dendrites? Energy is needed to

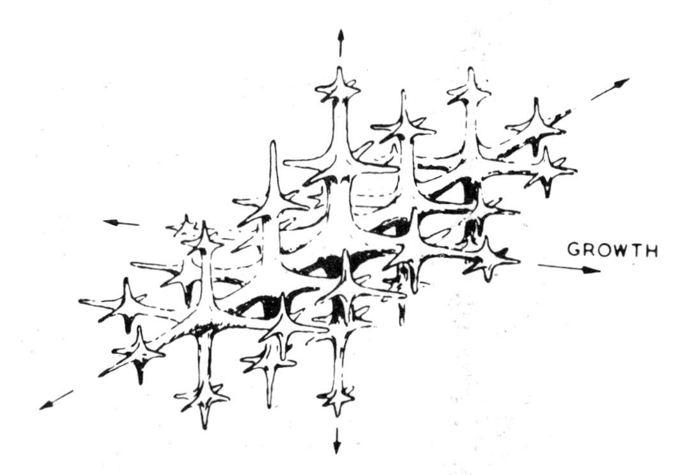

Figure 2.4 Growth of a metal dendrite (From Higgins, R. A., *Properties of Engineering Materials*, Hodder and Stoughton)

change a solid, at its melting point, to a liquid without any change in temperature occurring; this energy is called *latent heat*. Similarly, when a liquid at the fusion point (i.e. the melting point) changes to a solid, energy has to be removed, no change in temperature occurring during the change of state; this is the latent heat. Thus, when the liquid metal in the immediate vicinity of the metal crystal face solidifies, energy is released which warms up the liquid in front of that advancing crystal face. This slows, or stops, further growth in that direction. The result of this action is that spikes develop as the crystal grows in the directions in which the liquid is coolest. As these warm up in turn, so secondary, and then tertiary, spikes develop as the growth continues in these directions in which the liquid is coolest.

2.2 Block slip model

A simple theory to explain the elastic and plastic behaviour of metals is the '*block slip' theory*, where a metal is considered to be made of blocks of atoms which can move relative to each other. When a stress is applied to the metal, blocks of atoms become displaced (Figure 2.5). When the yield stress is reached there is a movement of large blocks of atoms as they slip past each other, the plane along which this movement occurs being the *slip plane*.

Metals are composed of many crystals. A crystal within a metal is just a region of orderly packed atoms. Such a region is generally referred to as a *grain*. The surfaces that divide the different regions of orderly packed atoms are termed *grain boundaries*. When plastic deformation occurs in a metal, movement occurs along slip planes and the result is rather like Figure 2.6. Slip occurs only in those planes which are at suitable angles to the applied stress.

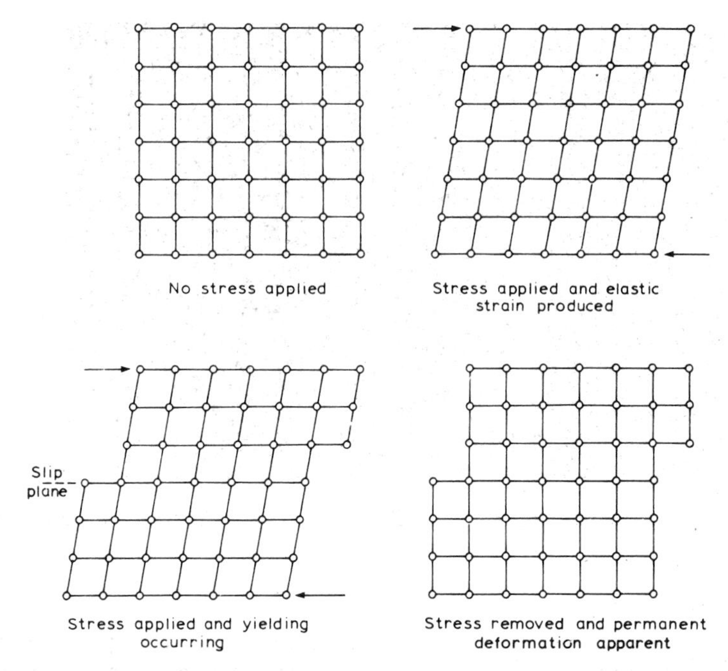

Figure 2.5 'Block slip' model showing plastic behaviour of metals under stress

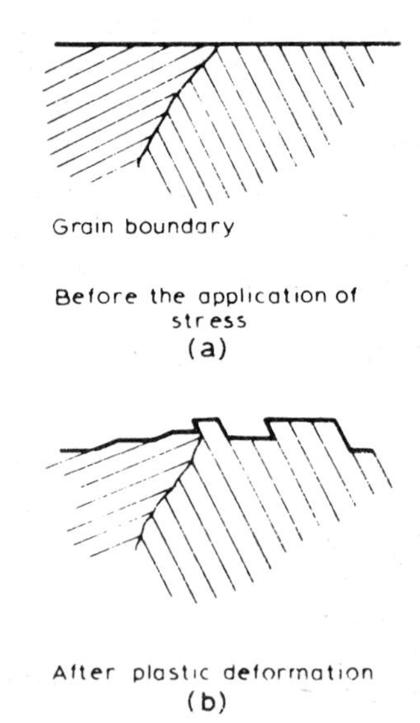

Figure 2.6 (a) Before the application of stress, (b) After plastic deformation

Figure 2.7 Slips steps in polycrystalline aluminium (Courtesy of the Open University)

The result is that the surface of the metal shows a series of steps due to the different movements of the various planes of atoms. These can be seen under a microscope (Figure 2.7). The slip lines do not cross over from one grain to another; the grain boundaries restrict the slip to within a grain. Thus the bigger the grains the more slippage that can occur; this would show itself as a greater plastic deformation. A fine grain structure should therefore have less slippage and so show less plastic deformation, i.e. be less ductile. A brittle material is thus one in which each little slip process is confined to a short run in the metal and not allowed to spread, a ductile material is one in which the slip process is not confined to a short run in the metal and does spread over a large part of the metal.

While this theory appears to give a plausible explanation of elastic and

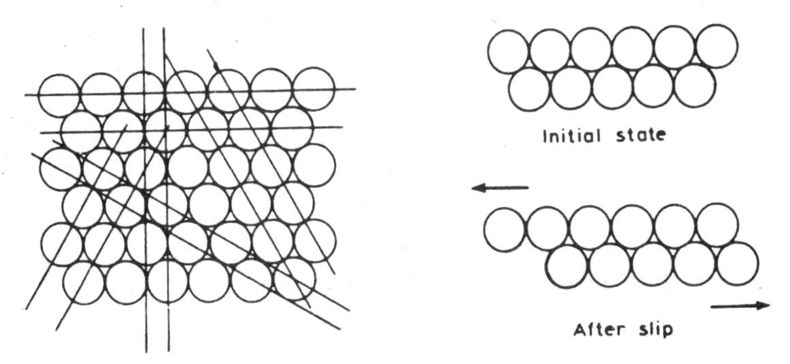

Figure 2.8 Some of the possible planes in a regularly-packed array of atoms

plastic behaviour there is one big disadvantage – calculations of the stress needed to displace all the atoms in one plane relative to those in the next plane by at least the 'width' of one atom, indicate a stress value considerably greater than the real results given by experiments. Real metals are not as strong as the theoretical model.

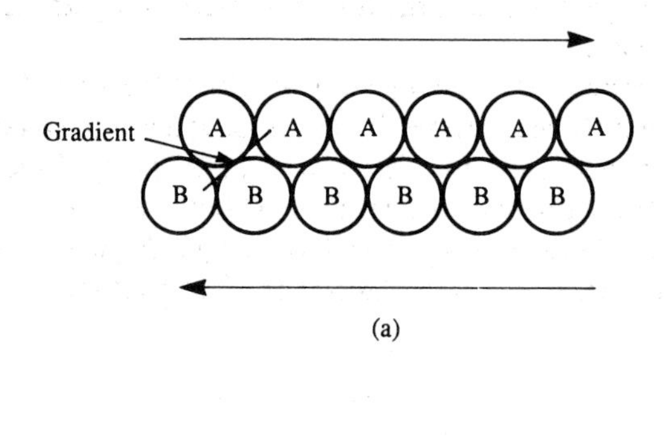

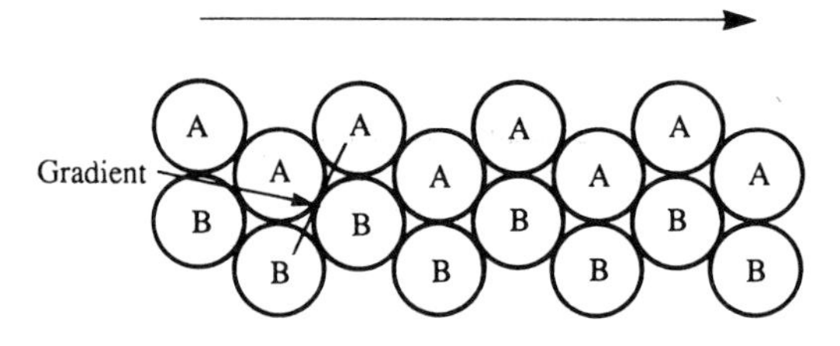

Figure 2.9 Ease of sliding with different planes, (a) Close-packed plane (b) Less closely-packed plane

Slip planes

The term 'slip' is used to describe the relative sliding of two parts of a crystal structure on either side of a plane which we call the slip plane. While there are many possible planes that can be considered to exist in crystalline structures

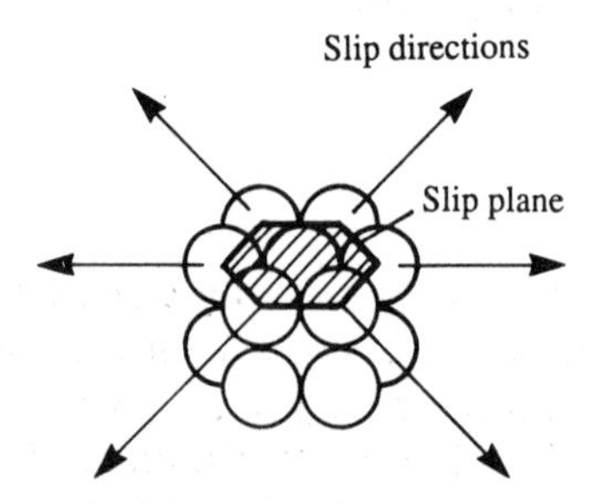

Figure 2.10 The slip plane and slip directions for a hexagonal close-packed structure

(see Figure 2.8), slip only occurs along some of them. These are the planes with the closest packing of atoms. Consider Figure 2.9 in which the ease of slip along a close-packed plane can be compared with that along a less closely-packed plane. If layer A is to be slid over layer B then a measure of the ease of sliding is the gradient line drawn on each figure. The close-packed plane involves a lower gradient than the less closely-packed plane and so appears to be more easily slid.

Figure 2.10 shows the hexagonal close-packed structure. It has just a single slip plane though there are three directions in which slip can occur. The face-centred cubic system has four slip planes, each having three slip directions. The body centred cubic system has more slip planes and directions than the other systems but because it is less well packed (see Figure 1.10) materials with this form of structure tend to be harder and less ductile.

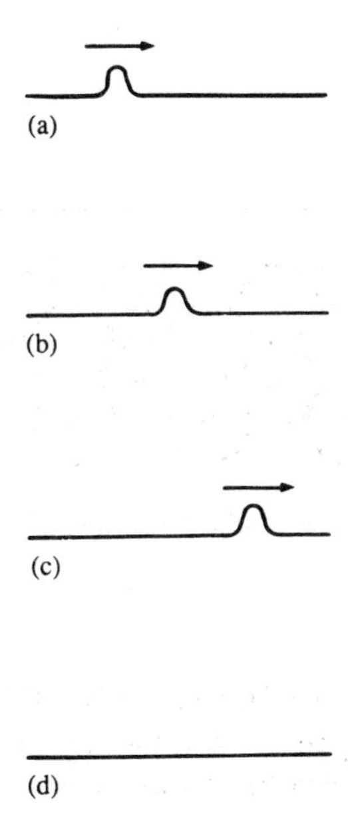

Figure 2.11 Movement of a ruck across a carpet

2.3 Dislocations

The 'block slip' model has atoms perfectly arranged in an orderly manner within the metal. If, however, we consider the arrangement to be imperfect then permanent deformations can be produced with much less stress. When you have a large carpet which is perfectly flat on the floor, it requires quite an effort to slide the entire carpet and make it move across the floor. But if there is a ruck in the carpet (Figure 2.11), then the carpet can be slid over the floor by

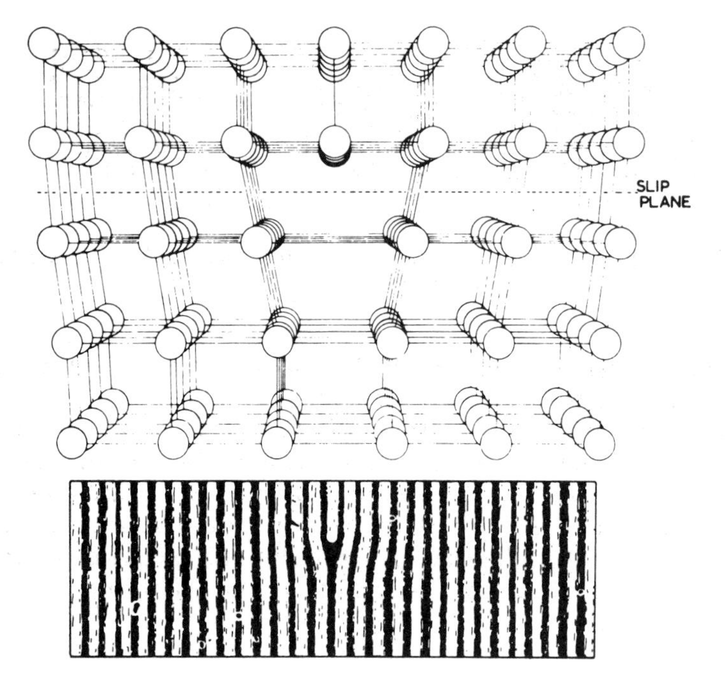

Figure 2.12 A 'ball-and-wire' model of a simple edge dislocation (From Higgins, R. A., *Properties of Engineering Materials*, Hodder and Stoughton)

pushing the ruck along a bit at a time and considerably less effort is required. This is the type of movement which is considered to take place within a metal, the 'ruck' in the crystal being a *dislocation* of atoms due to imperfect packing of the atoms within the metal. Figure 2.12 shows the type of arrangement of atoms that is considered to occur with what is called an *edge dislocation*. Figure 2.13 illustrates the movement that occurs when stress is applied and permanent deformation occurs. The dislocation moves through the array of

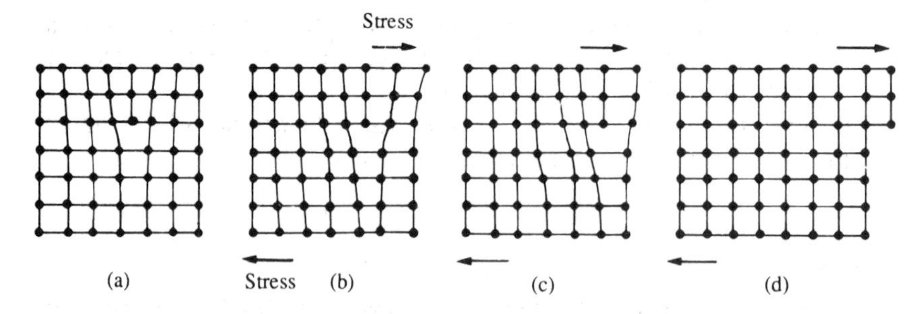

Figure 2.13 Movement of a dislocation through an atomic array under the action of stress

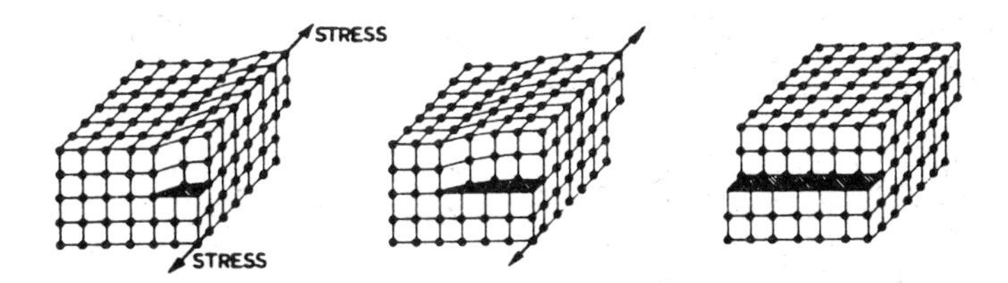

Figure 2.14 Principle of the screw dislocation (From Higgins, R. A., *Properties of Engineering Materials*, Hodder and Stoughton)

atoms without wholesale movement of planes of atoms past each other; it is a bit-by-bit process like the ruck in the carpet.

Figure 2.14 shows the form of a *screw dislocation* and its movement through the array of atoms under the action of stress. With an edge dislocation the line of dislocation is at right angles to the slip plane; with a screw dislocation the line of dislocation is parallel to the slip plane. In practice, dislocations are often neither straight lines, at right angles to, or parallel to, the slip plane but curved lines, which can, however, be considered to be a combination of edge and screw dislocations.

Crystal defects

The dislocations referred to above, i.e. edge and screw dislocations, are line defects in the crystal structure. *Line defects* are long in one direction while measuring only a few atomic diameters at right angles to their length. Another form of defect is a point defect. *Point defects* are only of the order of an atomic diameter in all directions. Such defects take the form of:

1 A vacancy in the crystal structure as a result of a missing atom (Figure 2.15a).
2 An atom displaced from its normal position to a position within the lattice of the other atoms (Figure 2.15b), this being called a self-interstitial defect, or onto the surface of the crystal.
3 A foreign atom, whether an impurity atom or a deliberate alloying addition, substituting for one of the crystal atoms (Figure 2.15c), this being called a substitutional atom defect.
4 A foreign atom occupying a vacant site within the crystal lattice (Figure 2.15d), this being called an interstitial atom defect.

Movement of dislocations

What happens when two dislocations come close to each other during their movement through a metal? As Figure 2.16 shows, the atoms on one side of the slip plane are in compression and on the other side in tension. When two

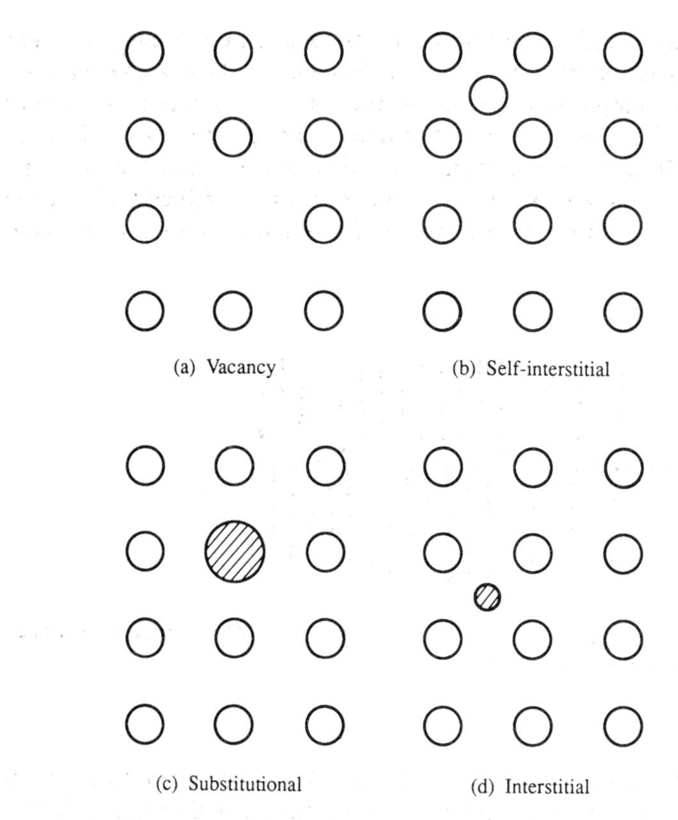

Figure 2.15 Point defects

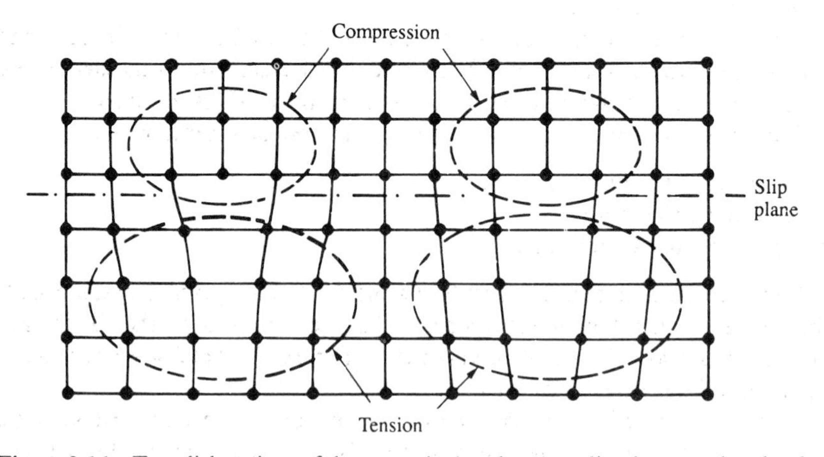

Figure 2.16 Two dislocations of the same sign on the same slip plane repel each other

dislocations come together the regions of compression can impinge on each other and so hinder the movement of the dislocations. If the movement is such as to bring the compression region against the tension region of another dislocation then it is possible for the two dislocations to annihilate each other (Figure 2.17). In general, the more dislocations a metal has, the more the dislocations get in the way of each other and so the more difficult it is for the dislocations to move through the metal. More stress is needed to cause yielding.

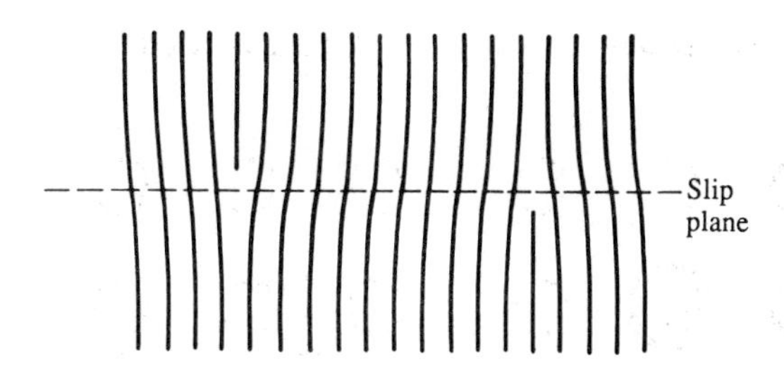

Figure 2.17 Two edge dislocations of opposite sign on the same slip plane can move together and annihilate each other

The movement of dislocations through a metal is also hindered by the grain boundaries. The more grain boundaries there are in a metal the more difficult it is to produce yielding of that metal. More grain boundaries occur when the grain size in a metal is small.

The movement of dislocations is hindered by anything that destroys the continuity of the atomic array. The presence of 'foreign' atoms can distort the atomic array of a metal and so hinder the movement of dislocations.

Thus possible ways of increasing the yield stress of a metal are by:

1 Increasing the number of dislocations.
2 Reducing the grain size.
3 Introducing 'foreign' atoms.

Work hardening occurs as a result of a material being plastically deformed, the result being a higher yield stress. This occurs because the dislocation density is increased by plastic deformation (Figure 2.18) and so there is more interaction between dislocations.

Dispersion hardening increases the yield stress of a material by producing a dispersion of fine particles throughout the material. These hinder the movement of dislocations, hence increasing the yield stress. One form of dispersion hardening is called *precipitation hardening*. With this the dispersion of fine particles is produced as a result of a specific form of heat treatment

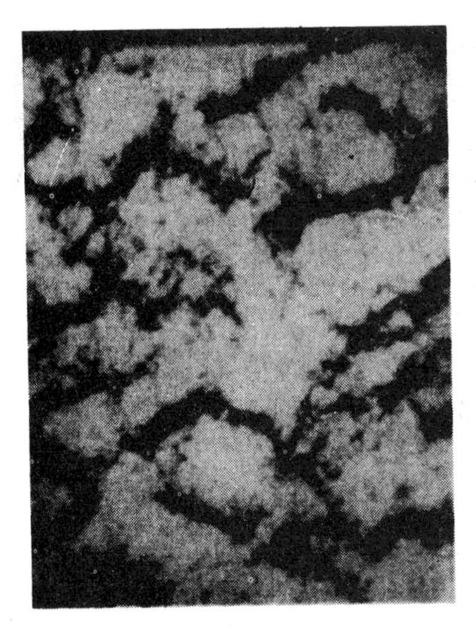

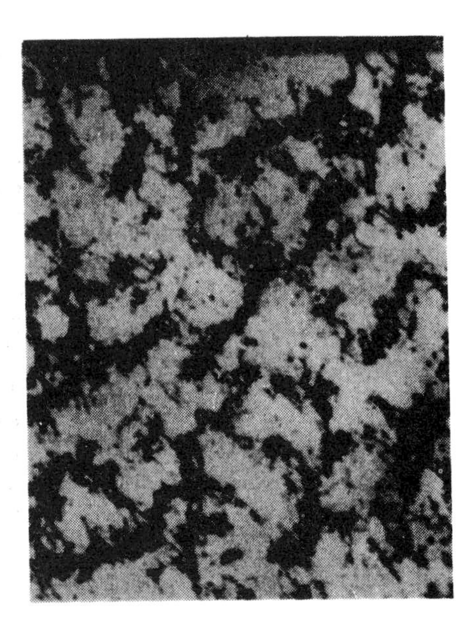

Figure 2.18 Dislocations in niobium under increasing strain. There is an increase in dislocation density (From Harris, B. and Bunsell, A. R., *The Structure and Properties on Engineering Materials*, Longman)

applied to the material to cause a precipitation to occur within the material.

The alloying of metals involves the introduction of foreign atoms into the crystal lattice. These produce interstitial and substitutional point defects which hinder the movement of dislocations, hence increasing the yield stress. This is referred to as *solution hardening*.

Dislocations can annihilate each other if they are of opposite sign and move together along the same slip plane, as in Figure 2.17. However, this does not occur too frequently at room temperature. At higher temperatures another annihilation mechanism can occur. This is because at higher temperatures diffusion of atoms can become significant. This leads to what is called

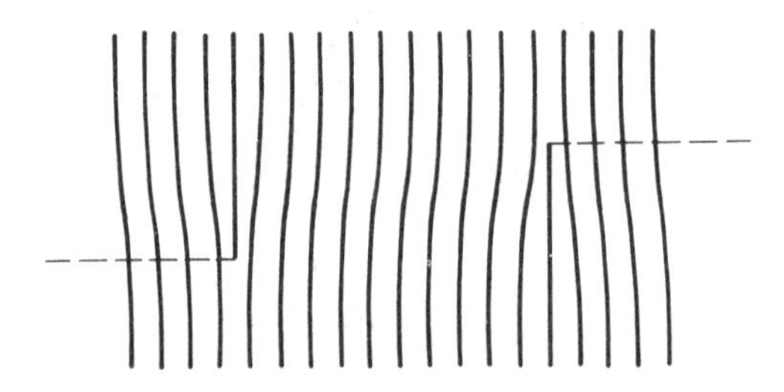

Figure 2.19 Two edge dislocations of opposite sign can climb to be on the same slip plane and then annihilate each other

dislocation climb when an edge dislocation moves in a direction at right angles to its slip plane. The result of such movement is that more edge dislocations of opposite sign can annihilate each other (Figure 2.19).

If a work hardened material is heated to about 0.3 to 0.4 times its melting temperature (in degrees kelvin), some diffusion occurs and dislocations become rearranged and the number reduced. At this temperature there is no change in grain size. The result of such changes is that residual stresses are released, *recovery* being said to occur, and there is a slight reduction in yield stress. At higher temperatures *recrystallization* occurs. New grains of low dislocation density are produced. The result is a marked decrease in yield stress. The heat treatment which allows recrystallization to occur with a consequent decrease in yield stress is called *annealing*.

The effects of diffusion and consequent annihilation of dislocations as a result of dislocation climb is particularly evident in the behaviour of metals when subjected to a load for a long period of time at a high temperature. The strain increases steadily with time. This effect is called *creep* and is dealt with in more detail in Chapter 8.

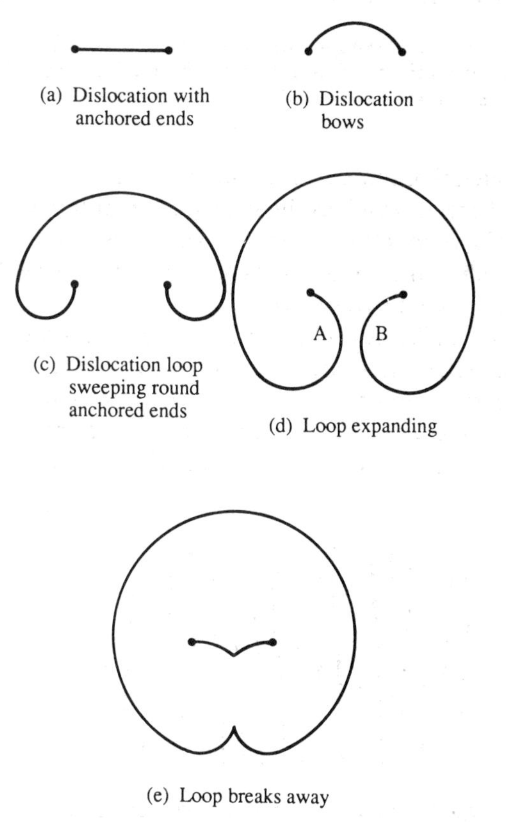

Figure 2.20 The Frank-Read source

Dislocation multiplication

The number of dislocations piercing a unit area of a polycrystalline material is of the order of 10 to a 100 million per square centimetre. These occur as a result of fabrication processes. However, if the material is subject to plastic deformation this dislocation density increases to as much as a million million per square centimetre. A possible mechanism for this multiplication is called the *Frank-Read source*, these being the names of the people who originally postulated such a source.

Consider the dislocation shown in Figure 2.20a. The movement of the ends of the dislocation are hindered, by other dislocations or possibly foreign atoms. Thus, when a stress is applied the ends of the dislocation cannot move. The result is that the dislocation bows out, as in Figure 2.20b. Initially the bowed dislocation has a large radius. However, as the stress increases the radius decreases until the minimum radius is reached when the dislocation forms a semi-circle. This is the condition of maximum stress. Beyond that point the radius decreases and the stress required to keep the dislocation expanding decreases. The dislocation forms a loop which grows by sweeping round the fixed ends until eventually the two sides meet to form a complete loop. When this occurs the portions of the dislocation A and B annihilate each other and the final result is an expanding dislocation loop which is free of the points anchoring the ends of the initial dislocation and the original pinned dislocation. This can continue multiplying.

Problems

1 Explain what is meant by the term grain.
2 Describe how metal crystals grow within a liquid metal.
3 Distinguish between elastic and plastic deformation of metals and give a simple explanation of such deformation in terms of the block slip model.
4 Explain why slip only occurs in particular directions in a crystal.
5 Explain the nature of point and line defects in crystals.
6 Describe how dislocations density, grain size and foreign atoms affect the mobility of dislocations within a metal and hence the yield stress of that metal.
7 Explain in terms of dislocations (a) work hardening, (b) dispersion hardening, (c) solution hardening, (d) recovery, (e) annealing.
8 Explain what is meant by a Frank-Read source of dislocations and explain how it multiplies dislocations.

3

Structure of alloys

3.1 Alloys

Brass is an alloy composed of copper and zinc. Bronze is an alloy of copper and tin. An *alloy* is a metallic material consisting of an intimate association of two or more elements. The everyday metallic objects around you will be made almost invariably from alloys rather than the pure metals themselves. Pure metals do not always have the appropriate combination of properties needed; alloys can however be designed to have them.

If you put sand in water, the sand does not react with the water but retains its density, as does the water. The sand in water is said to be a mixture, In a *mixture*, each component retains its own physical structure and properties. Sodium is a very reactive substance, which has to be stored under oil to stop it interacting with the oxygen in the air, and chlorine is a poisonous gas. Yet when these two substances interact, the product, sodium chloride, is eaten by you and me every day. The product is common salt. Sodium chloride is a compound. In a *compound* the components have interacted and the product has none of the properties of its constituents. Alloys are generally mixtures though some of the components in the mixture may interact to give compounds as well.

The term *binary alloy* is used to describe an alloy made up of just two components while the term *ternary alloy* is used when three components are involved.

Mixtures

If a pinch of common salt, sodium chloride, is dropped into water it will dissolve. The salt is said to be *soluble* in water, the resulting mixture being called a solution. If, however, sand is mixed with water the result is not a solution since the sand and the water can clearly be seen to be two separate entities. The sand is said to be *insoluble* in water.

When two liquids are mixed the result can be:

1 One liquid completely dissolves in the other, e.g. alcohol in water.
2 Each liquid is partially soluble in the other. Thus if a small amount of liquid A is mixed with liquid B a solution might be formed, but if more is added a limit of solubility is reached and the end result is a solution of a small amount of A in B and undissolved B.
3 Each liquid is completely insoluble in the other. Such a mixture of liquids will always separate out into two layers.

When liquid metals are mixed the result is that in most cases one liquid is completely dissolved in the other, a homogeneous solution being produced. With a solution it is not possible to identify the separate constituents.

When a liquid mixture solidifies a number of possibilities exist:

1 The two components separate out with each in the solid state maintaining its own separate identity and structure. The two components are said to be insoluble in each other in the solid state.
2 The two components remain completely mixed in the solid state. The two components are then said to be soluble in each other in the solid state, the components forming a *solid solution*.
3 On solidifying, the two components show a limited solubility in each other.
4 On solidifying, the elements may combine to form a compound referred to as an intermediate or intermetallic compound.

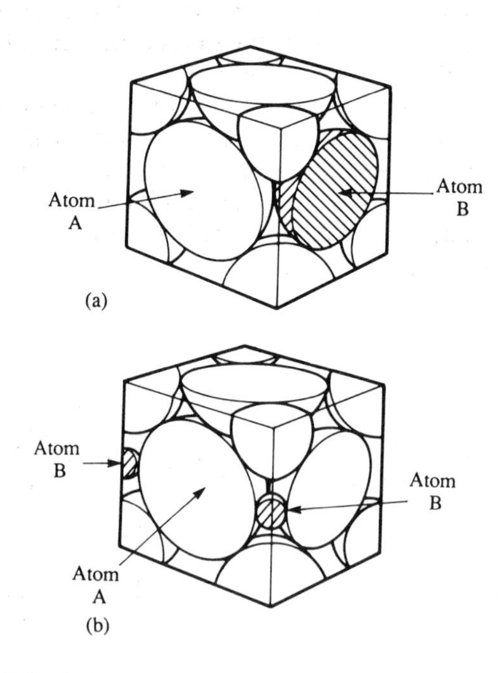

Figure 3.1 Solid solutions (a) A substitutional solid solution of B in A, (b) An interstitial solid solution of B in A

When liquid copper and liquid nickel are mixed the result is complete solubility in the liquid state. When the mixture solidifies a solid solution is produced. This has a face-centred cubic lattice. If the mixture was that of a small amount of nickel added to copper, then the resulting solid solution has a copper face-centred lattice with nickel atoms substituting for some of the copper atoms. Such a solid solution is said to be *substitutional* (Figure 3.1a). The substitution may be ordered with the atoms of the added metal always taking up the same fixed places in the lattice, or it can be disordered with the added atoms appearing virtually at random throughout the lattice.

With the copper-nickel solid solution the copper and nickel atoms are virtually the same size. This is necessary for a substitutional solid solution. Another form of solid solution can, however, occur when the sizes of the two atoms are considerably different. With an interstitial solid solution the added atoms are small enough to fit into the spaces between the atoms in the lattice (Figure 3.1b). Carbon can form an interstitial solid solution with the face-centred cubic form of iron.

Substitutional solid solutions can occur when the atoms of the two materials are either the same size or very similar. If they differ by more than about 15 per cent then the solubility becomes very limited. Copper atoms have a radius of 0.128 nm and nickel atoms 0.125 nm, hence a substitutional solid solution is feasible. The difference in size is about 2 per cent and complete solid solubility is possible. Silver atoms have a radius of 0.144 nm. Since this is about 13 per cent greater than the radius of copper atoms only a very limited substitutional solid solution is feasible for a copper–silver alloy.

Interstitial solutions can form if the atoms of the added material are small enough to fit into the spaces between the solvent atoms or only slightly distort the lattice. This tends to mean a radius of about 0.6 or less of the radius of the solvent atom. Carbon atoms have a radius of 0.077 nm, iron atoms a radius of 0.128 nm. An interstitial solid solution of carbon in iron is possible since the carbon atoms are about this critical size.

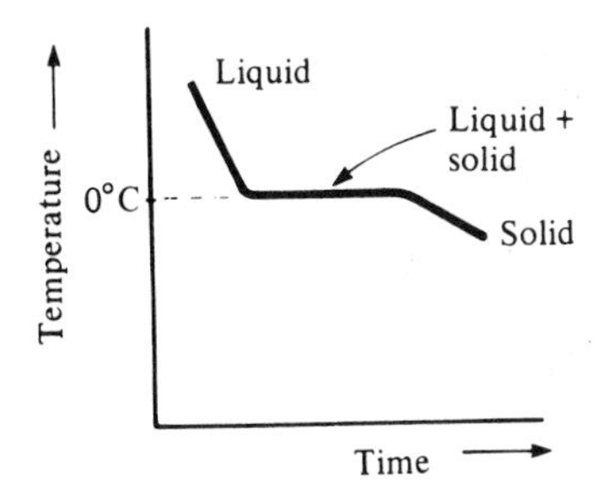

Figure 3.2 Cooling curve for water during solidification

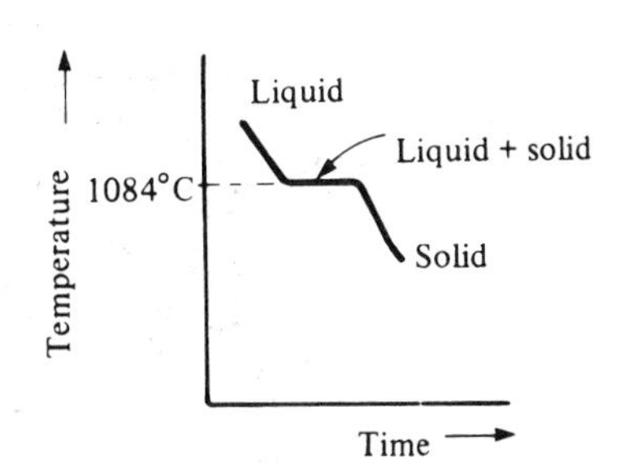

Figure 3.3 Cooling curve for copper during solidification

3.2 Thermal equilibrium diagram

When pure water is cooled to 0°C it changes state from liquid to solid, i.e. ice is formed. Figure 3.2 shows the type of graph that is produced if the temperature of the water is plotted against time during a temperature change from above 0°C to one below 0°C. Down to 0°C the water only exists in the liquid state. At 0°C solidification starts to occur and while solidification is occurring the temperature remains constant. Energy is still being extracted from the water but there is no change in temperature during this change of state. This energy is called *latent heat*. The *specific latent heat of fusion* is defined as the energy taken from, or given to, 1 kg of a substance when it changes from liquid to solid, or solid to liquid, without any change in temperature occurring.

All pure substances show the same type of behaviour as water when they change state. Figure 3.3 shows the cooling graph for copper when it changes state from liquid to solid.

During the transition of a pure substance from liquid to solid, or vice versa, the liquid and solid are both in existence. Thus for the water, while the latent heat is being extracted there is both liquid and ice present. Only when all the latent heat has been extracted is there only ice. Similarly with the copper, during the transition from liquid to solid at 1084°C, while the latent heat is being extracted both liquid and solid exist together.

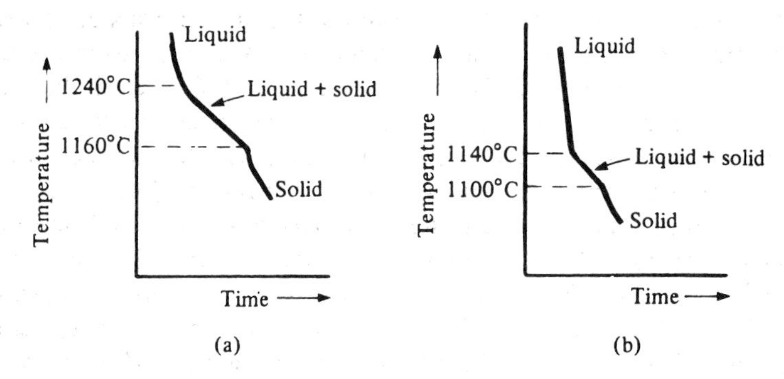

Figure 3.4 Cooling curves for copper–nickel alloys (a) 70 per cent copper–30 per cent nickel, (b) 90 per cent copper–10 per cent nickel

The cooling curves for an alloy do not show a constant temperature occurring during the change of state. Figure 3.4 shows cooling curves for two copper–nickel alloys. With an alloy, the temperature is not constant during solidification. The temperature range over which this solidification occurs depends on the relative proportions of the elements in the alloy. If the cooling curves are obtained for the entire range of copper–nickel alloys a composite diagram can be produced which shows the effect the relative proportions of the

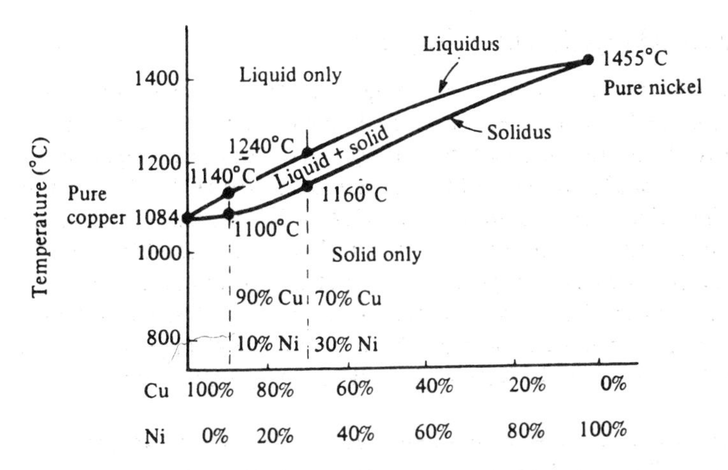

Figure 3.5 Equilibrium diagram for copper–nickel alloys

constituents have upon the temperatures at which solidification starts and that at which it is complete. Figure 3.5 shows such a diagram for copper–nickel alloys.

Thus for pure copper there is a single temperature point of 1084°C, indicating that the transition between liquid and solid takes place at a constant temperature. For 90 per cent copper–10 per cent nickel the transition between liquid and solid starts at 1140°C and terminates at 1100°C when all the alloy is solid. For 70 per cent copper–30 per cent nickel the transition between liquid and solid starts at 1240°C and terminates at 1160°C when all the alloy is solid.

The line drawn through the points at which each alloy in the group of alloys ceases to be in the liquid state and starts to solidify is called the *liquidus line*. The line drawn through the points at which each alloy in the group of alloys becomes completely solid is called the *solidus line*. These liquidus and solidus lines indicate the behaviour of each of the alloys in the group during solidification. The diagram in which these lines are shown is called the *thermal equilibrium diagram*.

The thermal equilibrium diagram is constructed from the results of a large number of experiments in which the cooling curves are determined for the whole range of alloys in the group. The diagram provides a forecast of the states that will be present when an alloy of a specific composition is heated or cooled to a specific temperature. The diagrams are obtained from cooling curves produced by very slow cooling of the alloys concerned. They are slow because time is required for equilibrium conditions to obtain at any particular temperature, hence the term thermal equilibrium diagram.

Phase

A *phase* is defined as a region in a material which has the same chemical composition and structure throughout. A piece of pure copper which is the face-centred cubic structure throughout; has but a single phase at that temperature. Molten copper does, however, represent a different phase in that the arrangement of the atoms in the liquid copper is different from that in the solid copper. A completely homogeneous substance at a particular temperature has only one phase at that temperature. If you take any piece of that homogeneous substance it will show the same composition and structure.

Liquid copper and liquid nickel are completely miscible, as are most liquid metals. The copper–nickel solution is completely homogeneous and thus at the temperature at which the two are liquid there is but one phase present. When the liquid alloy is cooled it solidifies. In the solid state the two metals are completely soluble in each other and so the solid state for this alloy has but one phase. In the case of the 70 per cent copper–30 per cent nickel alloy (Figure 3.4a), the liquid phase exists above 1240°C; between 1240°C and 1160°C there are two phases when both liquid and solid are present. Below 1160°C the copper–nickel alloy exists in just one phase as the two metals are completely soluble in each other and give a solid solution. The thermal equilibrium diagram given in Figure 3.5 for the range of copper–nickel alloys is thus a diagram showing the phase or phases present at any particular temperature of any composition of copper–nickel alloy.

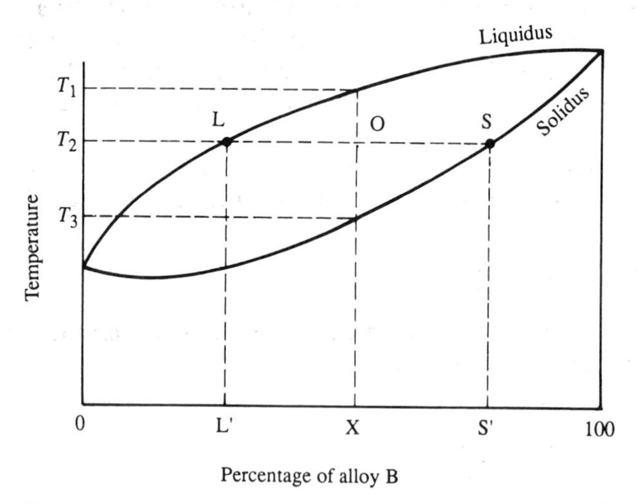

Figure 3.6 The lever rule

The lever rule

Consider an alloy of two components A and B with a composition of X per cent of B and (100 – X) per cent of A. If the thermal equilibrium diagram is of the form shown in Figure 3.6, then at a temperature T_1 the alloy will just be beginning to change from liquid to solid. At temperature T_2 the alloy will consist of a mixture of liquid and solid. To obtain the composition of this liquid and solid a horizontal, isothermal line is drawn at temperature T_2. This line is called a *tie line*. The intercepts of the tie line with the solidus and liquidus give the compositions of the liquid and solid in the mixture. Thus at temperature T_1 the composition of the liquid will be L′ and that of the solid S′. The weight of a piece of the alloy W_A must at this temperature T_1 be equal to the weights of the solid W_S and liquid W_L.

$$W_A = W_S + W_L$$

Similarly, the weight of B in the alloy must be equal to the weight of B in the solid plus the weight in the liquid. The weight of B in the alloy is $(X/100)W_A$, in the solid $(S'/100)W_S$ and in the liquid $(L'/100)W_L$.

$$\frac{XW_A}{100} = \frac{S'W_S}{100} + \frac{L'W_L}{100}$$

Thus by combining these two equations

$$W_A = W_S + Wa_L = \frac{S'W_S}{X} + \frac{L'W_L}{X}$$

$$W_L(X - L') = W_S(S' - X)$$

But $(X-L')$ is the same as the distance LO and $(S'-X)$ the distance OS. The tie line LS can be considered to be a simple beam resting on a pivot at O. The above equation then describes, according to the principle of moments, the condition for balance. Hence the equation is generally referred to as the *lever rule*.

The lever rule can be stated as: the amount of a particular phase multiplied by its lever arm is equal to the amount of the other phase multiplied by its lever arm, i.e.

$$W_L \times LO = W_S \times OS$$

We can use this equation to determine the fraction of an alloy that will be solid at a particular temperature since:

$$W_A = W_S + W_L$$

$$W_A = W_S + W_S \times \frac{OS}{LO}$$

$$\frac{W_S}{W_A} = \frac{LO}{LO + OS}$$

The fraction of the alloy that is solid is proportional to the lever arm LO. Similarly the fraction of the alloy that is liquid is proportional to the length of the lever arm OS.

Thus if a series of tie lines are drawn for different temperatures between T_1 and T_3, the fraction of the alloy that is solid increases from zero at T_1 to 100 per cent at T_3. During this the composition of the solid will change, following the composition plotted out by the solidus. The composition of the liquid also changes, following the composition plotted out by the liquidus. When the cooling rate is slow enough the solid formed early on will change its composition by diffusion so that at any temperature all the solid has the same composition and the solid alloy formed at temperature T_3 has a uniform composition.

Coring

Consider the solidification of a copper–nickel alloy, e.g. 70 per cent copper–30 per cent nickel. Figure 3.7 shows the relevant part of the thermal equilibrium diagram. When the liquid copper–nickel alloy cools to the liquidus temperature, small dendrites of copper–nickel solid solution form. Each dendrite will have the composition of 53 per cent copper–47 per cent nickel. This is the composition of the solid that can be in equilibrium with the liquid at the temperature concerned, the composition being obtained from the thermal equilibrium diagram by drawing a constant temperature line at this temperature and finding the intersection of the line with the solidus. As the overall composition of the liquid plus solid is 70 per cent copper–30 per cent nickel, the dendrites having a greater percentage of nickel mean that the remaining liquid must have a lower concentration of nickel than 30 per cent.

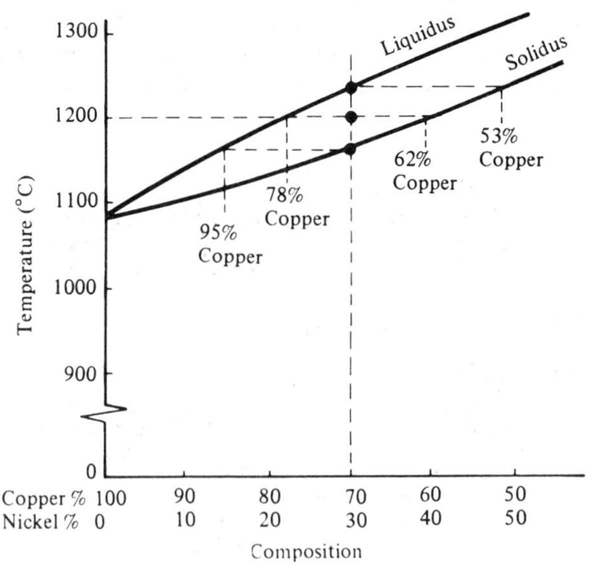

Figure 3.7 Thermal equilibrium diagram for copper-rich alloys of copper-nickel

As the alloy cools further, so the dendrites grow. At 1200°C the expected composition of the solid material would be 62 per cent copper–38 per cent nickel, the liquid having the composition 78 per cent copper–32 per cent nickel. So the percentage of copper has increased from the 53 per cent at the liquidus temperature, while the percentage of nickel has decreased. If the dendrite is to have a constant composition, movement of atoms within the solid will have to occur. The term *diffusion* is used for the migration of atoms. The nickel atoms will have to move outwards from the initial dendrite core and copper atoms will have to move inwards to the core; Figure 3.8 illustrates this process. This diffusion takes time, in fact the process of diffusion in a solid is very slow.

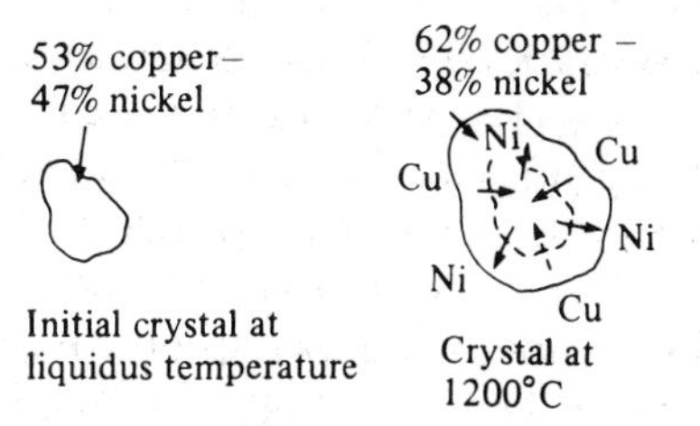

Figure 3.8 Diffusion during crystal growth

As the alloy cools further so the expected composition of the solid material changes until at the solidus temperature the composition becomes 70 per cent copper–30 per cent nickel, the last drop of the liquid having the composition 95 per cent copper–5 per cent nickel. For the solid to have this uniform composition there must have been a diffusion of copper atoms inwards to the core of the dendrite and a corresponding movement of nickel atoms outwards.

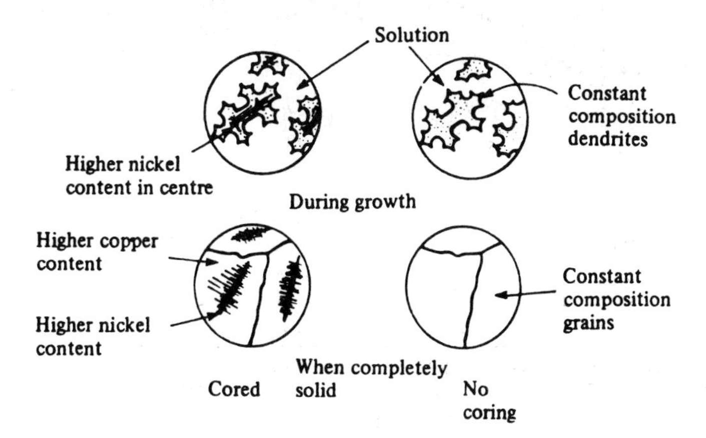

Figure 3.9 Coring with a 70 per cent copper–30 per cent nickel alloy

For this to happen, the entire process of cooling from the liquid must take place very slowly. This is what is meant by equilibrium conditions.

In the normal cooling of an alloy, in perhaps the production of a casting, the time taken for the transition from liquid to solid is relatively short and inadequate for sufficient diffusion to have occurred for constant composition solid to be achieved.

The outcome is that the earlier parts of the crystal growth have a higher percentage of nickel than the later growth parts. The earlier growth parts have, however, a lower percentage of copper than the later growth parts (Figure 3.9). This effect is called *coring*; the more rapid the cooling from the liquid, the more pronounced the coring, i.e. the greater the difference in composition between the earlier and later growth parts of crystals. Figure 3.10 shows the cored structure of the 70 per cent copper–30 per cent nickel alloy. The photograph shows the etched surface of the alloy; as the amount of etching that takes place is determined by the composition of the metal, the earlier and later parts of the dendritic growth are etched to different degrees and so show on the photograph as different degrees of light and dark. The effect of this is to show clearly the earlier parts of the dendritic growth within a crystal grain.

Magnification X 50

Figure 3.10 The cored structure in a 70 per cent copper–30 per cent nickel alloy (Courtesy of the Open University, TS 251/5, © 1973 Open University Press)

Coring can be eliminated after an alloy has solidified by heating it to a temperature just below that of the solidus and then holding it at that temperature for a sufficient time to allow diffusion to occur and a uniform composition to be achieved.

3.3 Forms of equilibrium diagrams for binary alloys

The form of the equilibrium diagram depends on the solubility conditions pertaining in both the solid and liquid states and whether any reactions occur during the liquid to solid transition or during the cooling of the solid.

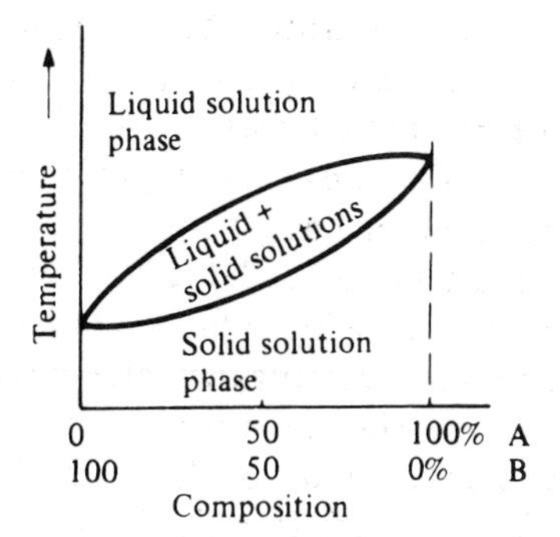

Figure 3.11 Equilibrium diagram for two metals that are completely soluble in each other in the liquid and solid states

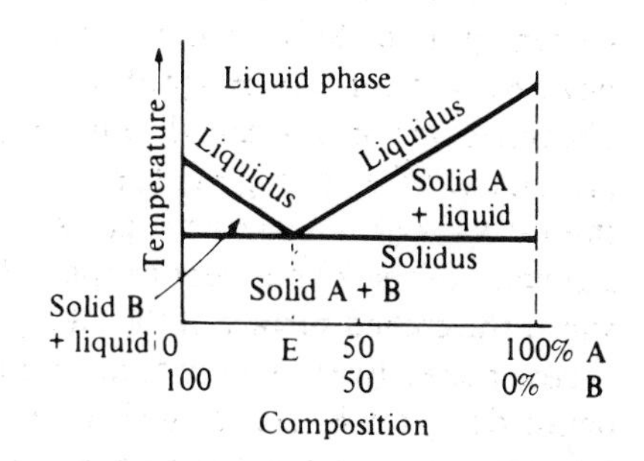

Figure 3.12 Equilibrium diagram for two metals that are completely soluble in each other in the liquid state and completely insoluble in each other in the solid state

Complete solubility in both liquid and solid states

Figure 3.11 shows the form of equilibrium diagram for an alloy involving two components which are soluble in each other in both the liquid and solid states. The equilibrium diagram for the copper–nickel alloy (Figure 3.5) is of this form.

Complete solubility in liquid state and complete insolubility in solid state

Figure 3.12 shows the type of equilibrium diagrams produced when the two alloy components are completely soluble in each other in the liquid state but completely insoluble in each other in the solid state. The solid alloy shows a mixture of crystals of the two metals concerned. Each of the two metals in the solid alloy retains its independent identity. At one particular composition, called the *eutectic composition*, the temperature at which solidification starts to occur is a minimum. At this temperature, called the *eutectic temperature*, the liquid changes to the solid state without any change in temperature (Figure 3.13). The solidification at the eutectic temperature, for the eutectic composition, has both the metals simultaneously coming out of the liquid. Both metals crystallize together. The resulting structure, known as the *eutectic structure*, is generally a laminar structure with layers of metal A alternating with layers of metal B (Figure 3.14).

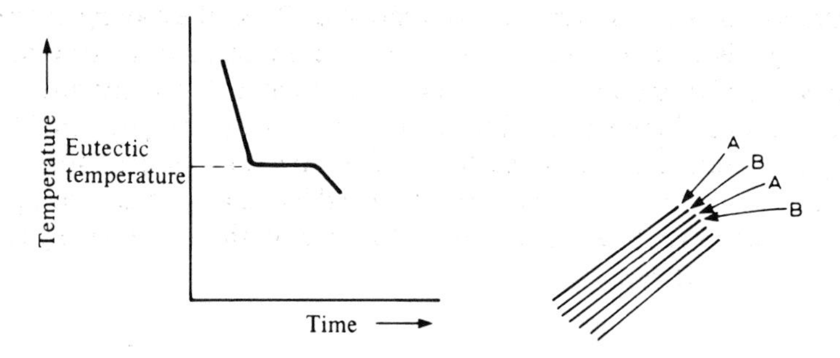

Figure 3.13 Cooling curve for the eutectic composition

Figure 3.14 The laminar structure of the eutectic

The properties of the eutectic can be summarized as:

1. Solidification takes place at a single fixed temperature.
2. The solidification takes place at the lowest temperature in that group of alloys.
3. The composition of the eutectic composition is a constant for that group of alloys.
4. It is a mixture, for an alloy made up from just two metals, of the two phases.
5. The solidified eutectic structure is generally a laminar structure.

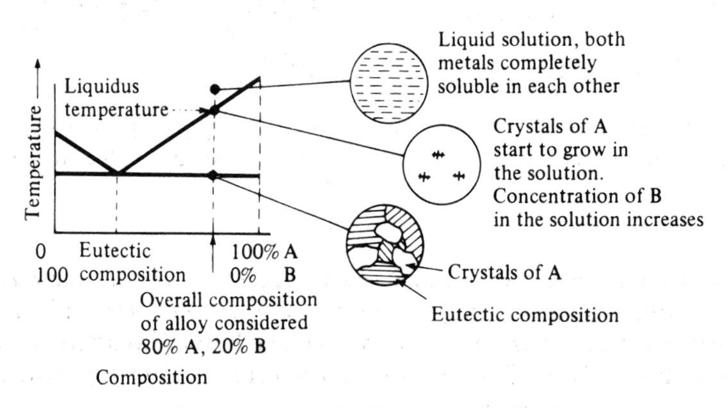

Figure 3.15 Cooling of an 80 per cent A–20 per cent B alloy

Figure 3.15 illustrates the sequence of events that occur when the 80 per cent A–20 per cent B liquid alloy is cooled. In the liquid state both metals are completely soluble in each other and the liquid alloy is thus completely homogeneous. When the liquid alloy is cooled to the liquidus temperature, crystals of metal A start to grow. This means that a metal A is being withdrawn from the liquid, the composition of the liquid must change to a lower concentration of A and a higher concentration of B. As the cooling proceeds and the crystals of A continue to grow so the liquid further decreases in concentration of A and increases in concentration of B. This continues until the concentrations in the liquid reach that of the eutectic composition. When this happens solidification of the liquid gives the eutectic structure. The resulting alloy has therefore crystals of A embedded in a structure having the composition and structure of the eutectic. Figure 3.16 shows the cooling curve for this sequence of events.

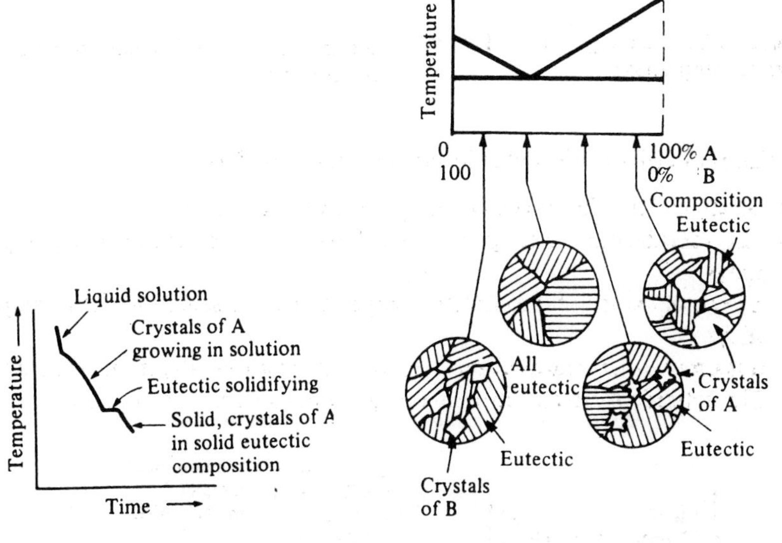

Figure 3.16 Cooling curve for an 80 per cent A–20 per cent B alloy

Figure 3.17 Composition of alloys that are insoluble in each other in the solid state

Apart from an alloy having the eutectic composition and structure when the alloy is entirely of eutectic composition, all the other alloy compositions in the alloy group show crystals of either metal A or B embedded in eutectic structure material (Figure 3.17). Thus for two metals that are completely insoluble in each other in the solid state:

1. The structure prior to the eutectic composition is of crystals of B in material of eutectic composition and structure.
2. At the eutectic structure the material is entirely eutectic in composition and structure.
3. The structure after the eutectic composition is of crystals of A in eutectic composition and structure material.

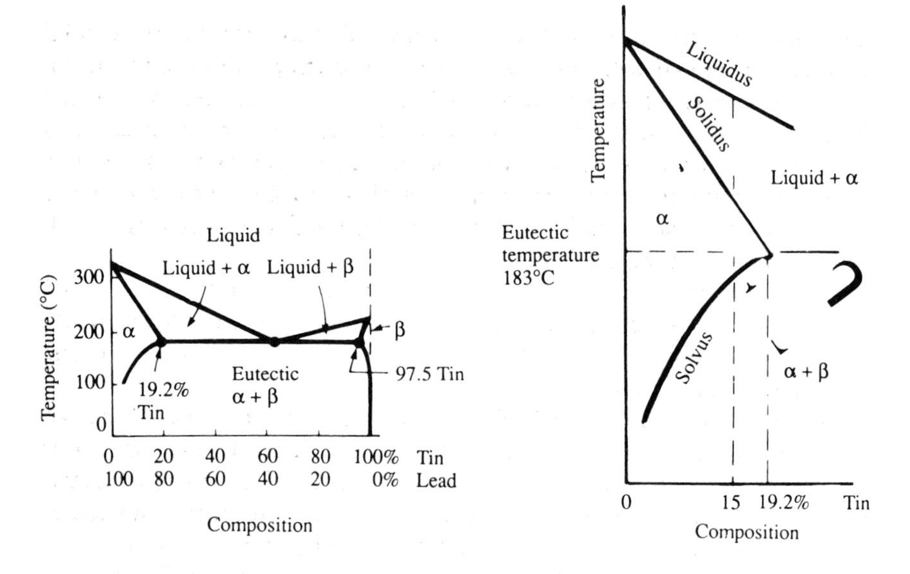

Figure 3.18 Equilibrium diagram for lead–tin alloys

Figure 3.19 The liquidus, solidus and solvus lines

Complete solubility liquid state and partial solubility in solid state

Many metals are neither completely soluble in each other in the solid state nor completely insoluble; each of the metals is soluble in the other to some limited extent. Lead–tin alloys are of this type. Figure 3.18 shows the equilibrium diagram for lead–tin alloys. The solidus line is that line, started at 0 per cent tin–100 per cent lead, between the (liquid+α) and the α areas, between the (liquid+α) and the (α+β) areas, between the (liquid+β) and the (α+β) areas, and between the (liquid+β) and the β areas. The α, the β, and the (α+β) areas all represent solid forms of the alloy. The transition across the line between α and (α+β) is thus a transition from one solid form to another solid form. Such a line is called the *solvus*. Figure 3.19 shows the early part of Figure 3.18 and the liquidus, solidus and solvus lines.

Consider an alloy with the composition 15 per cent–85 per cent lead. When this cools from the liquid state, where both metals are soluble in each other, to a temperature below the liquidus then crystals of the α phase start to grow. The α phase is a solid solution. Solidification becomes complete when the temperature has fallen to that of the solidus. At that point the solid consists entirely of crystals of the α phase. This solid solution consists of 15 per cent tin completely soluble in the 85 per cent lead at the temperature concerned. Further cooling results in no further change in the crystalline structure until the temperature has fallen to that of the solvus. At this temperature the solid solution is saturated with tin. Cooling below this temperature results in tin coming out of solution in another solid solution β. The more the alloy is cooled the greater the amount of tin that comes out of solution, until at room temperature most of the tin has come out of the solid solution. The result is largely solid solution crystals, the α phase, having a low concentration of tin in lead, mixed with small solid solution crystals, the β phase, having a high concentration of tin in lead.

At the eutectic temperature the maximum amount of tin that can be dissolved in lead in the solid state is 19.2 per cent (see Figure 3.19). Similarly the maximum amount of lead that can be dissolved in tin, at the eutectic temperature, is 2.5 per cent.

The eutectic composition is 61.9 per cent tin–38.1 per cent lead. When an alloy with this composition is cooled to the eutectic temperature the behaviour is the same as when cooling to the eutectic occurred for the two metals insoluble in each other in solid state (see Figure 3.15) except that, instead of pure metals separating out to give a laminar mixture of the metal crystals, there is a laminar mixture of crystals of the two solid solutions α and β. The α phase has the composition of 19.2 per cent tin–80.8 per cent lead, the β phase has the composition 97.5 per cent tin–2.5 per cent lead. Cooling below the eutectic temperature results in the α solid solution giving up tin, due to the decreasing solubility of the tin in the lead, and the β solid solution giving up lead, due to the decreasing solubility of the lead in the tin. The result at room temperature is a structure having a mixture of alpha and beta solid solution, the alpha solid solution having a high concentration of lead and the beta a high concentration of tin.

For alloys having a composition with between 19.2 per cent and 61.9 per cent tin, cooling from the liquid results in crystals of the α phase separating out when the temperature falls below that of the liquidus. When the temperature reaches that of the solidus, solidification is complete and the structure is that of crystals of the α solid solution in eutectic structure material. Further cooling results in α solid solution losing tin. The eutectic mixture has the α part of it losing tin and the β part losing lead. The result at room temperature is a structure having the α solid solution crystals with a high concentration of lead and very little tin and some β precipitate, and the eutectic structure a mixture of α with high lead concentration and β with high tin concentration.

For alloys having a composition with between 61.9 per cent and 97.5 per cent tin, cooling from the liquid results in crystals of the β phase separating out when the temperature falls below that of the liquidus. Otherwise the events are the same as those occurring for compositions between 19.2 per cent and 61.9 per cent tin. The result at room temperature is a structure having the β solid solution crystals with a high concentration of tin and very little lead and some α precipitate, and the eutectic structure a mixture of α with high lead concentration and β with high tin concentration.

For alloys having a composition with more than 97.5 per cent tin present, crystals of the β phase begin to grow when the temperature falls below that of the liquidus. When the temperature falls to that of the solidus, solidification becomes complete and the solid consists entirely of β solid solution crystals. Further cooling results in no further change in the structure until the temperature reaches that of the solvus. At this temperature the solid solution is saturated with lead and cooling below this temperature results in the lead coming out of solution. The result at room temperature, when most of the lead has come out of the solid solution, is β phase crystals having a high concentration of tin mixed with α phase crystals with a high concentration of lead.

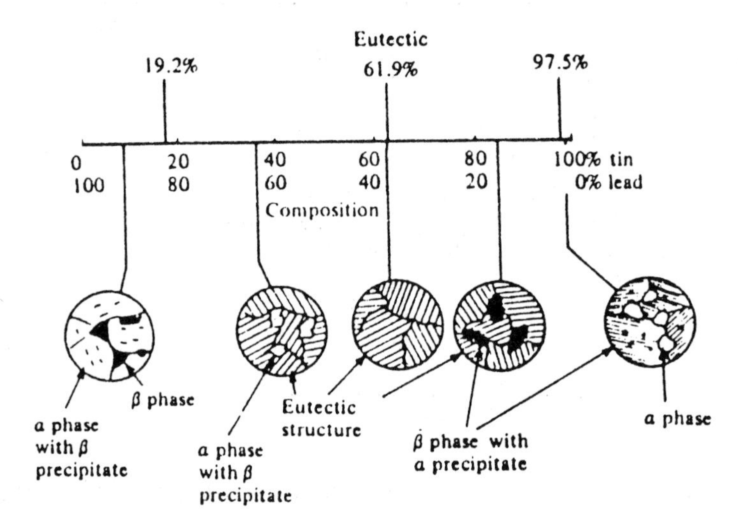

Figure 3.20 Lead–tin alloys

Figure 3.20 shows the types of structure that might be expected at room temperature for lead–tin alloys of different compositions.

Peritectic reaction

A peritectic reaction is said to occur if, during the cooling process from the liquid state, a reaction occurs between the solid that is first produced and the liquid in which it is forming, with the result that a different solid is produced. Figure 3.21 shows the form of a thermal equilibrium diagram when such a reaction occurs. At a composition X, solidification from the liquid starts at temperature T_1 and during cooling to T_p solid of phase α separates out from the liquid. At this temperature a reaction occurs between the solid and the remaining liquid and a new solid, β, is produced.

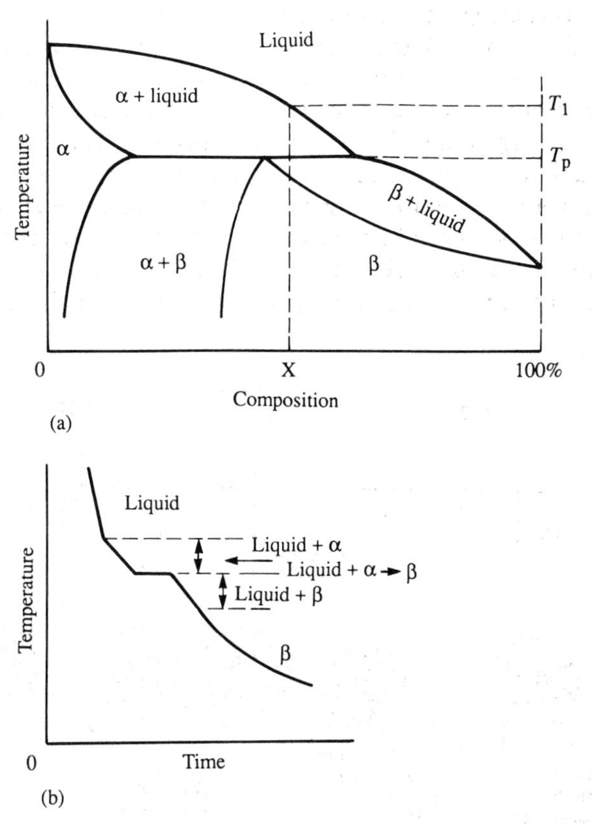

Figure 3.21 Peritectic reaction (a) Thermal equilibrium diagram, (b) Cooling curve for composition X

Eutectoid reaction

A eutectoid reaction is said to occur if, during the cooling process in the solid state, a reaction occurs and one phase of a solid is replaced by a mixture of two other phases. An example of this is shown in the iron–carbon thermal equilibrium diagram (see Figure 11.1, page 191).

3.4 Precipitation

If a solution of sodium chloride in water is cooled sufficiently, sodium chloride precipitates out of the solution. This occurs because the solubility of sodium chloride in water decreases as the temperature decreases. Thus very hot water may contain 37 g per 100 g of water. Cold water is saturated with about 36 g. When the hot solution cools down the surplus salt is precipitated out of the solution. Similar events can occur with solid solutions.

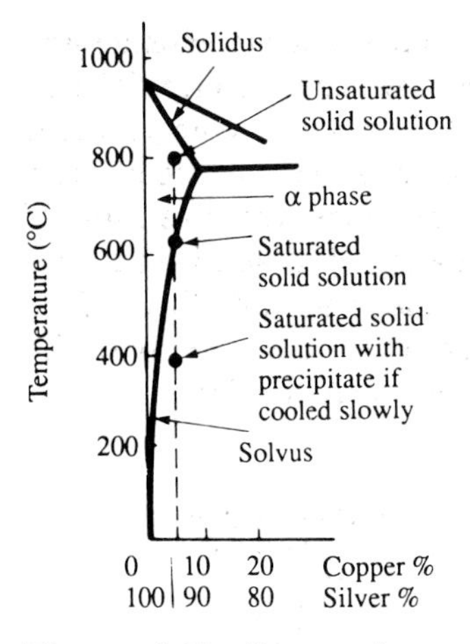

Figure 3.22 Copper–silver thermal equilibrium diagram

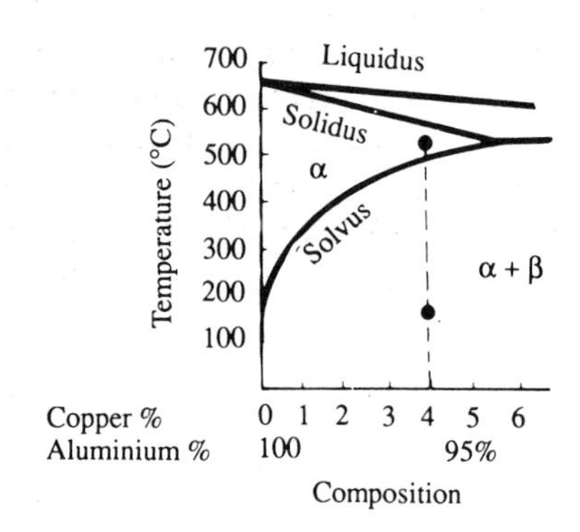

Figure 3.23 Aluminium–copper thermal equilibrium diagram

Figure 3.22 shows part of the copper–silver thermal equilibrium diagram. When the 5 per cent copper–95 per cent silver alloy is cooled from the liquid state to 800°C a solid solution is produced. At this temperature the solid solution is not saturated but cooling to the solvus temperature makes the solid solution saturated. If the cooling is continued, slowly, precipitation occurs. The result at room temperature is a solid solution containing a coarse precipitate.

The above discussion assumes that the cooling occurs very slowly. The formation of a precipitate requires the grouping together of atoms. This requires atoms to diffuse through the solid solution. Diffusion is a slow process; if the solid solution is cooled rapidly from 800°C, i.e. quenched, the precipitation may not occur. The solution becomes *supersaturated*, i.e. it contains more of the α phase than the equilibrium diagram predicts. The result of this rapid cooling is a solid solution, the α phase, at room temperature.

The supersaturated solid solution may be retained in this form at room temperature, but the situation is not very stable and a very fine precipitation may occur with the elapse of time. This precipitation may be increased if the solid is heated for some time (the temperature being significantly below the solvus temperature). The precipitate tends to be very minute particles dispersed throughout the solid. Such a fine dispersion gives a much stronger and harder alloy than when the alloy is cooled slowly from the α solid solution. This hardening process is called *precipitation hardening*. The term *natural ageing* is sometimes used for the hardening process that occurs due to precipitation at room temperature and the term *artificial ageing* when the precipitation occurs as a result of heating.

Figure 3.23 shows part of the thermal equilibrium diagram for aluminium–copper alloys. If the alloy with 4 per cent copper is heated to about 500°C and held at that temperature for a while, diffusion will occur and a homogeneous α solid solution will form. If the alloy is then quenched to about room temperature, supersaturation occurs. This quenched alloy is relatively soft. If now the alloy is heated to a temperature of about 165°C and held at this temperature for about ten hours, a fine precipitate is formed. Figure 3.24 shows the effects on the alloy structure and properties of these processes. The effect is to give an alloy with a higher tensile strength and harder.

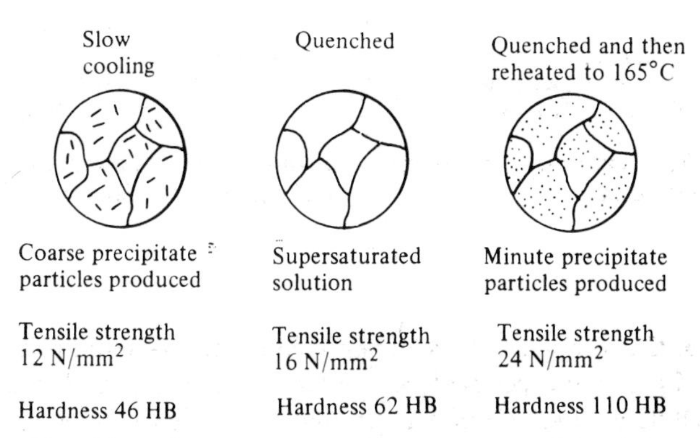

Figure 3.24 The effect of heat treatment for a 96 per cent aluminium–4 per cent copper alloy

Not all alloys can be treated in this way. Precipitation hardening can only occur, in a two-metal alloy, if one of the alloying elements has a high solubility at high temperatures and a low solubility at low temperatures, i.e. the solubility decreases as the temperature decreases. This means that the solvus line must slope as shown in Figure 3.25. Also the structure of the alloy at temperatures above the solvus line must be a single-phase solid solution. The

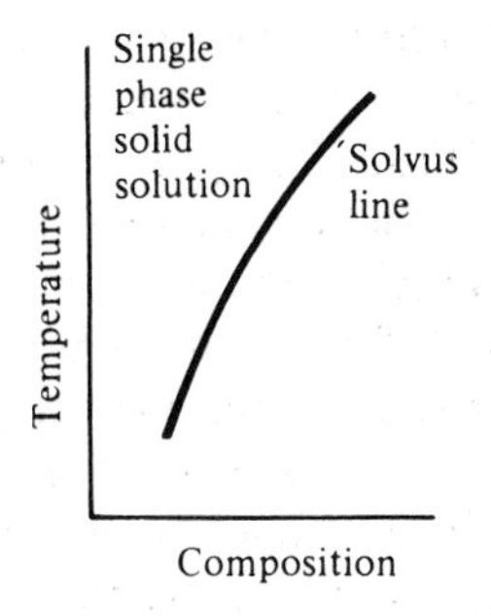

Figure 3.25 The required form of solvus line for precipitation hardening

alloy systems that have some alloy compositions that can be treated in this way are mainly non-ferrous, e.g. copper–aluminium and magnesium–aluminium.

Problems

1 Explain what is meant by a solid solution, distinguishing between substitutional and interstitial solid solutions.

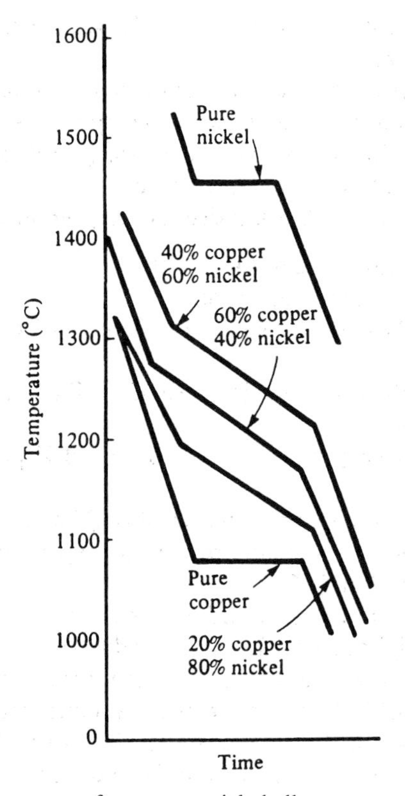

Figure 3.26 Cooling curves for copper–nickel alloys

2 Figure 3.26 shows the cooling curves for copper–nickel alloys. Use these to plot the copper–nickel thermal equilibrium diagram.

3 Use either Figure 3.5 or your answer to Problem 2 to determine the liquidus and solidus temperatures for a 50 per cent copper–50 per cent nickel alloy.

4 Germanium and silicon are completely soluble in each other in both the liquid and solid states. Plot the thermal equilibrium diagram for germanium–silicon alloys from the following data.

Alloy Germanium %	Silicon %	*Liquidus temperature* (°C)	*Solidus temperature* (°C)
100	0	958	958
80	20	1115	990
60	40	1227	1050
40	60	1315	1126
20	80	1370	1230
0	100	1430	1430

5 Explain what is meant by the liquidus, solidus and solvus lines on a thermal equilibrium diagram.
6 Describe the form of the thermal equilibrium diagrams that would be expected for alloys of two metals that are completely soluble in each other in the liquid state but in the solid state are (a) soluble, (b) completely insoluble, (c) partially soluble in each other.
7 The lead–tin thermal equilibrium diagram is given in Figure 3.18

   (a) What is the composition of the eutectic?
   (b) What is the eutectic temperature?
   (c) What will be the expected structure of a solid 40 per cent–60 per cent copper alloy?
   (d) What will be the expected structure of a solid 10 per cent tin–90 per cent copper alloy?
   (e) What will be the expected structure of a solid 90 per cent–10 per cent copper alloy?

8 For (a) a 40 per cent tin–60 per cent lead alloy, (b) an 80 per cent tin–20 per cent lead alloy, what are the phases present at temperatures of (i) 250°C, (ii) 200°C? (See Figure 3.18.)
9 Explain what is meant by coring and the conditions under which it occurs.
10 Explain how precipitation hardening is produced.
11 The relevant part of the aluminium–copper thermal equilibrium diagram is given in Figure 3.23. What type of microstructure would you expect for a 2 per cent copper–98 per cent aluminium alloy after it has been heated to 500°C, held at that temperature for a while, and then cooled (a) very slowly to room temperature, (b) very rapidly to room temperature?
12 Use the information given in the thermal equilibrium diagram for lead–tin alloys, Figure 3.18, for this question.

   (a) Sketch the cooling curves for liquid to solid transitions for (i) 20 per cent tin–80 per cent lead, (ii) 40 per cent tin–60 per cent lead, (iii) 60 per cent tin–40 per cent lead (iv) 80 per cent tin–20 per cent tin alloys.
   (b) Solder used for electrical work in the making of joints between wires has about 67 per cent lead–33 per cent tin. Why is this alloy composition chosen?
   (c) What is the lowest temperature at which lead–tin alloys are liquid?

4

Structure of non-metals

4.1 Polymer structure

The plastic washing-up bowl, the plastic measuring rule, the plastic cup – these are examples of the use of polymeric materials. The molecules in these plastics are very large. A molecule of oxygen consists of just two oxygen atoms joined together. A molecule in a plastic may have thousands of atoms all joined together in a long chain. The backbones of these long molecules are chains of carbon atoms. Carbon atoms are able to bond together strongly to produce long chains of carbon atoms to which other atoms can become attached.

The term *polymer* is used to indicate that a compound consists of many repeating structural units. The prefix 'poly' means many. Each structural unit in the compound is called a *monomer*. Thus the plastic polyethylene is a polymer which has as its monomer the substance ethylene. For many plastics the monomer can be determined by deleting the prefix 'poly' from the name of the polymer. Figure 4.1 shows the basic form of a polymer.

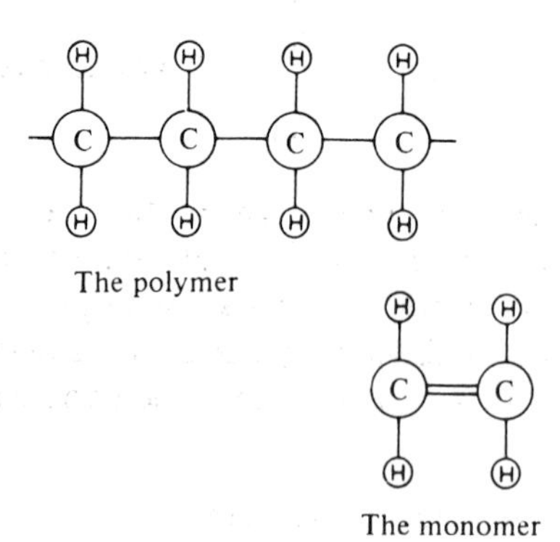

Figure 4.1 The polymer, polyethylene

The figure gives a two-dimensional representation of the polymer molecule. In fact the carbon atoms do not form a straight line because the carbon atom bonds have a directionality which means that the angle between the bonds between a carbon atom and its two neighbouring carbon atoms is 109°. The result is rather a zig-zag form of molecule (Figure 4.2).

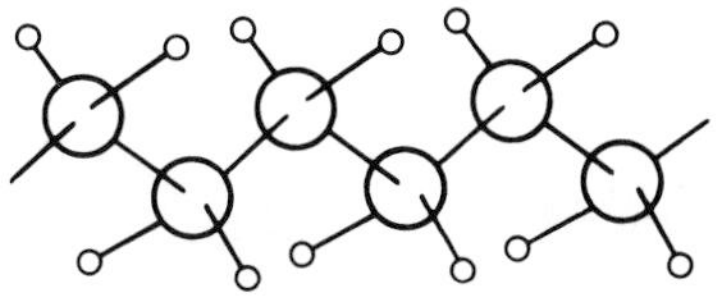

Figure 4.2 The polyethylene molecular chain

The molecular chains of polythene are said to be linear chains. Other possible forms of polymer chains are branched and cross-linked, as illustrated in Figure 4.3 (for simplicity only the carbon atoms are indicated). The term

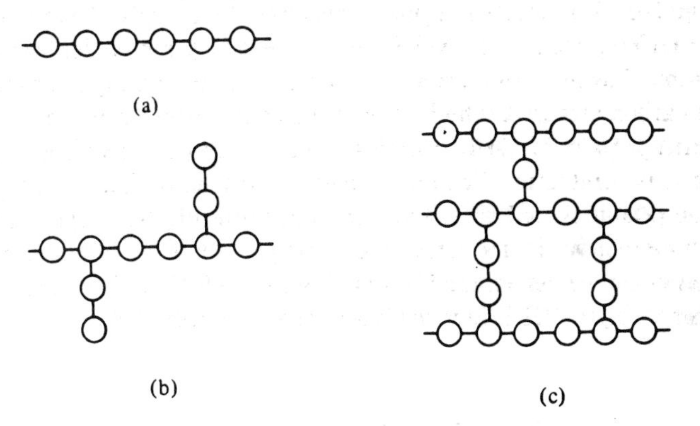

Figure 4.3 (a) Linear polymer chain, (b) Branched polymer chain, (c) Cross-linked polymer

homopolymer is used to describe those polymers that are made up of just one monomer: for instance, polyethylene is made up of only the monomer ethylene. Other types of polymers, *copolymers*, can be produced by combining two or more monomers in a single polymer chain. Figure 4.4 shows four possible types of structure of copolymers based on two monomers.

Thermoplastics, thermosets and elastomers

If you apply heat to a plastic washing-up bowl the material softens. Removal of the heat causes the material to harden again. Such a material is said to be

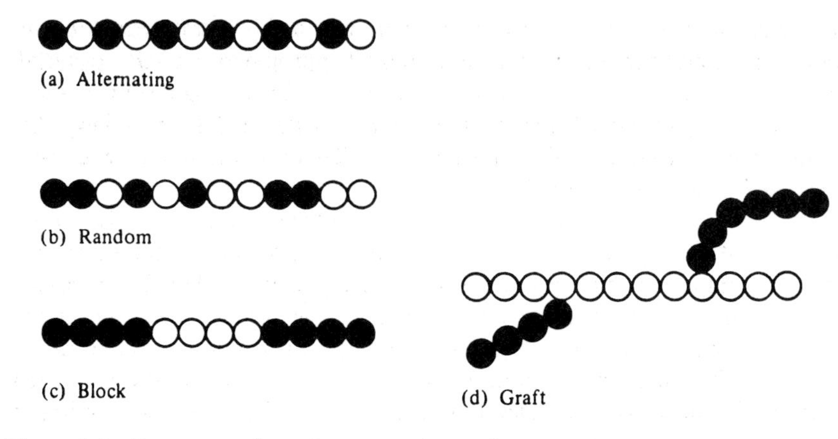

Figure 4.4 Structures of copolymers made up of two monomers

thermoplastic. The term implies that the material becomes 'plastic' when heat is applied.

If you applied heat to a plastic cup you might well find that the material did not soften but charred and decomposed. Such a material is said to be a *thermosetting plastic*.

Another type of polymer is the elastomers. Rubber is an elastomer. An *elastomer* is a polymer which by its structure allows considerable extensions which are reversible.

The thermoplastic, thermosetting and elastomer materials can be distinguished by their behaviour when forces are applied to them to cause stretching. Thermoplastic materials are generally flexible and relatively soft; if heated they become softer and more flexible. Thermosetting materials are rigid and hard and change little with an increase in temperature. Elastomers can be stretched to many times their initial length and still spring back to their original length when released. These different types of behaviours of polymers can be explained in terms of differences in the ways the long molecular chains are arranged inside the material.

Thermoplastics are linear or branched-chain polymers. Thermosets and elastomers are cross-linked polymers. Polyethylene has linear molecular chains and so is a thermoplastic material. It is easily stretched and is not rigid. Because the chains are independent of each other they can easily flow past each other and so the material has a relatively low melting point, no energy being needed to break bonds between chains. The absence of bonds between chains also means that, as none are broken when the material is heated, the removal of heat allows the material to revert to its initial harder state.

Because linear chains have no side branches or cross links with other chains they can move readily past each other. The presence of side branches reduces the ease with which chains can slide past each other and so the material is more rigid and has a higher strength. Thermoplastics with side branches are thus more rigid than linear chain thermoplastics. Polypropylene is such a material, being harder and more rigid than polyethylene. Another consequence of a material having branched chains is that, as they do not pack so readily together in the material as linear chains, the material will generally have a lower density than the linear chain material.

Thermosetting materials are cross-linked polymers with extensive cross-linking and are rigid. As energy is needed to break bonds before flow can occur, thermosetting materials have higher melting points than thermoplastic materials having linear or branched chains. Also the effect of heat is not reversible; when heat causes bonds to break, an irreversible change to the structure of the material is produced. Bakelite is an example of a thermosetting material. It can withstand temperatures up to 200°C, but most thermoplastics are not used above 100°C.

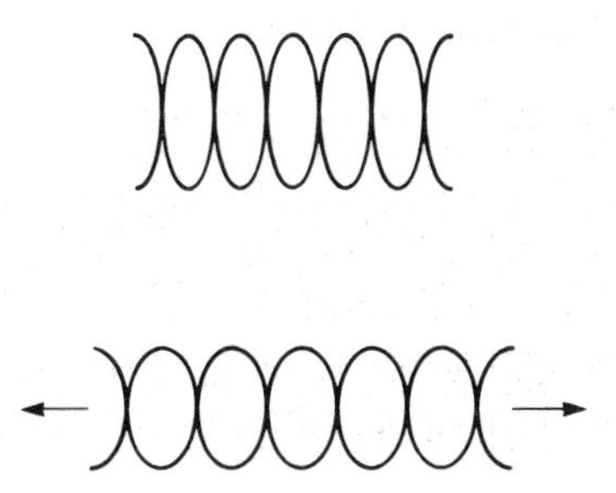

Figure 4.5 Simple model of an elastomer (a) Unstretched, (b) Stretched

Elastomers are highly cross-linked polymers. Between crosslinks the molecular chains are fairly free to move. A simple way of considering the elastomer structure is in the form of a concertina (Figure 4.5). A considerable extension of the concertina is possible without any of the crosslinks having to be broken.

Molecular structure

Polyethylene is, as shown in Figures 4.1 and 4.2, a linear chain with a backbone of carbon atoms, each carbon atom being linked to two hydrogen atoms. The length of the polymer chain is not fixed. In any one sample there will be a range of lengths. The average length can be controlled and thus polythenes with slightly differing properties can be produced. As the length of the molecule increases so does the tensile strength. This is because the longer a chain the more likely it is to become tangled and resist being moved. Since an increase in chain length means an increase in the number of carbon and hydrogen atoms there will be an increase in the weight of a polymer molecule.

```
           H
           |
       H — C — H
           |
       H — C — H
   H       |     H     H     H     H     H     H
   |       |     |     |     |     |     |     |
 — C  —    C  —  C  —  C  —  C  —  C  —  C  —  C —
   |       |     |     |     |     |     |     |
   H       H     H     H     H     H     |     H
                                     H — C — H
                                         |
                                     H — C — H
                                         |
                                     H — C — H
                                         |
                                     H — C — H
                                         |
                                         H
```

Figure 4.6 Branched polyethylene molecule

The properties of a sample of polythene thus depend on the molecular weight.

Polyethylene can also show branching (Figure 4.6) with branches occurring at intervals of every 25 to 100 carbon atoms along the chain, the frequency of the branches depending on the conditions under which the polyethylene was produced. Generally the branches are about two to four carbon atoms long. This tendency to form branches restricts the crystallinity possible with polyethylene.

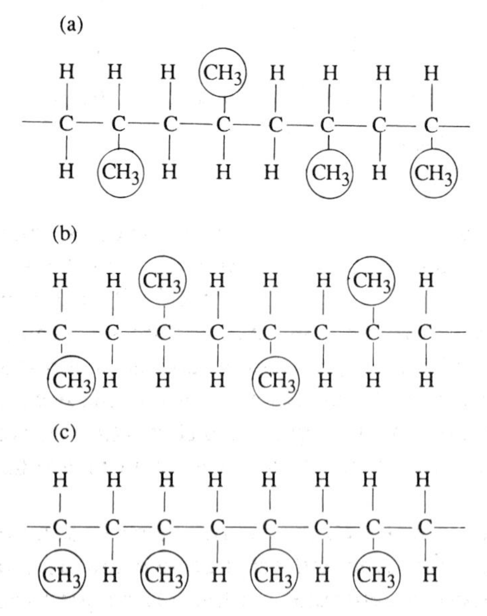

Figure 4.7 Polypropylene (a) Abactic form, (b) Syndiatactic form, (c) Isolactic form

Polypropylene differs from polyethylene only to the extent that every alternate carbon atom has one of its hydrogen atoms replaced by the methyl CH_3 group. This replacement can take a number of forms (Figure 4.7). With the *atactic* form it is random as to which side of the carbon atom the hydrogen atoms are replaced, with the *syndiatactic* form it alternates in a regular manner from one side to the other, while with the *isotactic* form all the atoms replaced are on the same side of the carbon chain. Commercial propylene is generally predominantly isotactic with small amounts of the other forms present.

Polyvinyl chloride differs from polyethylene only to the extent that every alternate carbon atom has one of its hydrogen atoms replaced by a chlorine atom (see Figure 4.15). Commercial polyvinyl chloride is largely atactic.

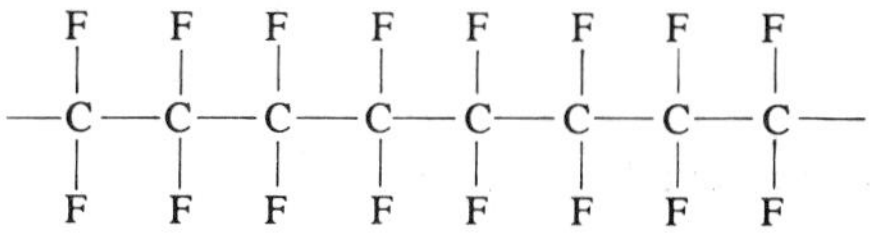

```
   F   F   F   F   F   F   F   F
   |   |   |   |   |   |   |   |
 —C—C—C—C—C—C—C—C—
   |   |   |   |   |   |   |   |
   F   F   F   F   F   F   F   F
```

Figure 4.8 Polytetrafluoroethylene

Polytetrafluoroethylene differs from polyethylene only to the extent that all the hydrogen atoms have been replaced by fluorine atoms (Figure 4.8). As the fluorine atoms are rather large the polymer chain can only accommodate them by twisting itself into a helix.

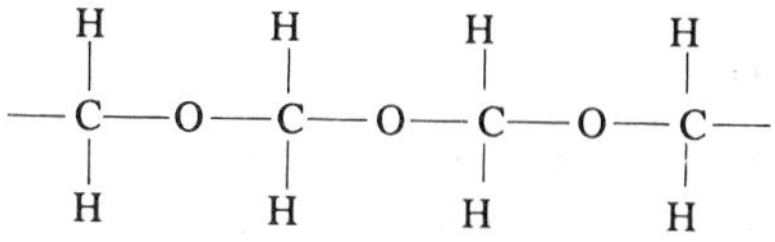

```
   H       H       H       H
   |       |       |       |
 —C—O—C—O—C—O—C—
   |       |       |       |
   H       H       H       H
```

Figure 4.9 Polyoxymethylene

The above polymers are based on the carbon chain backbone with changes being made to atoms linked to this chain. Another form of polymer is produced by making changes to the carbon chain itself. Thus polyoxymethylene is essentially a polythene chain with alternate carbon atoms replaced by oxygen atoms (Figure 4.9). Polyamides, i.e. nylons, consist of amide groups of atoms separated by lengths of (CH_2) chains. The lengths of these chains can be varied to give different forms of nylon. Figure 4.10 shows the form of nylon 6, the 6 refers to the number of carbon atoms in the chain before the chain repeats itself.

Another way of changing the chain structure is to combine two or more mers to give a copolymer. Ethylene and vinyl acetate can be combined to give copolymer, the properties depending on the relative proportions of the two constituents. Increasing the vinyl acetate component increases the flexibility of

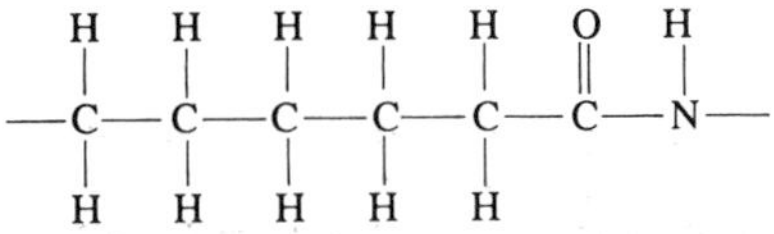

Figure 4.10 The basic unit of nylon 6

the product but decreases the upper temperature at which the product can be used. Large amounts of vinyl acetate produce a polymer with properties more like those of a rubber than a thermoplastic. The copolymer is referred to as EVA and has the basic structure of a linear chain with short branches. It is these short branches which reduce the crystallinity and so make the product more flexible.

Another way of modifying the properties of a polymer is to blend two or more polymers. Polystyrene has thus been mixed with rubbers to produce high impact polystyrene (HIPS). This overcomes the problem of brittleness that occurs with polystyrene alone.

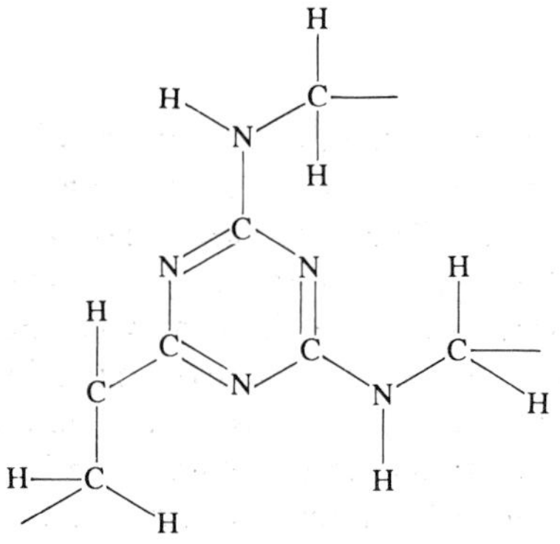

Figure 4.11 Melamine formaldehyde unit

The above have all been thermoplastics or thermoplastic rubber and have been linear or branched polymeric chains. Thermosets have a highly cross-linked structure. An example of this is melamine formaldehyde. Figure 4.11 shows the basic unit which is then repeated a large number of times by other units being connected to it in a random manner.

Elastomers are lightly cross-linked structures. The monomer from which natural rubber polymerizes is isoprene, C_5H_3. The monomer links up to form a long chain molecule with some 20,000 carbon atoms. The cross-linking between chains is produced as a result of a reaction with sulphur. The sulphur breaks some of the double bonds between carbon atoms in the chains to form sulphur cross-links (Figure 4.12).

(a)

```
  H    H   CH3   H    H    H   CH3   H
  |    |    |    |    |    |    |    |
 -C----C====C----C----C----C====C----C
  |              |    |              |
  H              H    H              H
```

(b)

```
  H    H   CH3   H    H    H   CH3   H
  |    |    |    |    |    |    |    |
 -C----C----C----C----C----C----C----C--
  |    |    |    |    |    |    |    |
  H    |    |    H    H    |    |    H
       S    S              S    S
  H    |    |    H    H    |    |    H
  |    |    |    |    |    |    |    |
 -C----C----C----C----C----C----C----C--
  |    |    |    |    |    |    |    |
  H    H   CH3   H    H    H   CH3   H
```

Figure 4.12 Natural rubber (a) An isoprene chain, (b) Two isoprene chains linked by sulphur

The isoprene long chain molecule can exist in two different forms. In one form, referred to as the *cis structure*, the CH_3 groups are all on the same side of the chain (as in Figure 4.12a). This concentration of the CH_3 grouped all on one side of the chain allows the chain to easily bend and coil in a direction which puts the CH_3 groups on the outside of the bend. In the other form, the *trans structure*, the CH_3 groups alternate between opposite sides of the chain. This has the result that the chain cannot easily bend since the CH_3 groups get in the way of each other. Cis-polyisoprene is natural rubber and shows a high degree of flexibility. Trans-polyisoprene is gutta-percha and is inflexible in comparison with the cis form.

Like thermoplastics the properties of elastomers can be changed by copolymerization and blending. Thus, randomly spaced units of styrenye and butadienye gives a linear chain, with virtually no branches. The elastomer, styrene-butadiene rubber (SBR) can have its properties varied by varying the ratio of styrenye to butadienye. This rubber finds a wide variety of uses, e.g. in car tyres.

Ethnic and non-ethnic thermoplastics

It is possible to consider thermoplastics as essentially forming two groups of materials, ethnic and non-ethnic. The *ethnic* family of materials is based on ethylene. A series of polymers having as its basis the polyethylene molecular chain can be derived by substituting other atoms or groups of atoms for some of the hydrogen atoms. Figure 4.13 shows some examples. The term *polyolefins* is used for those ethnic polymers based on polyethylene and polypropylene and the term *vinyls* for those based on vinyl chloride, vinyl acetate and various other vinyl compounds. *Non-ethnic* polymers include such materials as polyamides, polyacetals, polycarbonates and cellulosics. A fundamental difference between these and the ethnic polymers is that with ethnic polymers the backbone of the polymer chain is just carbon atoms while with non-ethnic ones this is not the case.

Generic structure		*Name of polymer*
X X | | —C—C— | | X X	X = H X = F	Polyethylene Polytetrafluorethylene
H X | | —C—C— | | H H	$X = CH_3$ X = Cl $X = C_6H_6$	Polypropylene Polyvinylchloride Polystyrene
H X | | —C—C— | | H Y	$X = CH_3$ $Y = COOCH_3$	Polymethyl methacrylate

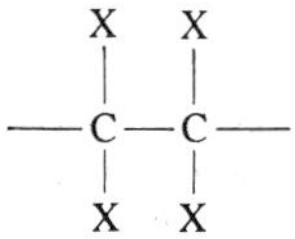

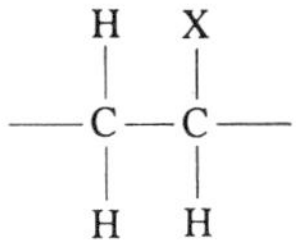

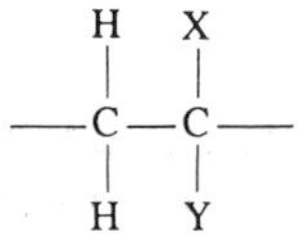

Figure 4.13 Ethnic thermoplastics

Crystallinity in polymers

A crystal can be considered to be an orderly packing-together of atoms. The molecular chains of a polymer may be completely tangled up in a solid with no order whatsoever. Such a material is said to be *amorphous*. There is, however, the possibility that the polymer molecules can be arranged in an orderly manner within a solid. Thus Figure 4.14 shows linear polymer molecules folded to give regions of order. The orderly parts of such polymeric materials are said to be *crystalline*. Because long molecules can easily become tangled up with each other, polymer materials are often only partially crystalline, i.e.

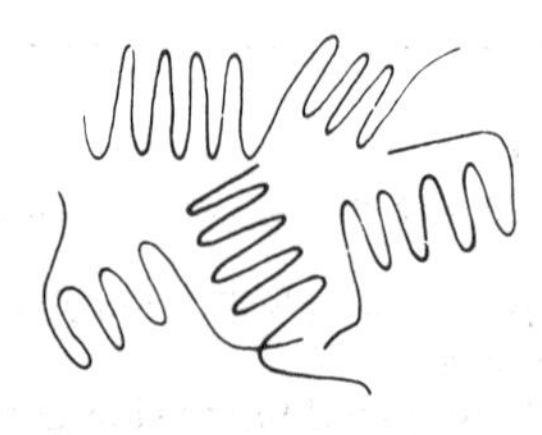

Figure 4.14 Folded linear polymer chains

parts of the material have orderly arrangements of molecules while other parts are disorderly.

Not all polymers can give rise to crystallinity. It is most likely to occur with simple linear chain molecules. Branched polymer chains are not easy to pack together in a regular manner, the branches get in the way. If the branches are completely regularly spaced along the chain then some crystallinity is possible; irregularly spaced branches make crystallinity improbable. Cross-linked polymers cannot be rearranged due to the links between chains and so crystallinity is not possible.

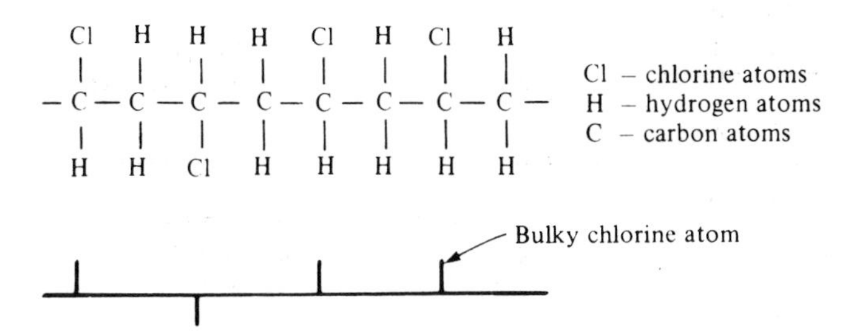

Figure 4.15 The basic form of a PVC molecule

There are many polymers based on the form of the polyethylene molecule. Despite being linear molecules, they do not always give rise to crystalline structures. PVC is essentially just the polyethylene molecule with some of the hydrogen atoms replaced with chlorine atoms (Figure 4.15). The molecule does not, however, give rise to crystalline structures. This is because the chlorine atoms are rather bulky and are not regularly spaced along the molecular chain. It is this lack of regularity which makes packing of the PVC molecular chains too difficult. Polypropylene has a molecule rather like that of polythene but with some of the hydrogen atoms replaced with CH_3 groups. These are, however, regularly spaced along the molecular chain and thus orderly packing is possible and so some degree of crystallinity. Table 4.1 shows the form of the molecular chains and the degree of crystallinity possible for some common polymers.

Table 4.1 *Chain structure of some common polymers*

Polymer	*Form of chain*	*Possible crystallinity (%)*
Polyethylene	Linear	95
	Branched	60
Polypropylene	Regularly spaced side groups on linear chain	60
Polyvinyl chloride	Irregularly spaced bulky chlorine atoms (Figure 4.8)	0
Polystyrene	Irregularly spaced bulky side groups	0

When an amorphous polymer is heated it shows no definite melting temperature but progressively becomes less rigid. The molecular arrangement in an amorphous material is all disorderly, just like that which occurs in a liquid. It is for this reason, i.e. no structural change occurring, that no sharp melting point occurs. For crystalline polymers there is an abrupt change at a particular temperature. Thus if the density of the polymer were being measured as a function of temperature, an abrupt change in density would be seen at a particular temperature. At this temperature the crystallinity of the polymer disappears, the structure changing from a relatively orderly one below the temperature to a disorderly one above it. The temperature at which the crystallinity disappears is defined as being the *melting point* of the polymer. Table 4.2 gives the melting points of some common polymers.

Table 4.2 *Melting points of common polymers*

Polymer	*Crystallinity (%)*	*Melting point (°C)*
Polyethylene	95	138
Polyethylene	60	115
Polypropylene	60	176
Polyvinyl chloride	0	212*
Polystyrene	0	–

*This is not a clear melting point but a noticeable softening region of temperature.

The degree of crystallinity of a polymer affects its mechanical properties. The more crystalline a polymer, the higher its tensile modulus. Thus the linear-chain form of polyethylene in its crystalline form has the molecules closely packed together. Greater forces of attraction e.g. van der Waal forces, can exist between the chains when they are closely packed. The result is a stiffer material, a material with a higher tensile modulus. The branched form of polyethylene has a lower crystallinity and thus a lower tensile modulus. This is because the lower degree of crystallinity means that the molecules are not so closely packed together, an orderly structure being easier to pack closely than a disorderly one (you can get more clothes in a drawer if you pack them in an orderly manner than if you just throw them in). The farther apart the molecular chains, the lower the forces of attraction between them and so the less stiff, i.e. more flexible, the material and hence the lower the tensile modulus. Table 4.3 gives the tensile modulus and tensile strength values for polyethylene with different degrees of crystallinity.

Glass transition temperature

PVC, without any additives and at room temperature, is a rather rigid material. It is often used in place of glass. But if it is heated to a temperature of about

Table 4.3 *Effect of crystallinity on polyethylene properties*

Polymer	*Crystallinity (%)*	*Tensile modulus ($GN\ m^{-2}$ or GPa)*	*Tensile strength ($MN\ m^{-2}$ or MPa)*
Polyethylene	95	21 to 38	0.4 to 1.3
Polyethylene	60	7 to 16	0.1 to 0.3

87°C a change occurs, the PVC becomes flexible and rubbery. The PVC below this temperature gives only a moderate elongation before breaking, above this temperature it stretches a considerable amount and behaves rather like a strip of rubber.

Polythene is a flexible material at room temperature and will give considerable extensions before breaking. If, however, the polythene is cooled to below about −120°C it becomes a rigid material.

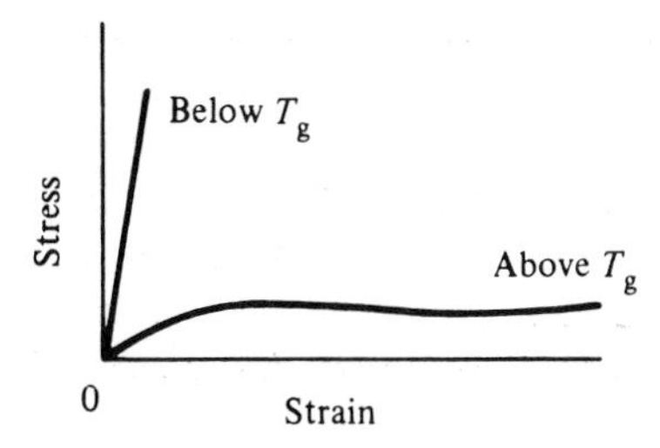

Figure 4.16 Stress/strain graphs for a polymer below and above its glass transition temperature

The temperature at which a polymer changes from a rigid to a flexible material is called the *glass transition temperature*. The material is considered to be changing from a glass-like material to a rubber-like material. At glass transition temperature the tensile modulus shows an abrupt change from a high value for the glass-like material to a low value for the rubber-like material. Figure 4.16 shows the form of the stress/strain graph for a polymer both below and above this transition temperature. Below the transition temperature the material has the type of stress/strain graph characteristic of a relatively brittle substance; above the transition temperature the graph is more like that of a rubber-like material in that the polymer may be stretched to many times its original length.

Below the glass transition temperature only very limited molecular motion is possible; above the glass transition temperature quite a large amount of motion is possible. The extent to which motion is possible at any particular temperature obviously depends on the structure of the polymer molecules. Thus linear chain molecules tend to have lower glass transition temperatures than molecules with bulky side groups or side chains and these, in turn, have lower values than cross-linked polymers. The greater the degree of cross-linking the higher the glass transition temperature.

Table 4.4 shows some typical values of glass transition temperatures.

Table 4.4 *Glass transition temperatures*

Material		T_g(°C)
Thermoplastic:	Polyethylene, low density	−90
	Polypropylene	−27
	Polyvinyl chloride	+80
	Polystyrene	+100
Thermoset:	Phenol formaldehyde	Decomposes first
	Urea formaldehyde	Decomposes first
Elastomer:	Natural rubber	−73
	Butadiene styrene rubber	−58
	Polyurethane	−48

In compounding a plastic, other materials are added to the polymer. These can affect the glass transition temperature. Thus an additive referred to as a plasticizer (see later in this chapter) depresses the glass transition temperature by coming between the polymer molecules and weakening the forces between them.

Heat deflection temperature

For polymeric material a temperature referred to as the *heat deflection temperature* is frequently given in specifications. This is defined as the temperature at which a bar yields under a specified load, when loaded in a specified manner. Slightly different temperatures are given depending on the load and manner of loading specified. The temperatures given by this method indicate the upper temperature at which the material can be used before softening becomes too high. Table 4.5 gives the heat deflection temperatures from tests carried out to BS 2782:M.102C.

Table 4.5 *Heat deflection temperatures*

Material	*Heat deflection temperature* °C
Polyethylene, high density	115
Polypropylene	145
Polyvinyl chloride	78
Polystyrene	95
Acetal	170
Nylon 66	180

4.2 Mechanical properties of polymers

For polymers the variation of the tensile modulus with temperature is of considerable significance in that, for instance, it determines whether the plastic spoon which is stiff at room temperature remains stiff when it is used to stir hot coffee.

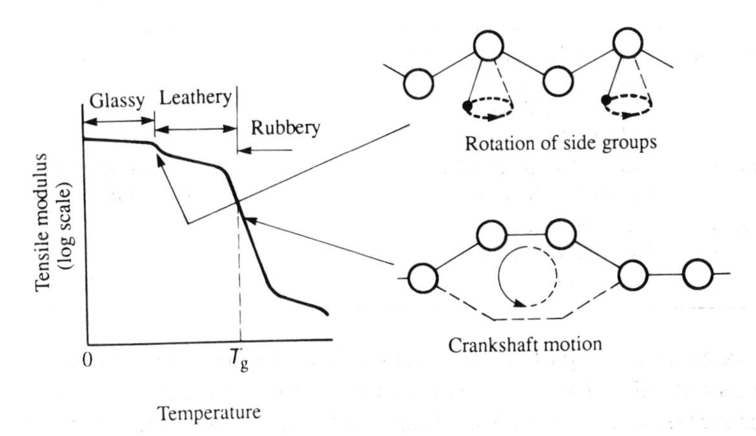

Figure 4.17 Variation of tensile modulus with temperature for an amorphous polymer

The overall way in which the tensile modulus of amorphous polymers changes with time is the same for all such polymers, the differences between polymers being just the actual temperatures at which the changes occur. Figure 4.17 shows the general variation that occurs with temperature – note that the modulus axis of the graph is on a log scale. The material shows a high modulus in what is termed the glassy region. When the temperature is sufficiently increased this modulus begins to decrease and the polymer is said to be leathery. Prior to this the polymer chains were frozen into a tightly-packed, tangled structure and no chain or part of the molecule could easily be moved. The decrease in modulus occurs when rotation of side groups on molecular chains starts to occur. At the glass transition temperature, T_g, complete segments of the polymer chain are able to move, e.g. in a crankshaft motion. This produces a marked drop in tensile modulus and the material is said to be rubbery. At yet higher temperatures large-scale motion of chains becomes possible and the polymer can be considered to have become a liquid.

Time is needed for rotation or motion of polymer chains. A polymer that, at some particular temperature, may be rubbery with a slow application of stress, may be quite glass-like with a faster application of stress. The effects of increasing the rate of application of stress is to make the polymer more brittle and have a higher tensile modulus.

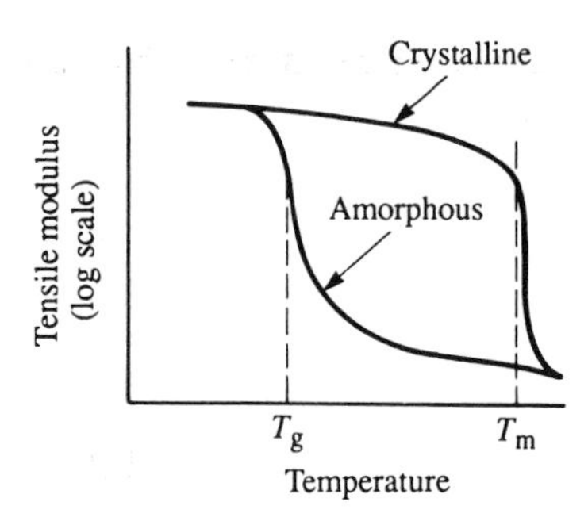

Figure 4.18 Effect of crystallinity on modulus–temperature graph

The effect on the tensile modulus of temperature for a partially crystalline polymer is shown in Figure 4.18. At the glass transition temperature the amorphous part of the polymer becomes rubbery but the crystalline part tends to dominate and the material retains much of its stiffness. At the melting point when the crystalline structure disappears the material becomes amorphous and there is a sharp drop in the modulus.

Heavily cross-linked polymers, i.e. thermosets, show very little change in tensile modulus with temperature and often decompose before reaching their glass transition temperature.

Elastomers, i.e. lightly cross-linked polymers, under normal conditions are amorphous. They have glass transition temperatures below room temperature and so are normally used about T_g and in the rubbery region of Figure 4.17. If the temperature is dropped to below T_g, elastomers become a stiff, glassy, solid with a high tensile modulus.

Temperature and polymer use

Amorphous polymers tend to be used below their glass transition temperature T_g. They are, however, formed and shaped at temperatures above the glass transition temperature when they are in a soft condition. Crystalline polymers are used up to their melting temperature T_m. They can be hot-formed and shaped at temperatures above T_m, or cold-formed and shaped at temperatures between T_g and T_m.

Polythene is a crystalline polymer, with a melting point of 138°C and a glass transition temperature of −120°C for the form that gives 95 per cent crystallinity. The maximum service temperature of polythene items is about 125°C, i.e. just below the melting point. The form of polythene with 60 per cent crystallinity has a melting point of 115°C and a maximum service temperature of about 95°C.

Polystyrene is an amorphous polymer with a glass transition temperature of

100°C. It has a maximum service temperature of about 80°C, i.e. just below the glass transition temperature.

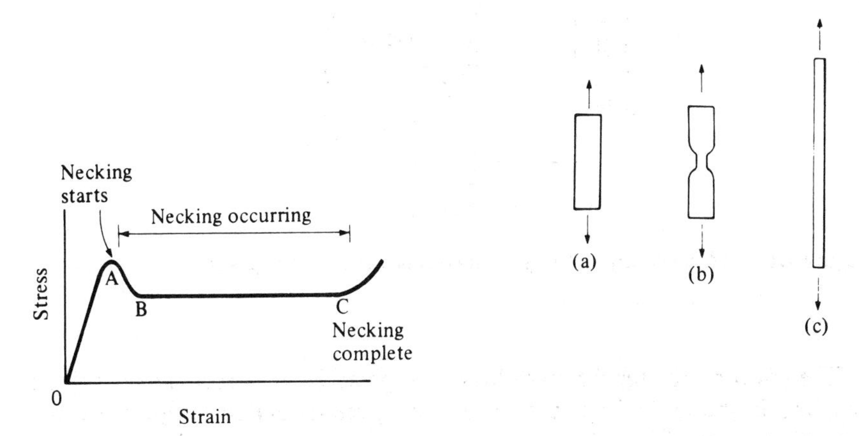

Figure 4.19 Typical stress/strain graph for a crystalline polymer

Figure 4.20 Necking with a polymer (a) Prior to point 'A' and necking starting, (b) Necking starting, (c) The entire material at the necked stage

Orientation

Figure 4.19 shows a typical stress–strain graph for a crystalline polymer, e.g. polythene. When the stress reaches point 'A' the material shows a sudden large reduction in cross-sectional area at some point (Figure 4.20). After this initial necking a considerable increase in strain takes place at essentially a constant stress, as the necked area gradually spreads along the entire length of the material. When the entire piece of material has reached the necked stage an increase in stress is needed to increase the strain further.

The above sequence of events can be explained by considering the orientation of the polymer chains. Prior to necking starting, the polymer chains are folded to give regions of order in the material (as in Figure 4.14). When necking starts the polymer chains unfold to give a material with the chains lying along the direction of the forces stretching the material. As the necking spreads along the material so more of the chains unfold and line up. Eventually when the entire material is at the necked stage all the chains have lined up. Material that has reached this stage shows a different behaviour and is said to be *cold drawn*. It is completely crystalline, i.e. all the chains are packed in a very orderly manner.

The above sequence of events only tends to occur if the material is stretched slowly and sufficient time elapses for the molecular chains to slide past each other. If a high strain rate is used, the material is likely to break without becoming completely orientated. You can try pulling a strip of polythene for yourself and see the changes. (A strip cut from a polythene bag can be used and you can pull it with your hands.)

With crystalline or semi-crystalline polymers, orientation is produced at temperatures both below and above the glass transition temperature. With amorphous polymers orientation is produced at temperatures above the glass transition temperature. Below that temperature an amorphous polymer is too brittle and breaks. The effect of the chains becoming orientated is to give a harder, stronger material. The effect can be considered to be similar to work hardening with metals.

In order to improve the strength of polymer fibres, e.g. polyester fibres, they are put through a drawing operation to orientate the polymer chains. Stretching a polymer film causes orientation of the polymer chains in the direction of the stretching forces. The result is an increase in strength and stiffness in the stretching direction. Such stretching is referred to as uniaxial stretching and the effect as *uniaxial orientation*. The material is, however, weak and has a tendency to split if forces are applied in directions other than those of the stretching forces. For the polyester fibres this does not matter as the forces will be applied along the length of the fibre; with film this could be a serious defect. The problem can be overcome by using a biaxial stretching process in which the film is stretched in two directions at right angles to each other. The film has then *biaxial orientation*.

Rolling through compression rollers is similar in effect to the drawing operation and results in a uniaxial orientated product. Extrusion has a similar effect.

Orientation can be obtained by both cold and hot working processes. Hot working involves working at temperatures just below the glass transition temperature. Under such conditions orientation can be produced without any internal stresses being developed. Cold working, at lower temperature than those of hot working, requires more energy to produce orientation but also results in internal stresses being produced.

If orientated polymers are heated to above their glass transition temperature they lose their orientation. On cooling they are no longer orientated and are in the same state as they were before the orientation process occurred. This effect is made use of with shrinkable films. The polymer film is stretched, and so made longer even when the stretching force is removed. If it is then wrapped around some package and heated, the film contracts back to its initial, prestretched state. The result is a plastic film tightly fitting the package. The film is said to show *elastic memory*.

Changing polymer properties

With metals heat treatment is a method that can be used to change the properties of the material. With polymeric materials there are few instances where heat treatments can be used. The methods used to change the properties of polymeric materials involve changing the chemical reactions used to produce the material. The following are some of the outcomes of such changes.

1 *Increasing the length of chain for a linear polymer* – this increases the tensile strength and stiffness since longer chains more readily become tangled and so cannot so easily be moved.
2 *Introducing large side groups into a linear chain* – this increases the tensile strength and stiffness since the side groups inhibit chain motion.
3 *Producing branches on a linear chain* – this increases the tensile strength and stiffness since the branches inhibit chain motion.
4 *Introducing large groups into the chain* – these reduce the ability of the chain to flex and so increase the rigidity.
5 *Cross-linking chains* – the greater the degree of cross-linking between chains the more chain motion is inhibited and so the more rigid the material.
6 *Introducing liquids between chains* – the addition of plasticers, liquids which fill some of the space between polymer chains, makes it easier for the chains to move and so increases flexibility.
7 *Making some of the material crystalline* – with linear chains a degree of crystallinity is possible. This degree can be controlled. The greater the degree of crystallinity the more dense the material and the higher its tensile strength and modulus.
8 *Including fillers* – the properties of polymeric materials can be affected by the introduction of fillers. Thus, for example, the tensile modulus and strength can be increased by incorporating glass fibres (see Chapter 16 for a discussion of composites). Graphite can improve frictional characteristics etc.
9 *Orientation* – stretching or applying shear stresses during processing can result in polymeric molecules becoming aligned in a particular direction. The properties in that direction will be markedly different to those in a transverse direction.
10 *Copolymerization* – combining two or more monomers in a single polymer chain will change polymer properties, the properties being determined by the ratio of the components.
11 *Blending* – mixing two or more polymers together to form a material will affect the properties, the result depending on the proportions of the polymers blended.

4.3 Additives

The term *plastic* is used to describe a material based on polymers but also including other substances which are added to the polymers in order that the resulting material has the required properties. The following are some of the main types of additives:

1 *Fillers* to modify the mechanical properties, e.g. wood flour, cork dust, chalk, etc. to reduce brittleness and increase the tensile modulus. They also have the effect of reducing the overall cost of the plastic.

2 *Reinforcement* e.g. glass fibres or spheres, to improve the tensile modulus and strength (see Chapter 6 for details of composites).
3 *Plasticizers* to enable molecular chains to slide more easily past each other, hence making the plastic more flexible.
4 *Stabilizers* to enable the material to resist degradation better.
5 *Flame retardants* to improve the fire resistance properties.
6 *Lubricants and heat stabilizers* to assist the processing of the material.
7 *Pigments* and *dyes* to give colour to the material.

Plasticization

An important group of additives are called *plasticizers*. Their primary purpose is to enable the molecular polymer chains to slide more easily past each other.

Internal plasticization involves modifying the polymer chain by the introduction into the chain of bulky side groups. An example of this is the plasticization of polyvinyl chloride by the inclusion of some 15 per cent of vinyl acetate in the polymer chains. These bulky side groups have the effect of forcing the polymer chains farther apart, thus reducing the attractive forces between the chains and so permitting easier flow of chains past each other.

A more common method of plasticization is called *external plasticization*. It involves a plasticizer being added to the polymer, after the chains have been produced. This plasticizer may be a liquid which disperses throughout the plastic, filling up the spaces between the chains. The effect of the liquid is the same as adding a lubricant between two metal surfaces, the polymer chains slide more easily past each other. The effect of the plasticizer is to weaken the attractive forces that exist between the polymer chains. The plasticizer decreases the crystallinity of polymers as it tends to hinder the formation of orderly arrays of polymer chains. The plasticizer also reduces the glass transition temperature. The effect on the mechanical properties is to reduce the tensile strength and increase the flexibility. The effects of plasticizer on the mechanical properties of PVC are given in Table 4.6.

Table 4.6 *The effect of plasticizer on PVC properties*

	Tensile strength ($MN\ m^{-2}$ or MPa)	*Elongation (%)*
No plasticizer	52 to 58	2 to 40
Low pasticizer	28 to 42	200 to 250
High plasticizer	14 to 21	350 to 450

4.4 Ceramics

The term ceramics covers a wide range of materials, e.g. brick, earthenware pots, clay, glasses and refractory materials. Ceramics are formed from a combination of one or more metals with a non-metallic element such as oxygen, nitrogen or carbon. Ceramics are usually hard and brittle, good electrical and thermal insulators and have good resistance to chemical attack. They tend to have a low thermal shock resistance because of their low thermal conductivity and a low thermal expansivity; think of the effect of pouring a very hot liquid into a drinking glass.

Examples of ceramics are silica (an oxide of silicon), magnesium oxide, aluminium oxide, silicon nitride, boron nitride and silicon carbide. The structure of the ceramic can be amorphous, crystalline or a mixture of crystalline and amorphous phases. In some cases a layered structure can be produced, like that of graphite (see Figure 1.13). The bonding between the atoms may be ionic or covalent and where a layered structure occurs there may be van der Waal bonding between the layers.

Structure of ceramics

To illustrate the possible structures of ceramics consider first a simple structure, magnesium oxide. This forms a crystalline structure similar to that of sodium chloride, i.e. common salt (see Figure 1.8). The bonding between magnesium and oxygen atoms in the structure is ionic, each magnesium atom losing two electrons to become negatively charged. Figure 4.21 shows the cubic form of structure taken by the magnesium oxide. The oxygen ions are comparatively large compared with the magnesium ions and so the structure can be considered as a matrix of oxygen ions with the smaller magnesium ions tucked into the spaces between the oxygen ions.

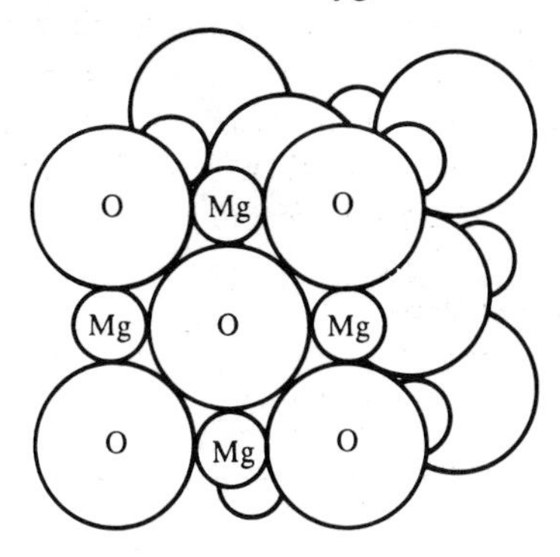

Figure 4.21 Part of the crystal structure of magnesium oxide

Alumina, i.e. aluminium oxide, has a more complex structure. The bonds between the aluminium and the oxygen are a mixture of ionic and covalent. Magnesium oxide has an equal number of magnesium and oxygen atoms. With

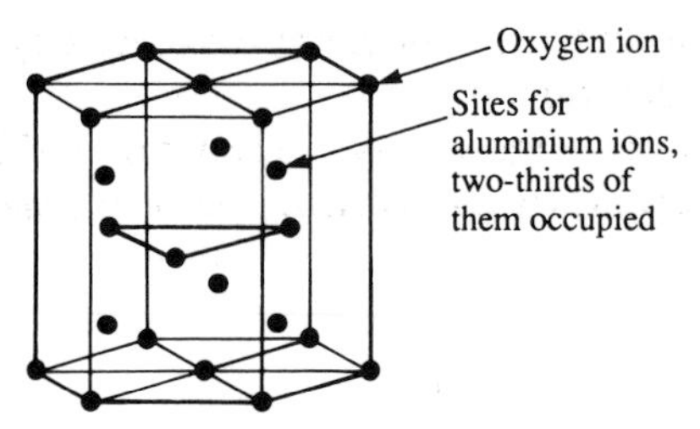

Figure 4.22 Part of the crystal structure of aluminium oxide

aluminium oxide this is not the case, there being two aluminium atoms to three oxygen atoms. As with magnesium oxide the oxygen atoms are much bigger in the structure than the metal, i.e. aluminium, atoms. The structure can thus be considered to be an hexagonal close-packed structure of oxygen ions with interstices between them partially occupied by aluminium ions. The occupation is only partial, in fact two-thirds, because there are the same number of

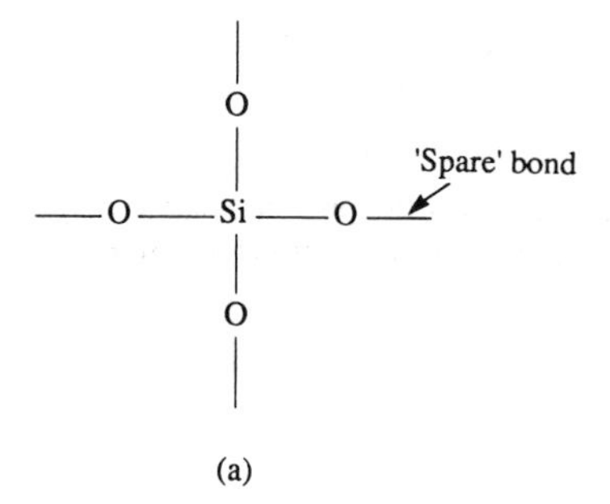

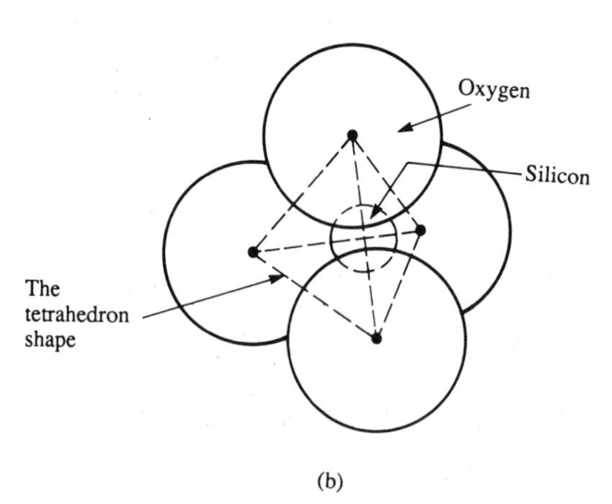

Figure 4.23 (a) The silicon and oxygen bond, (b) The tetrahedron-shaped silicon–oxygen structure

interstices as oxygen ions but less aluminium ions than oxygen ions. Figure 4.22 shows the form of the structure.

Silica forms the basis of a large variety of ceramics. A silicon atom forms covalent bonds with three oxygen atoms to give a tetrahedron-shaped structure

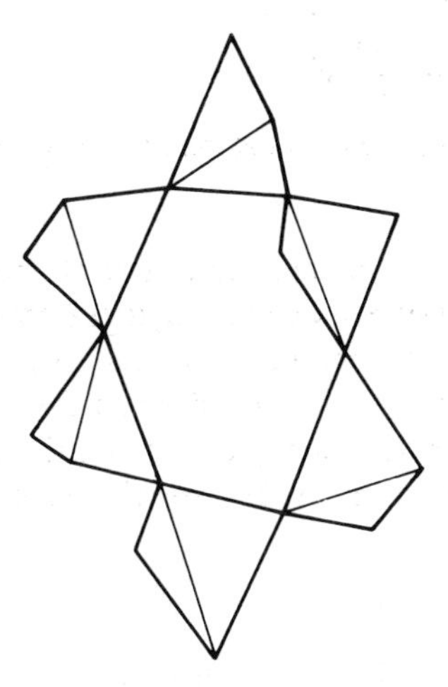

Figure 4.24 The linking of silicon–oxygen tetrahedra to give a three-dimensional structure, this pattern being repeated in a regular manner

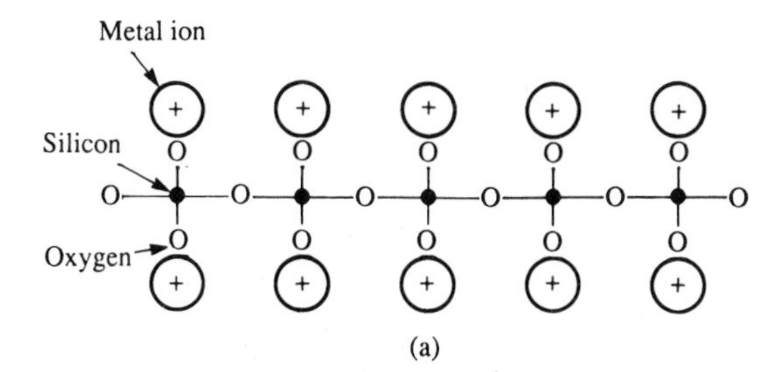

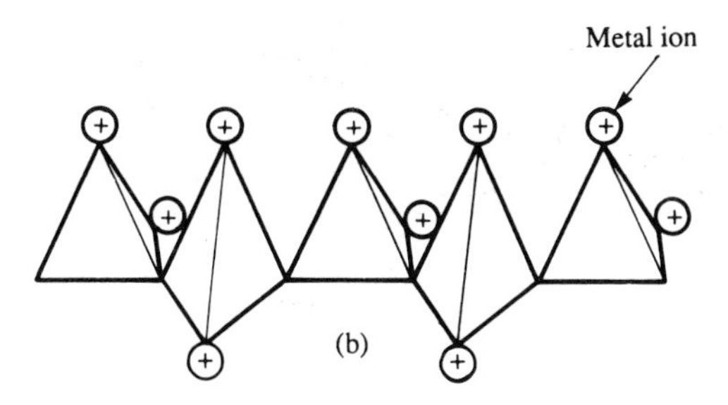

Figure 4.25 The linking up of silicon–oxygen tetrahedra with metal ions to give chains (a) Two-dimensional representation, (b) Three-dimensional representation

(Figure 4.23). This form of structure leaves the oxygen atoms with 'spare' bonds in which they can link up with other silicon-oxygen tetrahedra or metal ions. This bonding can be ionic.

Crystalline silica can be described as a number of tetrahedra joined corner to corner, i.e. an array linked through the 'spare' bonds. A three-dimensional structure occurs (Figure 4.24). Quartz is such a silica structure.

Other structures can be produced by linking tetrahedra in chains and then linking adjacent chains by metallic ions (Figure 4.25). The links in the chains are covalent bonds while the links between adjacent chains are ionic bonds. Another chain structure involves a double chain of tetrahedra with links between these double chains being by metallic ions (Figure 4.26). Asbestos is an example of such a material.

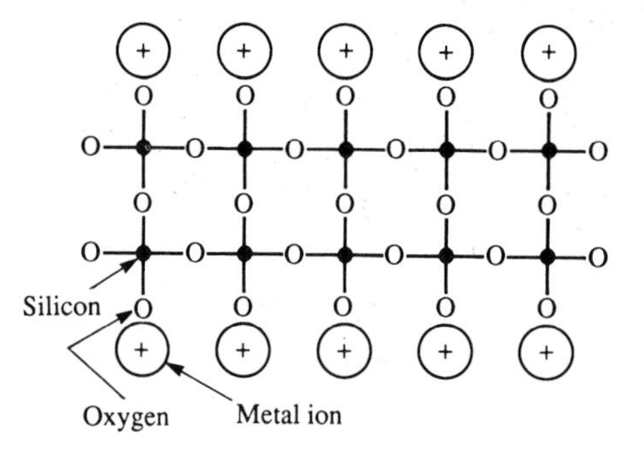

Figure 4.26 Two-dimensional representation of silicon–oxygen tetrahedra forming double chains

The tetrahedra can link up to give sheets instead of chains (Figure 4.27). Such a sheet is the basis of many minerals. Aluminium hydroxide is also a layer structure and such a layer can form a 'laminate' with the silicate sheet to produce a mineral called kaolin.

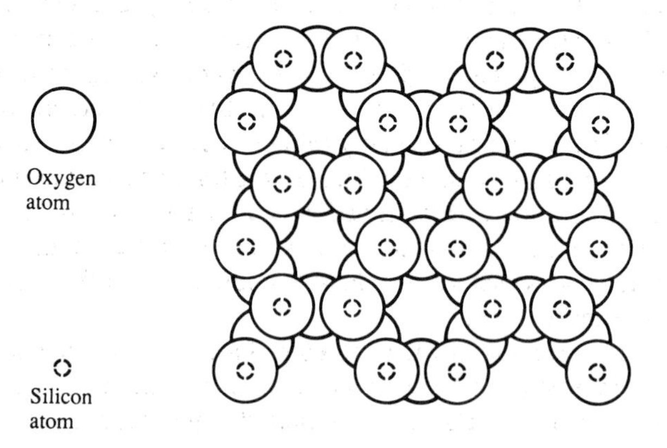

Figure 4.27 A silicate sheet

Glasses

If silica in the molten state is cooled very slowly it crystallizes at the freezing point. However, if the molten silica is cooled more rapidly it is unable to get all its atoms into the orderly arrangement required of a crystal and the resulting solid is a disorderly arrangement required of a crystal and the resulting solid is a disorderly arrangement which is called a glass. Figure 4.28 shows such a structure (compare this with the orderly structure of crystalline silica represented in Figure 4.24). The temperature at which molten silica turns into a glass is called the glass transition temperature T_g and this depends on the rate of cooling from the molten material.

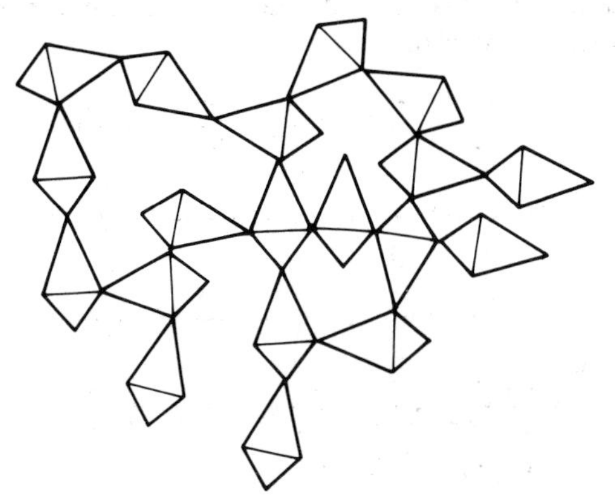

Figure 4.28 The linking up of silicon–oxygen tetrahedra to give a glass

A glass can be considered, in many respects, to be a supercooled liquid. When liquids are cooled and solidify there is a marked change in density when the liquid changes state and latent heat is extracted. This is, however, not the case with glass. Like any other liquid it contracts when the temperature is reduced, due to the rearrangement of atoms leading to closer packing, but instead of the abrupt change in volume when solidification occurs and atoms are much more closely packed, glass just goes on contracting in the same way as if it were a liquid. At a temperature, lower than the normal solidification temperature of the material when crystalline, when the closer packing of the atoms has to virtually cease and further contractions occur due to just a reduction in the thermal vibrations of the atoms there is a change in volume for a glass. This change is smaller than that occurring when a liquid crystallizes. The temperature at which such a change occurs is called the glass transition temperature (Figure 4.29).

Silica is an important constituent of general usage glasses. In addition to the silica other oxides, such as those of sodium and calcium, are added to change the characteristics of the glass, e.g. reduce the melting point or change the colour.

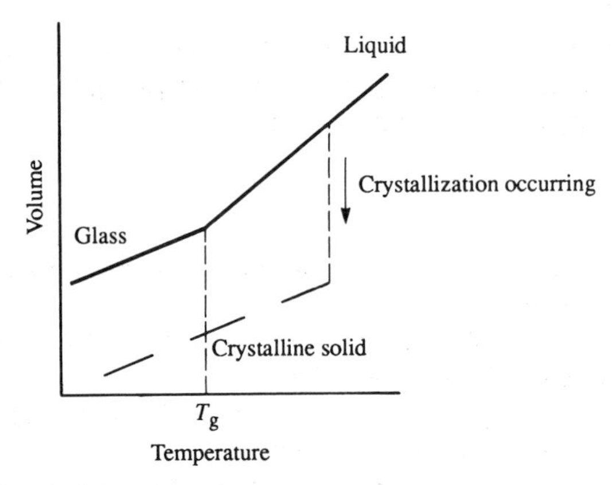

Figure 4.29 The glass transition temperature

Deforming ceramics

Ionic crystals have, like metal crystals, slip planes. In metals these are the closest packed planes with the slip direction being in the closest packed direction. This is also true of ionic crystals, but there are the additional provisos that the slip cannot bring ions of the same sign into juxtaposition and can only occur when it results in ions of the same element replacing each other. Figure 4.30 shows a slip plane in an ionic crystal which is feasible under the above conditions.

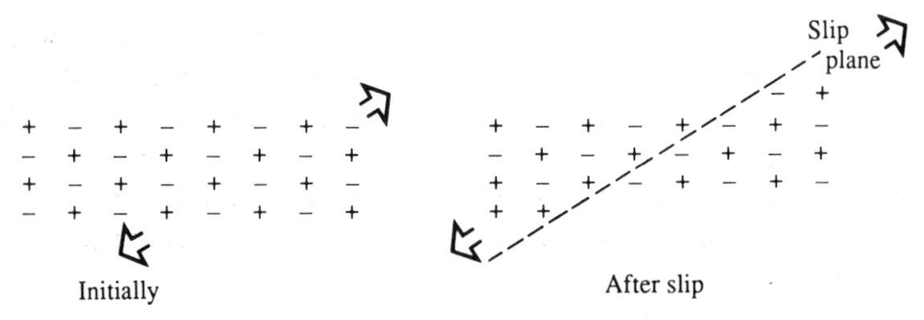

Figure 4.30 Slip with an ionic crystal

These conditions mean that very few planes in an ionic crystal permit slip. Because of this crystalline ceramics are brittle materials. There are just insufficient slip planes for such materials to be ductile.

However, some ceramic materials have a layered structure with ionic or covalent bonds between atoms in a layer but weaker van der Waal bonds between the layers. Because of this such structures permit one layer to be slid relatively easily over another.

Glasses have a disorderly arrangement of atoms and as a result do not have slip planes like crystalline materials with their regular, repetitive arrangement of atoms. Glasses are three-dimensional bonded arrays of atoms and thus the movement of an atom or group of atoms through a glass is barely feasible. Because of this, glasses are brittle materials.

Problems

1 How does the structure of an amorphous polymer differ from that of a crystalline polymer?
2 How does the form of the polymer molecular chain determine the degree of crystallinity possible with the polymer?
3 Explain how the crystallinity of a polymer affects its properties.
4 What is the glass transition temperature?
5 PVC has a glass transition temperature of 87°C. How would its properties below this temperature differ from those above it?
6 A polypropylene article was designed for use at room temperature. What difference might be expected in its behaviour if used at about −15°C. The glass transition temperature for polypropylene is −10°C.
7 When a piece of polythene is pulled it starts necking at one point. Further pulling results in no further reduction in the cross-section of the material at the necked region but a spread of the necked region along the entire length of the material. Why doesn't the material just break at the initial necked section instead of the necking continuing?
8 Why are the properties of a cold drawn polymer different from those of the undrawn polymer?
9 Why are polyester fibres cold drawn before use?
10 Polypropylene is a crystalline polymer with a glass transition temperature of −10°C and a melting point of 176°C. What would be its normal maximum service temperature? At what temperatures would the polymer be hot formed?
11 Explain what is meant by internal and external plasticization.
12 What is the effect of a plasticizer on the mechanical properties of a polymer?
13 Describe the basic properties of ceramics.
14 Explain the effect of the rate of cooling from the molten state on the resulting structure of silica.
15 Describe how silica can form the basis of a range of ceramics with differing properties.
16 Describe the structure of a glass.

PART TWO

Properties of materials

5

Basic properties

5.1 Stress/strain graphs

The behaviour of materials subject to tensile and compressive forces can be described in terms of their stress/strain behaviour, stress being the applied

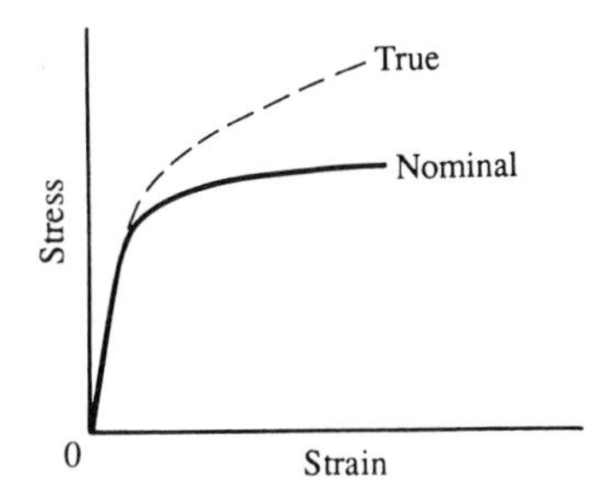

Figure 5.1 Effect of different definitions of stress

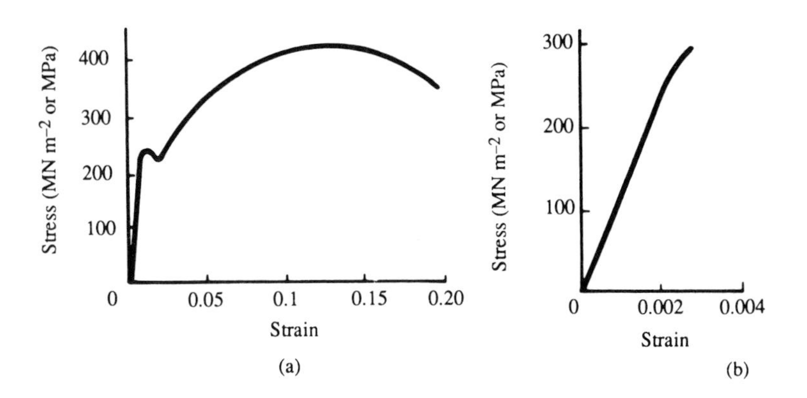

Figure 5.2 Stress/strain graphs for (a) Mild steel, (b) Cast iron

force per unit area and strain the extension, or contraction, per unit original length of material. The area referred to in the stress definition is the area at right-angles to the line of action of the force. In stress/strain graphs it is usually the area before any forces are applied, no account thus being taken of the reduction in area that occurs when the material is stretched or of the increase in area if it is compressed. The true stress at any particular extension is thus different to the nominal stress defined in this way (Figure 5.1).

Figure 5.2 shows stress/strain graphs for typical samples of mild steel and cast iron. For both materials there is a part of the graph where the strain is directly proportional to the stress and for this region a *modulus of elasticity* is defined as

$$\text{modulus of elasticity } E = \frac{\text{stress}}{\text{strain}}$$

This modulus is often referred to as Young's modulus and is a measure of the ease of stretching or compressing a material i.e. the stiffness. The higher the value of the modulus the more stress is needed to produce a given extension, the steeper being the gradient of the stress–strain graph. Some typical values for metals are given in Table 5.1.

Table 5.1 *Typical modulus of elasticity values*

Material	*Modulus of elasticity/GN m^{-2} or GPa*
Mild steel	220
Cast iron	150
Brass	120
Aluminium alloy	70

Thus, aluminium alloy is less stiff and easier to extend than mild steel. For most materials the modulus of elasticity is the same in tension as in compression.

The maximum tensile stress that a material can withstand is known as the *tensile strength*. In the case of the mild steel in Figure 5.2a, the tensile strength is about 400 MN m^{-2} (MPa), for the cast iron in Figure 5.2b about 300 MN m^{-2} (MPa). Some typical values for metals are given in Table 5.2.

Table 5.2 *Typical values of tensile strength*

Material	*Tensile strength/MN m^{-2} or MPa*
Mild steel	400
Cast iron	140–300
Brass	120–400
Aluminium alloy	146–600

The maximum stress which materials can withstand in compression is known as the *compressive strength*, which is not usually the same as the tensile strength.

When a material has a stress applied to it a strain results. When the stress is removed the strain can vanish and is said to have been *elastic* strain. Up to some limiting stress, most materials will exhibit elastic strain, the limiting stress being called the *elastic limit*. Stresses above this limit give rise to a permanent set, i.e. the material does not return to its original length when the stress is removed. Above this limiting stress the material is said to exhibit a combination of elastic and plastic behaviour, a combination because, generally, not all the strain is retained as a permanent set.

For some materials the difference between the elastic limit stress and the stress at which failure occurs is very small, as with the cast iron in Figure 5.2b. The length of the sample after breaking is not much different from the initial length of the sample, very little permanent set having occurred. Such materials are said to be *brittle*, in contrast to a material which suffers a considerable amount of plastic strain before breaking, which is called a *ductile* material, e.g. the mild steel in Figure 5.2a. A measure of ductility is the *percentage elongation*

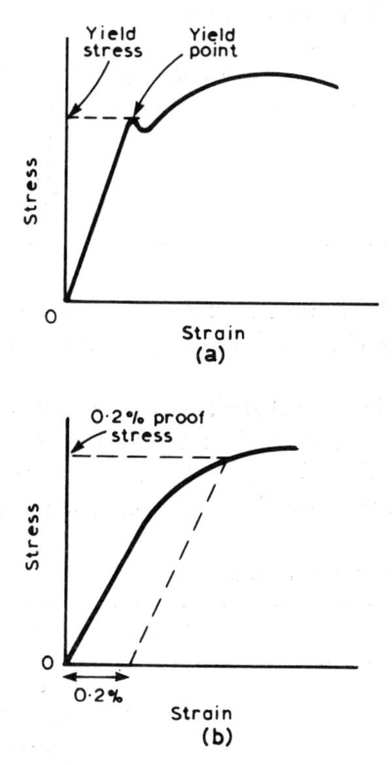

Figure 5.3 Yield and proof stress

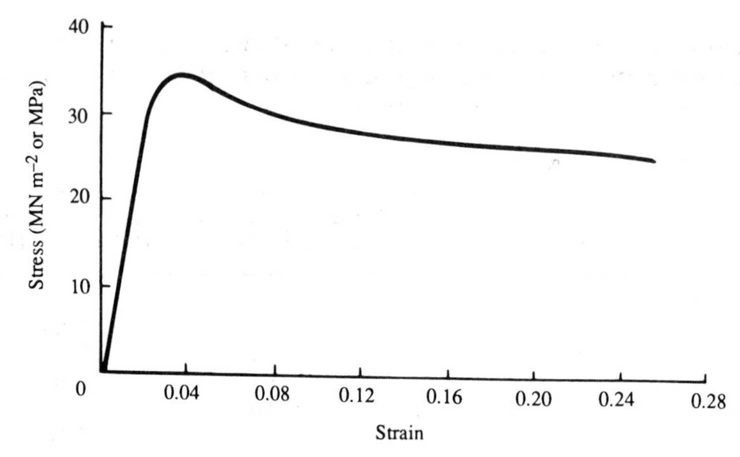

Figure 5.4 Stress/strain graph for ABS polymer, Novodur PK grade (Courtesy of Bayer UK Ltd)

of the sample after breaking; a very ductile material can have a percentage elongation at breaking of 40 per cent, a brittle material perhaps just a few per cent.

$$\text{Percentage elongation} = \frac{(\text{final length} - \text{initial length})}{\text{initial length}} \times 100\%$$

With some materials, e.g. the mild steel in Figure 5.2a, there is a noticeable dip in the stress/strain graph at some specific stress, where the strain increases without any increase in load. The material is said to have yielded and the stress at which this occurs is called the *yield stress*. Some materials, such as aluminium alloys, do not show a noticeable yield stress point, and it is usual here to specify a 0.2 per cent *proof stress* where 0.2 is the percentage off-set (Figure 5.3). Some typical values of yield and proof stress for metals are given in Table 5.3.

In general all the above discussions concerning the stress/strain graphs for metals can be applied to those for plastics. Figure 5.4 shows a stress/strain graph for ABS polymer. The stress/strain relationship for plastics depends on

Table 5.3 *Typical values of yield and proof stress*

Material	*Yield stress/ $MN\ m^{-2}$ or MPa*	*0.2% Proof stress/ $MN\ m^{-2}$ or MPa*
Mild steel	230	–
Stainless steel	620	–
Brass, drawn	–	200
Manganese bronze	–	250
Copper	–	240

the rate at which the strain is applied, unlike metals, where the strain rate is not usually a significant factor. Some typical values of modulus of elasticity and tensile strength for plastics are in Table 5.4.

Table 5.4 *Properties of plastics*

Material (room temp.)	*Modulus of elasticity/ GN m^{-2} or GPa*	*Tensile strength/ MN m^{-2} or MPa*
Thermoplastics		
Polyvinyl chloride (PVC)	2.5–4.0	35–60
Acrylonitrile-butadiene-styrene (ABS)	2.0–3.0	25–50
Polycarbonate (PC)	2.0–3.0	55–65
Polytetrafluoroethylene (PTFE)	0.3–0.6	15–35
Cellulose acetate (CA)	0.5–2.8	13–62
Thermosets		
Phenol formaldehyde (PH)	5.2–6.0	50–55
Polyester (unsaturated) (UP)	2.0–4.4	40–90

The stress/strain properties of plastics change quite significantly when the temperature changes (Figure 5.5). Both the modulus of elasticity and the tensile strength decrease with an increase in temperature. Plastics are generally mixtures of polymers with other materials, these additions can markedly affect the stress–strain graph.

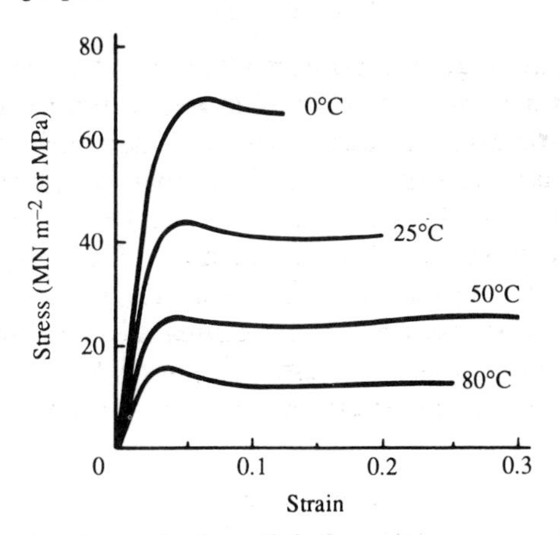

Figure 5.5 Stress/strain graphs for cellulose acetate

Elastomers are a group of materials that produce very large strains. Some typical data for elastomers are given in Table 5.5.

As will be seen from the form of the stress/strain graph in Figure 5.6 the modulus of elasticity is not so useful a quantity for elastomers, as such a modulus can refer to only a very small portion of the stress/strain graph. A

Table 5.5 *Properties of elastomers*

Material	*Tensile strength/ MN m^{-2} or MPa*	*Maximum elongation (%)*
Natural rubber	30	700
Neoprene	28	600
Nitrile	28	550

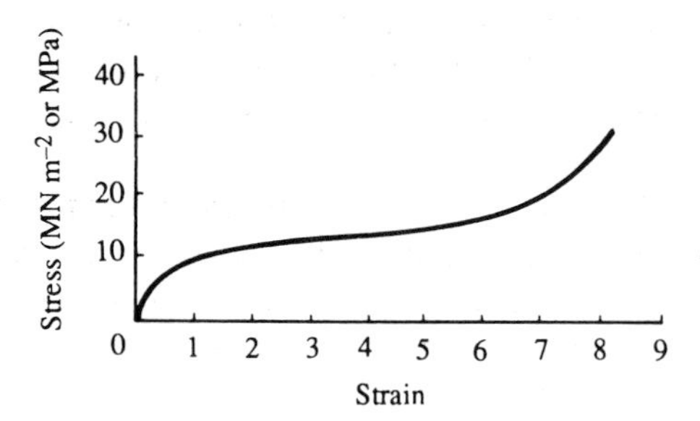

Figure 5.6 Stress/strain graph for rubber

typical modulus would be about 30 MN m^{-2} (MPa). Because for many polymeric materials there is no initial straight line part of the stress/strain graph and a value for the tensile modulus of elasticity cannot be arrived at, a property called the secant modulus is sometimes quoted. The *secant modulus* is obtained by dividing the stress at a value of 0.2 per cent strain by the strain (Figure 5.7).

Another property sometimes quoted for a material is the flexural strength.

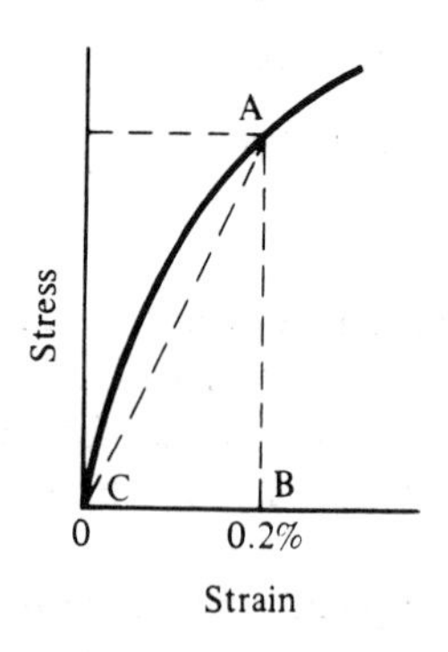

Figure 5.7 The secant modulus is AB/BC, i.e. the slope of the line AC

The *flexural strength* is the outer fibre stress developed when a material is loaded as a simply supported beam and deflected to a specified value of strain.

The tensile test

In a tensile test, measurements are made of the force required to extend a standard size test piece at a constant rate, the elongation of a specified gauge length of the test piece being measured by some form of extensometer. In order to eliminate any variations in tensile test data due to differences in the shapes of test pieces, standard shapes are adopted. Figure 5.8 shows the forms of two standard test pieces, one being a flat test piece and the other a round test piece.

Table 5.6 gives the dimensions for metal test pieces. For the tensile test data for the same material to give essentially the same stress–strain graph, regardless of the length of the test piece used, it is vital that these standard dimensions be adhered to. An important feature of the dimensions is the

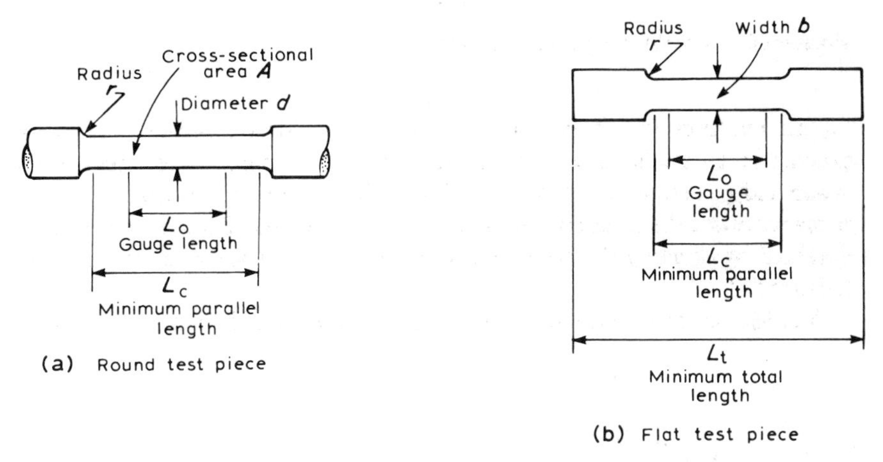

Figure 5.8 Tensile test pieces (a) Round test piece (b) Flat test piece

radius given for the shoulders of the test pieces. Variations in the radii can affect markedly the tensile test data. For example, small radii can cause localized stress concentrations which can result in the test piece failing prematurely. The surface finish of the test piece is important for the same reason.

The round test pieces are said to be proportional test pieces, for which the relationship between the gauge length L_0 and the cross-sectional area A of the gauge length is given by

$$L_0 = 5.65\sqrt{A}$$

Table 5.6 *Dimensions of standard test pieces*

Flat test pieces

M (mm)	L_0 (mm)	r (mm)	L_f (mm)
50	200	25	450
20	200	25	450
25	100	25	300
20	80	25	250
12.5	50	25	200
6	25	12	100
3	12.5	6	50

Note: $L_c = L_0 + \frac{b}{2}$

Round test pieces

d (mm)	A (mm^2)	L_0 (mm)	L_c (mm)	r (mm)
22.56	400	113	124	23.5
15.96	200	80	88	15
13.82	150	69	76	13
11.28	100	56	62	10
10	78.5	50	55	9
7.98	50	40	44	8
5.64	25.0	28	31	6
5	19.6	25	28	6
3.99	12.5	20	22	4

Since $A = \frac{1}{4}\pi d^2$, then to a reasonable approximation

$$L_0 = 5d$$

The reason for the specification of a relationship between the gauge length and the cross-sectional area is in order to give reproducible test results for the same material when different size test pieces are used. For example, when a ductile material is being extended in the plastic region the cross-sectional area does not reduce uniformly but necking occurs (Figure 5.9). The effect of this is to cause most of the further plastic deformation to occur in the necked region where the cross-sectional area is least and the stress consequently the greatest. The percentage elongation can thus differ markedly for different gauge lengths encompassing this necked portion of the test piece, since for the smaller gauge length a *greater proportion* of the gauge length is necked, and so with high elongation, than with the large gauge length. The same percentage elongation is, however, given if the test piece is a proportional test piece and follows the relationship given above.

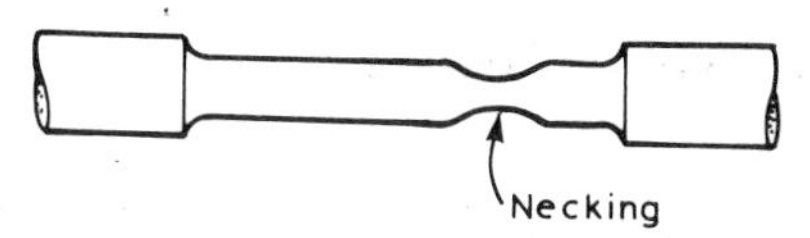

Figure 5.9 Necking of a tensile test piece

5.2 Hardness

The *hardness* of a material can be defined in terms of the ability of a material to resist indentation or scratching. There is no absolute scale of hardness and hardness values are expressed in terms of a scale associated with a particular test. Though some relationships exist between results on one scale and on another, care has to be taken in making comparisons between measurements on different scales since the tests associated with the scales are measuring different things.

The most common form of hardness tests for metals involves standard indentors being pressed into the surface of the material concerned. Measurements associated with the indentation are then taken as a measure of the hardness of the surface. The Brinell test, the Vickers test and the Rockwell test are the main forms of such tests.

With the *Brinell test* a hardened steel ball is pressed for a time of 10 to 15 s into the surface of the material by a standard force (Figure 5.10). After the load and the ball have been removed, the diameter of the indentation is measured. The Brinell hardness number (HB) is obtained by dividing the size of the applied force by the spherical surface area of the indentation:

$$\mathrm{HB} = \frac{\text{applied force}}{\text{spherical surface area of indentation}}$$

with the area being in mm^2 and the force in kgf (1 kgf = 9.8 N). This area can be obtained by calculation from the values of the diameter D of the ball used

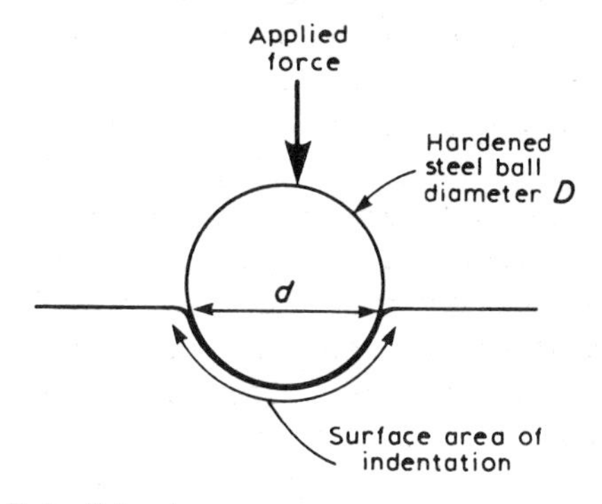

Figure 5.10 The Brinell hardness test

and the diameter d of the indentation.

$$\text{Surface area} = \tfrac{1}{2}\pi D(D - \sqrt{(D^2 - d^2)})$$

Alternatively, tables can be used which give hardness values for different diameter indentations. For example, with a 10 mm ball and a 3000 kg load applied to it, tables give for an impression of diameter 6 mm a hardness of 95 HB, for 5 mm a hardness 143 HB, for 4 mm a hardness 229 HB, for 3 mm a hardness 415 HB.

The diameter of the ball used and the value of the applied force F are chosen, for the British Standard, to give F/D^2 values of 1, 5, 10 or 30, the diameters of the balls being 1, 2, 5 or 10 mm. In principle, the same value of F/D^2 will give the same hardness value, regardless of the diameter of the ball used.

The Brinell test cannot be used with very soft or very hard materials. In one case the indentation becomes equal to the diameter of the ball and in the other case there is no or little indentation on which measurements can be made. The thickness of the material being tested should be at least ten times the depth of the indentation if the results are not to be affected by the thickness of the sample.

The *Vickers test* uses a diamond indenter (Figure 5.11) which is pressed for 10 to 15 s into the surface of the material under test. The result is a square-shaped impression. After the load and indenter are removed, the diagonals of the indentation are measured. The Vickers hardness number (HV) is obtained by dividing the size of the force applied by the surface area of the indentation. The surface area can be calculated from the diagonal values, the indentation being assumed to be a right pyramid with a square base and a vertex angle of 136°. More usually, however, tables are used which relate the lengths of the diagonals to hardness values. The Vickers test has the advantage over the Brinell test of the increased accuracy that is possible in determining the diagonals of the square as opposed to the diameter of a circle. Otherwise it has the same limitations as the Brinell test.

The *Rockwell test* uses either a diamond cone or a hardened steel ball as the indenter. A force is applied to press the indenter in contact with the surface. A

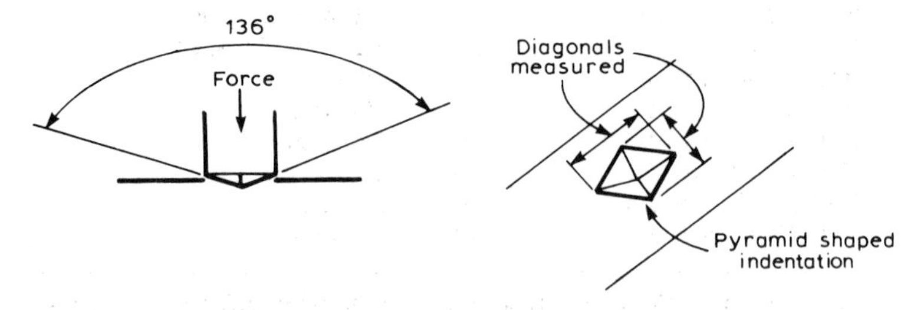

Figure 5.11 The basis of the Vickers hardness test

further force is then applied and causes an increase in depth of the indenter penetration into the material. When the additional force is removed there is some reduction in the depth of the indenter due to the deformation of the material being to some extent elastic. The difference in the depth of the indenter after the initial force is removed and the depth before it was applied is determined. This is the permanent increase in penetration e due to the additional force. The Rockwell hardness number (HR) is

$$\mathrm{HR} = E - e$$

where E is a constant determined by the form of the indenter. For the diamond cone indenter E is 100, for the steel ball E is 130.

There are a number of Rockwell scales, the scale being determined by the indenter and the additional force used. Table 5.7 indicates the scales and the types of materials for which each is typically used. In any reference to the results of a Rockwell test, the scale letter must be quoted. The B and the C scales are probably the most commonly used. For the most commonly used

Table 5.7 *Rockwell hardness scales*

Scale	*Indenter*	*Additional load (kg)*	*Typical applications*
A	Diamond	60	Extremely hard materials, e.g. tool steels
B	Ball 1.588 mm dia.	100	Soft materials, e.g. Cu alloys, Al alloys
C	Diamond	150	Hard materials, e.g. steels, hard cast irons
D	Diamond	100	Medium case hardened materials
E	Ball 3.175 mm dia.	100	Soft materials, e.g. Al alloys, Mg alloys, bearing metals
F	Ball 1.588 mm dia.	60	As E, the smaller ball being more appropriate where inhomogeneities exist
G	Ball 1.588 mm dia.	150	Malleable irons, gun metals, bronzes
H	Ball 3.175 mm dia.	60	Aluminium, lead, zinc
K	Ball 3.175 mm dia.	150	Al and Mg alloys
L	Ball 6.350 mm dia.	60	Plastics
M	Ball 6.350 mm dia.	100	Plastics
P	Ball 6.350 mm dia.	150	Plastics
R	Ball 12.70 mm dia.	60	Plastics
S	Ball 12.70 mm dia.	100	Plastics
V	Ball 12.70 mm dia.	150	Plastics

Note: the diameters of the balls arise from standard sizes in inches, 1.588 mm being 1/16 in, 3.175 mm being 1/8 in, 6.350 mm being 1/4 in, and 12.70 mm being 1/2 in.

Table 5.8 *Rockwell superficial hardness scales*

Scale	*Indenter*	*Additional load (kg)*
15-N	Diamond	15
30-N	Diamond	30
45-N	Diamond	45
15-T	Ball 1.588 mm dia.	15
30-T	Ball 1.588 mm dia.	30
45-T	Ball 1.588 mm dia.	45

Note: the N scales are used for materials that if thick enough would have been tested on the C scale, the T-scales for those on the B scale.

indenters, the size of the indentation is rather small. Localized variations of structure, composition and roughness can thus affect the results. The Rockwell test is however more suitable for workshop or 'on site' use as it is less affected by surface conditions than the Brinell or Vickers tests, which require flat and polished surfaces to permit accurate measurements.

The Rockwell test described above cannot be used with thin sheet. However, a variation of the test, called the *Rockwell superficial hardness test*, can be used with thin sheet. Smaller forces are used and the depth of indentation is measured with a more sensitive device. Table 5.8 indicates the scales given by this test.

The Brinell, Vickers and Rockwell tests can be used with plastics. The Rockwell test with its measurement of penetration depth rather surface area of indentation is more widely used. Scale R is a commonly used scale., e.g. nylon 6 (Durethan 30S) has a Rockwell hardness on scale R of 120.

Another measure that is used with plastics is the *softness number*. This involves an indenter, a ball of diameter 2.38 mm, being pressed against the plastic by an initial force of 0.294 N for 5 s and then an additional force of 5.25 N for 30 s. The difference between the two penetration depths is measured. This difference expressed in units of 0.01 mm becomes the softness value. Thus a penetration depth difference of 0.05 mm is a softness number of 5. The test is carried out at a temperature of $23 \pm 1\,^{\circ}\mathrm{C}$. Another form of this test is the *Shore durometer*. This involves an indenter in the form of either a truncated cone with a flat end or a spherically-ended cone.

One form of hardness test, called the *Moh scale*, is based on assessing the resistance of a material to being scratched. Ten materials are used to establish a scale, the arrangement being that each one in the list will scratch the one preceding it but not the one that succeeds it. The list is

1 Talc
2 Gypsum

3 Calcite
4 Fluorspar
5 Apatite
6 Felspar
7 Quartz
8 Topaz
9 Corundum
10 Diamond

Ten styli of the materials are used for the test. The hardness number of a material is one number less than that of the substance that just scratches it.

There are no simple theoretical relationships between the hardness values given by the Brinell, Vickers and Rockwell tests, though some approximate, experimentally derived relationships have been obtained. Different relationships, however, hold for different metals. Table 5.9 shows the conversions that can be used between the different tests for steels. Up to a hardness value of about 300, the Vickers and Brinell values are almost identical.

Table 5.9 *Hardness scale comparisons for steels*

Brinell value	*Vickers value*	*Rockwell values* B	*Rockwell values* C
112	114	66	
121	121	70	
131	137	74	
140	148	78	
153	162	82	
166	175	86	4
174	182	88	7
183	192	90	9
192	202	92	12
202	213	94	14
210	222	96	17
228	240	98	20
248	248	102	24
262	263	103	26
285	287	105	30
302	305	107	32
321	327	108	34
341	350	109	36
370	392		40
390	412		42
410	435		44
431	459		46
452	485		48
475	510		50
500	545		52

There is an approximate relationship between hardness values and tensile strengths. Thus for annealed steels the tensile strength in MN/mm^2 (MPa) is about 3.54 times the Brinell hardness value, and for quenched and tempered steels 3.24 times the Brinell hardness value. For brass the factor is about 5.6 and for aluminium alloys about 4.2.

5.3 Impact tests

Impact tests are designed to give a measure of the response of a material to a sudden high rate of loading. There are two main forms of such test, the *Izod* and the *Charpy* tests. In both tests a heavy pendulum swings through an arc and strikes a notched test piece; Figure 5.12 shows the basic arrangement for the Izod test. After breaking the test piece, the pendulum continues its swing but as some energy has been used to break the test piece the pendulum does not swing back up to the same height as that from which it started. The height to which the pendulum swings is thus a measure of the energy needed to break the test piece. The energy lost is mgh, where m is the mass of the pendulum, g the acceleration due to gravity and h the difference in height between the starting height of the pendulum and its maximum height after breaking the test piece. This energy is quoted for standard size test pieces and forms of notch.

Figure 5.13 shows the basic standard form of Izod test piece for a metal, Figure 5.14 that for a plastic. The test pieces are, in the case of metals, either 10 mm square or 11.4 mm diameter and for plastics either 12.7 mm square or 12.7 mm by 6.4 to 12.7 mm, depending on the thickness of the material concerned. With metals the pendulum strikes the test piece with a speed of between 3 and 4 m/s and with plastics the speed is 2.44 m/s. The Charpy test differs from the Izod test in the mounting of the test piece, Figure 5.15 showing the mounting (compare this with that shown for the Izod test in figure

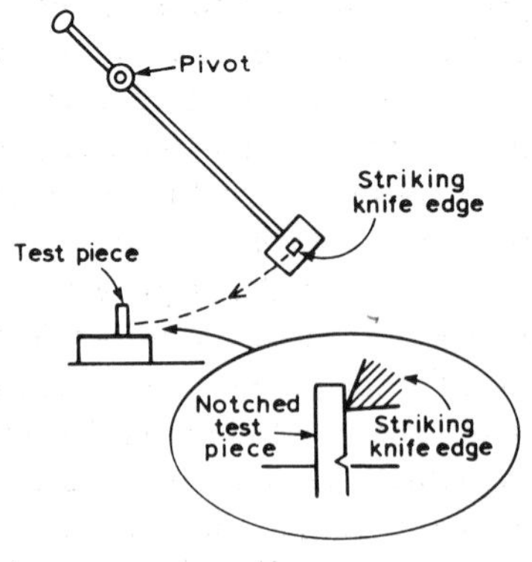

Figure 5.12 The Izod test

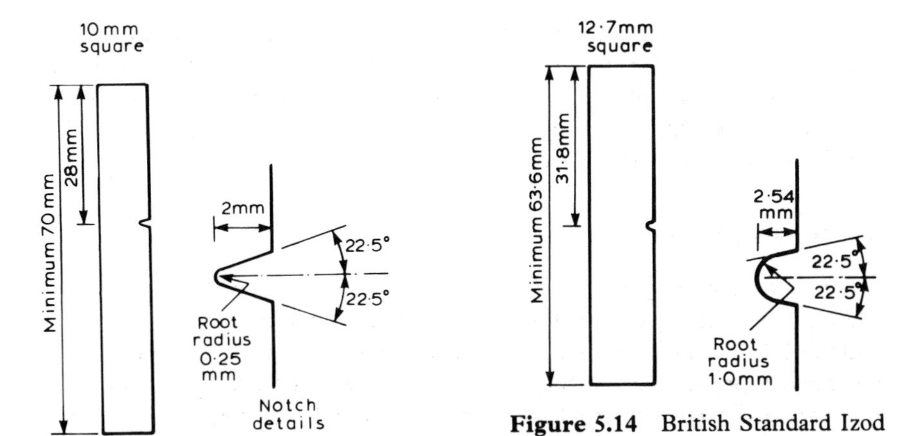

Figure 5.13 British Standard Izod test piece for a metal

Figure 5.14 British Standard Izod test piece for a plastic

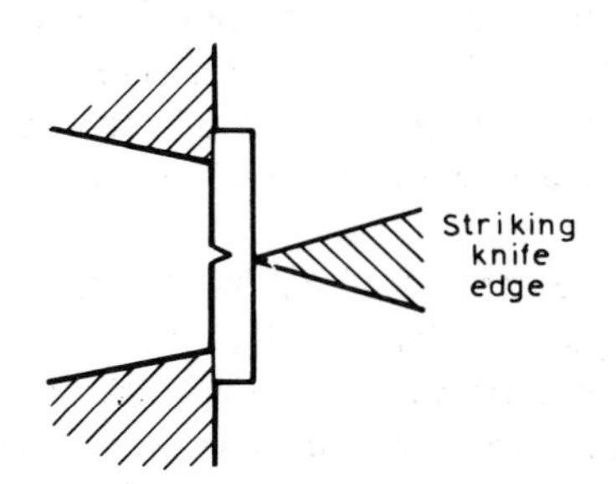

Figure 5.15 Form of the Charpy test piece (plan view)

5.12). Figure 5.16 shows the basic standard forms of Charpy test piece for a metal. The test pieces for plastics are in either the notched or un-notched state. With metals the pendulum strikes the test piece with a speed of between 3 and 5.5 m/s and with plastics between 2.9 and 3.8 m/s.

The results of impact tests need to specify the type of test and the form of notch used. In the case of metals, the results are expressed as the amount of energy absorbed by the test piece when it breaks. In the case of plastics, the results are often expressed as absorbed energy divided by either the cross-sectional area of the un-notched test piece or the cross-sectional area behind the notch in the case of notched test pieces. Table 5.10 shows typical values of impact strengths for metals and Table 5.11 for plastics.

Another test that is sometimes used with plastics involves a weight falling onto a disc of the material under test. The disc for the British Standard test is 57 to 64 mm in diameter, or a square of 57 to 64 mm side, resting on an annular support of 50.8 mm inner diameter and maximum external diameter 57.2 mm.

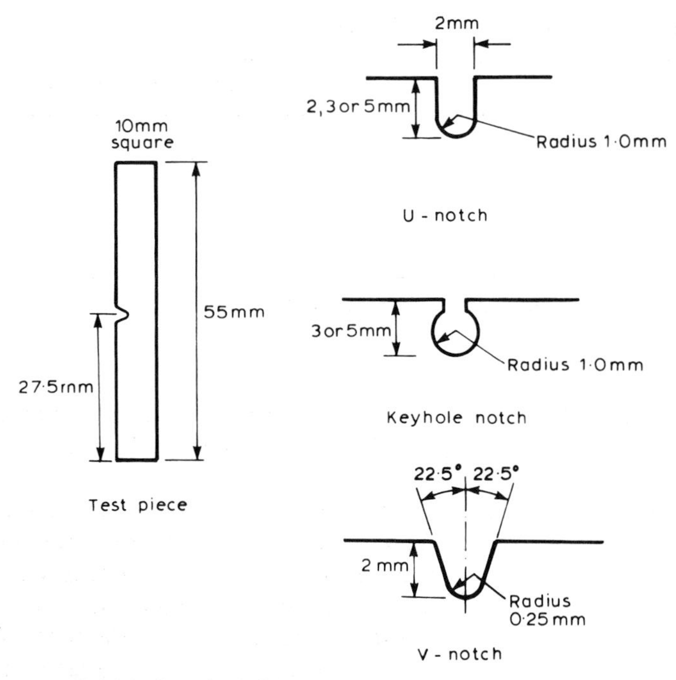

Figure 5.16 British Standard Charpy test piece for a metal

Table 5.10 *Impact strengths for metals*

Material	*Charpy V Impact strength/J*
Aluminium, commercial pure, annealed	30
Aluminium–1.5% Mn alloy, annealed	80
hard	34
Copper, oxygen free HC, annealed	70
Cartridge brass (70% Cu, 30% Zn), annealed	88
¾ hard	21
Cupronickel (70% Cu, 30% Ni), annealed	157
Magnesium–3% Al, 1% Zn alloy, annealed	8
Nickel alloy, Monel, annealed	290
Titanium–5% Al, 2.5% Sn, annealed	24
Grey cast iron	3
Malleable cast iron, Blackheart, annealed	15
Austenitic stainless steel, annealed	217
Carbon steel, 0.2% carbon, as rolled	50

Table 5.11 *Impact strengths for plastics*

Material	*Impact strength*/kJ m^-2*
Polythene, high-density	30
Nylon 6.6	5
PVC, unplasticized	3
Polystyrene	2
ABS	25

*Notch-tip radius 0.25 mm, depth 2.75 mm, Izod test

The thickness of the test piece is 1.52 mm in the case of moulded plastics. The impact strength is the energy needed to fracture half of a large number of samples.

One use of impact tests is to determine whether heat treatment of metals has been successfully carried out. A comparatively small change in heat treatment can lead to quite noticeable changes in impact test results. The changes can be considerably more pronounced than changes in other mechanical properties, e.g. percentage elongation or tensile strength. Figure 5.17 shows the effect on the Izod impact test results for cold worked mild steel annealed to different temperatures. The use of an impact test could then indicate whether annealing has been carried out to the correct temperature.

The properties of metals change with temperature. For example, a 0.2 per cent carbon steel undergoes a gradual transition from a ductile to a brittle material at a temperature of about room temperature (Figure 5.18). At about −25°C the material is a brittle material with a Charpy V-notch impact energy of only about 4 J, whereas at about 100°C it is ductile with an impact energy of about 120 J. This type of change from ductile to a brittle material can be

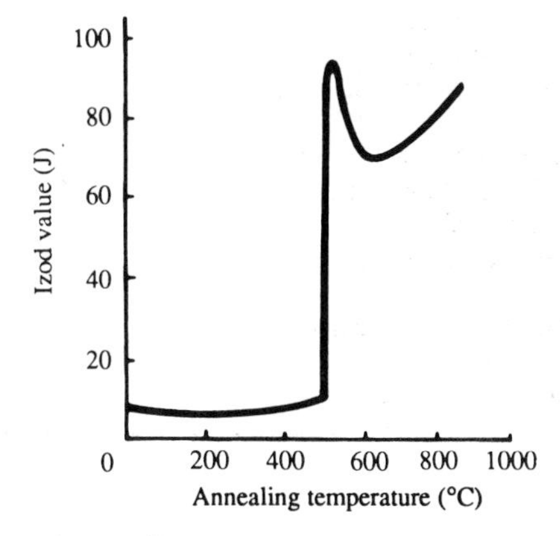

Figure 5.17 Effect of annealing temperatures on Izod test values

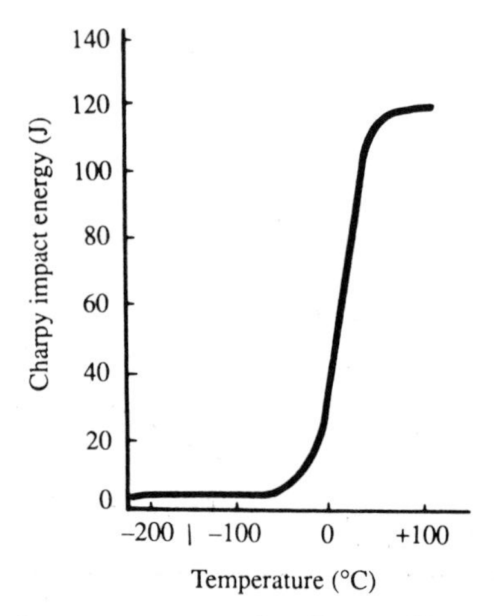

Figure 5.18 Effect of temperature on the Charpy V-notch impact energies for a 0.2 per cent carbon steel

charted by impact test results and the behaviour of the material at the various temperatures predicted.

At room temperature plastics can be grouped into three categories with regard to their impact properties:

1 Essentially brittle materials even when unnotched, e.g. polystyrene and dry glass-filled 66 nylon.
2 (a) Tough when unnotched and brittle when bluntly notched, e.g. polyvinyl chloride and acetal copolymer; (b) tough unnotched and brittle when sharply notched, e.g. high density polythene and dry nylon 66.
3 Essentially tough under all conditions, e.g. low density polythene and wet nylon 66.

The impact properties of plastics vary quite significantly with temperature, changing in many cases from brittle to tough at some particular transition temperature. Figure 5.19 shows the impact strength versus temperature for propylene homopolymer. This material at room temperature is classified as category 2(b) above, i.e. tough when unnotched and brittle when bluntly notched.

5.4 Bend tests

The bend test is a simple test of ductility. It involves bending a sample of the material through some angle and determining whether the material is unbroken and free from cracks after such a bend. Figure 5.20 shows one way of conducting a bend test. The results of a bend test are specified in terms of the angle of bend (Figure 5.21). Thus a material might be specified as withstanding a 90° transverse bend over a radius equal to its own thickness.

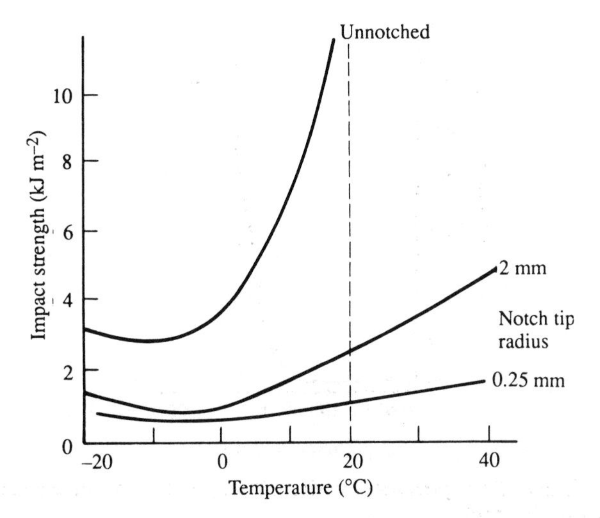

Figure 5.19 Impact strength versus temperature for a propylene homopolymer

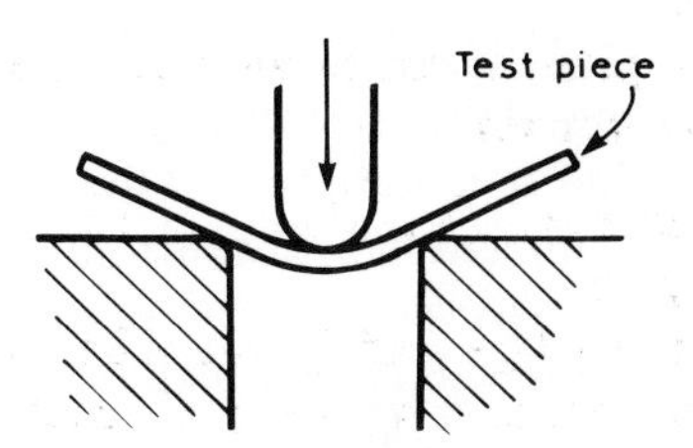

Figure 5.20 A bend test

Figure 5.21 The angle of bend

5.5 Thermal properties

The three thermal properties generally relevant to the choice of a solid material in engineering are a consequence of a heated material 'absorbing' heat and increasing its temperature, expanding and transmitting heat. The three properties are described by:

1 *Specific heat capacity* – the specific heat capacity is a measure of the amount of heat needed to raise the temperature of unit mass of the material by one degree. It is defined by the equation

$$\text{Specific heat capacity} = \frac{\text{amount of heat}}{(\text{mass}) \times (\text{change in temperature})}$$

2 *Thermal expansion* – the linear expansivity or coefficient of linear expansion is a measure of the amount by which a unit length of the material will expand when the temperature rises by one degree. It is defined by the equation

$$\text{Linear expansivity} = \frac{\text{change in length}}{(\text{original length}) \times (\text{change in temperature})}$$

3 *Thermal conductivity* – this is a measure of the rate at which heat will flow through the material; the higher the conductivity, the greater the rate at which heat will flow through the material. Thermal conductivity is defined by the equation

$$\text{Rate of flow of heat} = \text{thermal conductivity} \times \text{area} \times \text{temperature gradient in direction perpendicular to area}$$

Table 5.12 gives some typical values of these properties at about 20°C. The values should only to considered as indicative of the types of values that might be encountered at normal ambient temperatures since they are considerably affected by impurities and temperature.

The main characteristic of metals is their high thermal conductivity, while polymers and ceramics have much lower values. Thus if heat insulation is

Table 5.12 *Thermal properties*

Material	*Specific heat capacity* ($J\ kg^{-1}\ K^{-1}$)	*Linear-expansivity* ($K^{-1} \times 10^{-6}$)	*Thermal conductivity* ($W\ m^{-1}\ K^{-1}$)
Metals			
Aluminium	920	24	230
Copper	385	18	380
Mild steel	480	11	54
Stainless steel	510	16	16
Polymers			
Bakelite (thermoset)	1600	80	0.23
Nylon 66 (thermoplastic)	2000	100	0.025
PVC (thermoplastic)	1000	700	0.0019
Ceramics			
Alumina	750	8	2
Silica, fused	800	0.05	0.1
Glass	800	8	1

required polymers and ceramics are feasible, while if good heat conduction is required metals are appropriate. Polymers have a high linear expansivity and thus an increase in temperature can cause quite marked changes in dimensions of polymeric products. Unlike metals, the expansion of polymers is not usually linear with temperature and the linear expansivity increases with increasing temperature.

5.6 Electrical properties

The main electrical properties that are likely to be of relevance in the choice of materials for a general engineering application are:

1 *Electrical resistivity and conductivity* – the electrical resistivity is a measure of the electrical resistance of the material, being defined by the equation

$$\text{Resistivity} = \frac{(\text{resistance}) \times (\text{cross-section area})}{\text{length}}$$

The electrical conductivity is a measure of electrical conductance of the material; the bigger the conductance, the greater the current for a particular potential difference. It is defined by the following equation, being the reciprocal of resistivity:

$$\text{Conductivity} = \frac{\text{length}}{(\text{resistance}) \times (\text{cross-sectional area})}$$

$$\text{Conductance} = \frac{1}{\text{resistance}}$$

Table 5.13 shows conductivity values at 20°C for some typical materials. Metals are good electrical conductors, while polymers and ceramics are generally insulators.

Table 5.13 *Electrical conductivity values for typical materials*

Material	*Conductivity/ohm^{-1} m^{-1}*
Metals	
Aluminium	40,000
Copper	64,000
Iron	11,000
Mild steel	6,600
Polymers	
Acrylic	Less than 10^{-14}
Nylon	10^{-9} to 10^{-14}
PVC	10^{-13} to 10^{-14}
Ceramics	
Alumina	10^{-10} to 10^{-13}
Glass	10^{-10} to 10^{-12}

Table 5.14 *Dielectric strength values for typical materials*

Material	*Dielectric strength (V/m × 10^8)*
Polymers	
Acrylic	2.0
Nylon	1.7
PVC	5.5
Ceramics	
Alumina	1.2
Glass	0.2

2 *Dielectric strength* – this is a measure of the highest potential difference an insulating material can withstand without electrical breakdown, being defined by the equation

$$\text{Dielectric strength} = \frac{\text{breakdown voltage}}{\text{insulator thickness}}$$

A high value of dielectric strength is a vital requirement for an electrical insulator. The values given above, however, must only be taken as indicative of the types of values that can occur, since the dielectric strength is markedly affected by the presence of flaws and very small holes in the material, and also moisture and other contaminants on the surface of the material. Table 5.14 shows typical dielectric strength values.

Problems

1 To use the data given for grey iron castings below to sketch the form of the stress/strain graph in both tension and compression.

Grey iron castings		
(Courtesy of BCIRA)		
Tensile strength	180 N/mm^{-2}	
0.01% proof stress	50 N/mm^{-2}	
0.1% proof stress	117 N/mm^{-2}	
Phosphorus (%)	‹0.4	0.4–1.0
Total strain at failure (%)	0.7	0.5
Elastic strain at failure (%)	0.17	0.17
Total-minus-elastic strain (%) at failure	0.54	0.33
Compressive strength	672 N/mm^{-2}	
0.01% proof stress	100 N/mm^{-2}	
0.1% proof stress	234 N/mm^{-2}	
Modulus of elasticity		

Tension 109 GN/m^{-2}
Compression 109 GN/m^{-2}
Hardness With ‹0.4% phosphorus 150–183 HB

2 From the stress/strain graph for cast iron (Figure 5.2b), determine (a) the modulus of elasticity and (b) the tensile strength of the sample.
3 From the stress/strain graph for ABS (Figure 5.4) determine (a) the modulus of elasticity and (b) the tensile strength of the sample.
4 Sketch the forms that would be taken by the stress/strain graphs for (a) strong brittle materials, (b) strong ductile materials, (c) weak ductile materials.
5 Why do some materials have a yield stress quoted whereas others have a proof stress quoted?
6 For the data given below for brass: (a) Calculate the maximum force that can be experienced by a sample of T1 brass which has a cross-sectional area of 50 mm^2. (b) Calculate the force needed to reach the 0.2 per cent yield stress for the above T1 sample. (Data courtesy of Delta Extruded Metals Co. Ltd.)

Delta alloy	*Condition*	*0.2% proof stress/MN m^{-2}*	*Tensile strength/ MN m^{-2}*	*Hardness (HV)*	*Modulus of elasticity/ GN m^{-2}*
T1	As extruded	110–140	370–420	90–110	100
	Drawn 6–80 mm	200–260	430–530	140–160	100
	Drawn ›80 mm	170–230	380–460	130–150	100
T2	As extruded	110–140	370–420	90–110	100
	Drawn 6–80 mm	200–260	430–530	140–160	100
	Drawn ›80 mm	170–230	380–460	130–150	100
T12	As extruded 10–25 mm	110–140	430–460	90–110	100
	Drawn 6 mm	200–260	460–530	140–160	100
TR1	Extruded	140	330	70–80	100
	Drawn 6–50 mm	220–290	370–450	110–130	100
	Drawn ›50 mm	190–260	340–420	100–120	100
	Hard	430–480	510–570	150–160	100

7 Describe the basic principles of hardness measurement.
8 A steel is said to have a Charpy V-notch impact value for 10 × 10 mm specimens at 0°C of 27 J. Explain the significance of this data.
9 The following are Izod impact energies at different temperatures for samples of annealed cartridge brass (70 per cent copper–30 per cent zinc). What can be deduced from the results?

Temperature/°C	+27	−78	−197
Impact energy/KJ	88	92	108

10 The following are Charpy V-notch impact energies for annealed titanium at different temperatures. What can be deduced from the results?

Temperature/°C	+27	−78	−196
Impact energy/J	24	19	15

11 The following are Charpy impact strengths for nylon 6.6 at different temperatures. What can be deduced from the results?

Temperature/°C	−23	−33	−43	−63
Impact srength/KJm^{-2}	24	13	11	8

12 The impact strength of samples of nylon 6, at a temperature of 22°C, is found to be 3 kJ m^{-2} in the as moulded condition, but 25 kJ m^{-2} when the sample has gained 2.5 per cent in weight through water absorption. What can be deduced from the results?

13 Compare and contrast for metals and polymeric materials (a) thermal properties and (b) electrical properties.

6

Fracture

6.1 Types of fracture

When a ductile material has a gradually increasing stress applied to it it behaves elastically up to a limiting stress and then beyond that stress plastic deformation occurs. In the case of a tensile stress this deformation takes the form of necking (Figure 6.1a). As the stress is increased, the cross-sectional area of the material becomes considerably reduced until at some stress failure occurs. The fracture shows a typical cone and cup formation, which results because under the action of the increasing stress small internal cracks form which gradually grow in size until there is an internal, almost horizontal, crack. The final fracture occurs when the material shears at an angle of 45° to the axis of the direct stress. This type of failure is known as a *ductile fracture*. A characteristic of a ductile fracture is that the surfaces of the fractured material are dull or fibrous.

Materials can fail in a ductile manner in compression. Such failures show a characteristic bulge and a series of axial cracks around the edge of the material (Figure 6.1b).

Another form of failure is known as *brittle fracture*. If you drop a china cup and it breaks it is possible to pick up the pieces and stick them together again and have something which still looks like a cup. The china cup has failed by a brittle fracture. If you had dropped a tin mug then it might have deformed and shown a dent. If you could have broken the tin mug it would not have been possible to stick the pieces together and have something that looked like the original tin mug. If somebody drives a car into a wall the metal car wings are most likely to show a ductile type fracture like the tin mug. With a brittle failure the material fractures before any significant plastic deformation has occurred.

Figure 6.2a shows the possible forms of a brittle tensile failure with metals. The surfaces of the fractured material appear bright and granular due to the reflection of light from individual crystal surfaces. This is because the fracture has taken place by the separation of grains within the material along specific crystal planes of atoms. Such planes are called clevage planes.

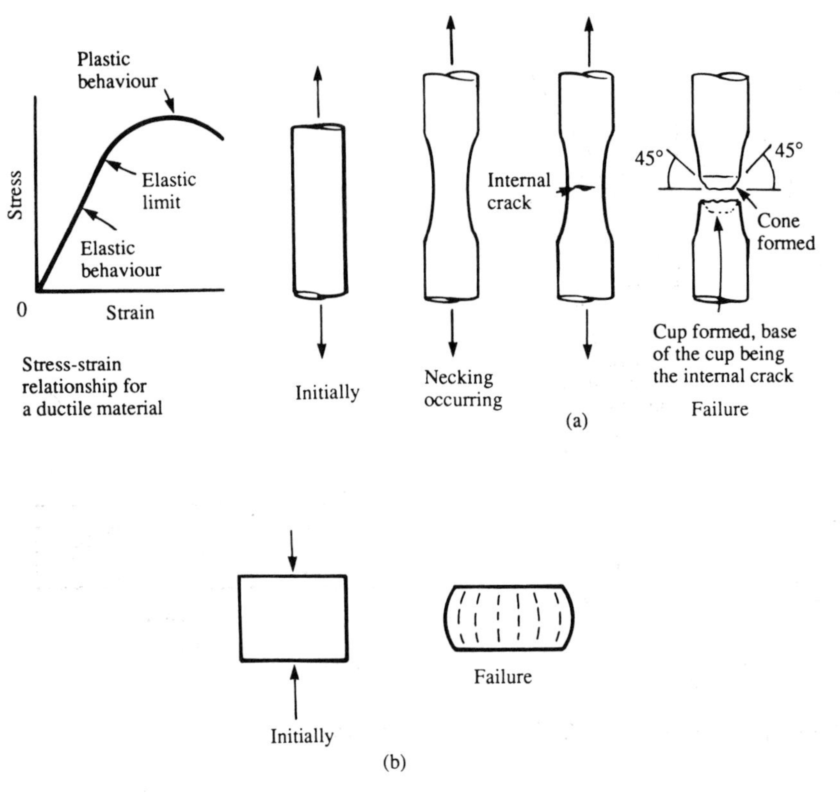

Figure 6.1 Ductile failure (a) Tensile, (b) Compressive

Materials can fail in a brittle manner in compression. Figure 6.2b shows the possible forms of a brittle compressive failure.

Factors affecting fracture

If you want to break a piece of material, one way is to make a small notch in the surface of the material and then apply a force. The presence of a notch or any sudden change in section of a piece of material can very significantly change the stress at which fracture occurs. The notch or sudden change in section produces what are called *stress concentrations*. They disturb the assumed stress distribution and produce local concentrations of stress.

The amount by which the stress is raised depends on the depth of the notch, or change in section, and the radius of the tip of the notch. The greater the depth of the notch and/or the smaller the radius of the tip of the notch, the greater the amount by which the stress is raised.

A crack in a brittle material will have a quite pointed tip and hence a small radius. Such a crack thus produces a large increase in stress at its tip. One way

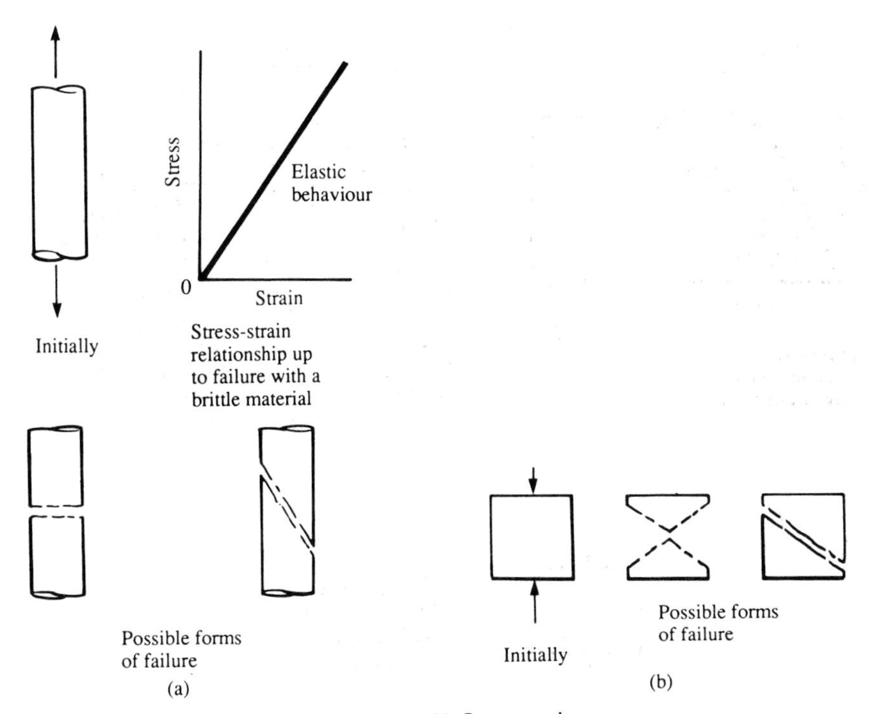

Figure 6.2 Brittle failure (a) Tensile, (b) Compressive

of arresting the progress of such a crack is to drill a hole at the end of the crack to increase its radius and so reduce the stress concentration.

An approximate relationship that has been derived for the stress at the end of a notch is:

$$\text{Stress at end of notch} = \text{applied stress} \times [1+2\sqrt{(L/r)}]$$

where L is the length of the notch and r the radius, of the tip of the notch. The increase in stress due to the notch is thus:

$$\text{Increase in stress} = 2\sqrt{(L/r)}$$

A crack in a ductile material is less likely to lead to failure than in a brittle material because a high stress concentration at the end of a notch leads to plastic flow and so an increase in the radius of the tip of the notch. The result is a decrease in the stress concentration.

Another factor which can affect the behaviour of a material is the speed of loading. A sharp blow to the material may lead to a fracture where the same stress applied more slowly would not. With a very high rate of application of stress there may be insufficient time for plastic deformation of the material to occur and so what was, under normal conditions, a ductile material behaves as though it were brittle. If a material has a notch or change in section and is subject to a sudden impact, the dual effects of the stress concentration due to the notch and the material behaving as though brittle can result in failure.

The Charpy and Izod tests referred to in Chapter 5 give a measure of the behaviour of a notched sample of material when subject to a sudden impact load. The results are expressed in terms of the energy needed to break a standard size test specimen; the smaller the energy needed to break the specimen, the easier it will be for failure to occur in service. The smaller energies are associated with materials which are termed brittle; ductile materials need higher energies for fracture to occur.

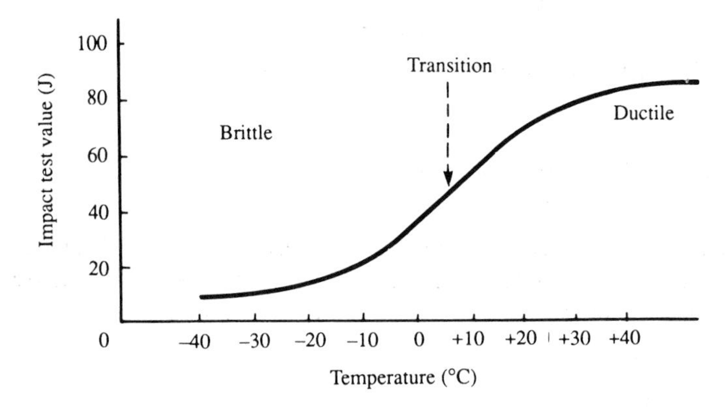

Figure 6.3 Ductile/brittle transition

The temperature of a material when it is subject to stress can affect its behaviour, many metals which are ductile at high temperatures being brittle at low temperatures. Figure 6.3 shows how the impact test results for a steel change with temperature. At room temperature and above the steel behaves as a ductile material; below 0°C it behaves as a brittle material. The *transition* temperature at which the change from ductile to brittle behaviour occurs is thus of importance in determining how a material will behave in service.

The transition temperature with a steel is affected by the alloying elements in the steel. Manganese and nickel reduce the transition temperature. Thus for low-temperature work a steel with these alloying elements should be preferred. Carbon, nitrogen and phosphorus increase the transition temperature.

Elastic strain energy

Energy is needed to extend a piece of material. There is a transfer of energy from some source to the material. Figure 6.4 shows the linear part of a force/extension graph for a material. To cause the material to extend and have an extension x a force is applied which gradually increases from zero to F. The average force is $F/2$ and thus the energy transferred to the material is given by

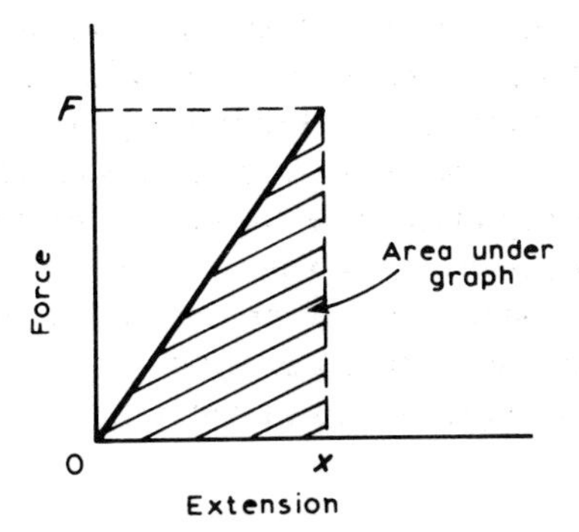

Figure 6.4 Elastic strain energy

Energy transferred = force × distance
$= \frac{1}{2}Fx$

This is the area under the graph between zero extension and extension x.

The volume of material being extended is AL, where A is the cross-sectional area and L the length. Hence the energy per unit volume is

$$\frac{\text{energy}}{\text{volume}} = \frac{Fx}{2AL}$$
$$= \frac{1}{2}Fx/AL$$

But F/A is the stress and x/L the strain, hence:

Energy per unit volume = ½ stress × strain

Units: stress, N/m^2; strain, no units; energy per unit volume, J/m^3 or N/m^2.

If the material is stretched only within its elastic limit, then when the stress is removed, the material springs back to its original dimensions. The energy used to extend the material is surrendered by the material. A stretched material thus has a store of energy, called *strain energy*. For example, a catapult relies on energy being slowly stored in the rubber of the catapult which is suddenly released and given to a projectile.

6.2 Ductile fracture

With ductile fracture the sequence of events is considered to be as follows: following elastic strain the material becomes plastically deformed and a neck occurs; within the neck small cavities or voids are formed; these then link up to form a crack which spreads across the material in a direction at right-angles to the applied tensile stress, and then finally the crack propagates to the material surface by shearing in a direction which is approximately at 45° to the applied stress to give a fracture in the form of a cup and cone. The small cavities or voids develop at impurities or other discontinuities in the material and, are said to require nucleation. The more nuclei available to trigger off the development of cavities or voids the less the material extends before fracture occurs and so the less ductile the material. Thus increasing the purity of a material increases its ductility.

6.3 Brittle fracture

When brittle fracture occurs the fracture surface appear bright and granular due to the reflection of light from individual crystal surfaces. This is because grains within the material have cleaved along planes of atoms. We can consider the sequence of events leading to a brittle fracture to be as follows: the bonds between the atoms in the material are elastically strained and the material acquires strain energy, then at some critical stress the bonds between atoms break and a crack propagates through the material to give fracture.

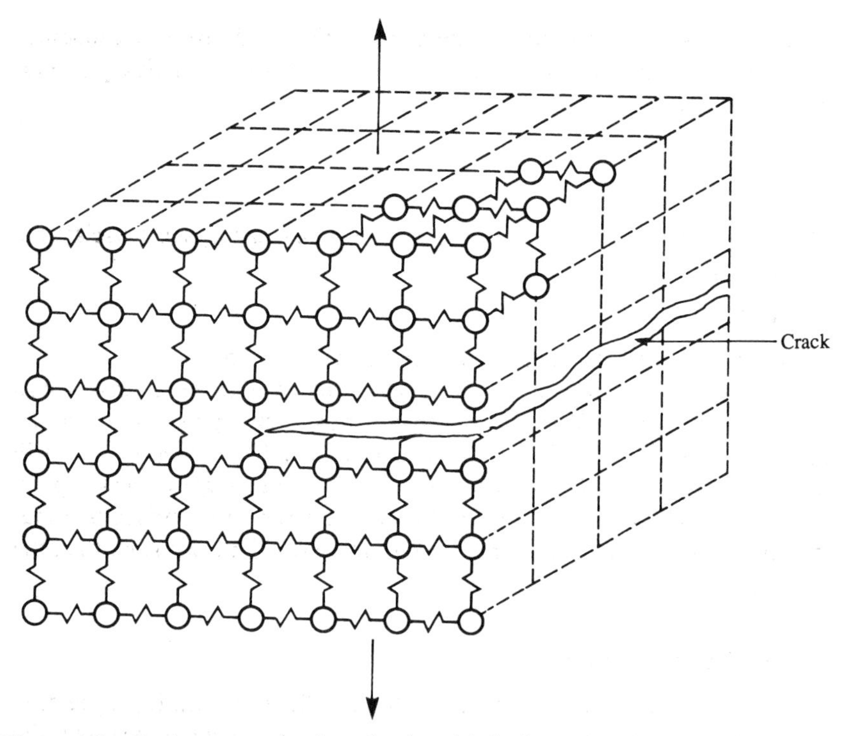

Figure 6.5 Crack propagation in a simple cubic lattice

In 1920, A. A. Griffith advanced the theory that all materials contain small cracks but that a crack will not propagate until a particular stress is reached, the value of the stress depending on the length of the crack. To illustrate the reasoning he employed, consider a simple cubic lattice within which there is a crack (Figure 6.5). We can think of each atom in the crystal being rather like a sphere linked to neighbouring atoms by springs, the springs representing the bonds between the atoms. When the material is stretched springs are stretched and elastic strain energy is stored in the material. For a crack to propagate further bonds must be broken. When bonds break there will be a release of elastic strain energy. But the breaking of bonds creates new surfaces. Energy is necessary to create a surface (see Chapter 1). The crack can thus be considered to propagate when the released elastic strain energy is just sufficient to provide the surface energy necessary for the creation of the new surfaces. The crack will not propagate if the released elastic strain energy would not be large enough to supply the required surface energy.

The condition for crack propagation derived by Griffiths is

$$\sigma = \sqrt{\left(\frac{2\gamma E}{\pi c}\right)}$$

where σ is the stress, γ the surface energy per unit area, E the tensile modulus and $2c$ the crack length which will propagate at the stress σ. This equation is often approximated to:

$$\sigma \simeq \sqrt{\left(\frac{\gamma E}{c}\right)}$$

Thus for a given material, the crack length which will propagate is related to the stress by

$$c \propto 1/\sigma^2$$

The bigger the stress the smaller the critical crack length.

If you stick a pin in the rubber of a deflated balloon all that happens is that a pin hole is produced; the hole does not result in a crack propagating, so the stress is low. If you stick a pin in the rubber of an inflated balloon there will probably be a bang as the crack started by the hole propagates very rapidly through the rubber. In the inflated balloon situation, the rubber is under sufficient stress for the critical crack length to be very small, smaller than the size of the pinhole.

Modified Griffith equation

The Griffith equation (see above) gives the condition for crack propagation as when the released elastic strain energy is just large enough to supply the required surface energy. Such an equation is reasonably valid when a material fails in a completely brittle manner. However, though this is the case with ceramics and brittle plastics, metals generally show some plastic deformation. Some plastic deformation even occurs with what can be termed brittle metals. Energy is required for plastic deformation. Thus the Griffith equation has been modified to include a term to take account of plastic deformation.

$$\sigma = \sqrt{\left(\frac{(2\gamma+\gamma_p)E}{\pi c}\right)}$$

where γ_p is the energy required for plastic deformation per unit area of crack. In most metals γ_p is greater than γ.

6.4 Fracture toughness

Fracture toughness can be defined as being a measure of the resistance of a material to fracture, i.e. a measure of the ability of a material to resist crack propagation.

One measure of the fracture toughness of a material is the *elastic strain energy release rate* (G) at the tip of a crack. When the rate of release of elastic strain energy at the tip of a crack reaches a high enough value then the crack propagates. This critical value is given the symbol G_c. Thus in the case of a brittle material, crack propagation occurs when

$$G_c = 2\gamma$$

where γ is the energy required to product unit area of new surface (the 2 is because there are two surface areas produced by a single crack). Where there is also plastic deformation, crack propagation occurs when

$$G_c = 2\gamma + \gamma_p$$

The Griffith equation, or its modified form, thus becomes:

$$\sigma = \sqrt{\left(\frac{G_c E}{\pi c}\right)}$$

The smaller the value of G_c the less tough a material is. The units of G_c are kJ/m^2.

Another measure of the fracture toughness of a material is the *stress intensity factor* (K). The stress intensity factor is a measure of the stresses at the tip of a crack, being related to the elastic strain energy release rate G by

$$K^2 = GE$$

Crack propagation occurs when the stress intensity factor reaches a high enough value. This critical value is given the symbol K_c. Since

$$K_c^2 = G_c E$$

then

$$\sigma = \frac{K_c}{\sqrt{(\pi c)}}$$

The smaller the value of K_c the less tough a material is. The units of K_c are generally MN/m$^{3/2}$.

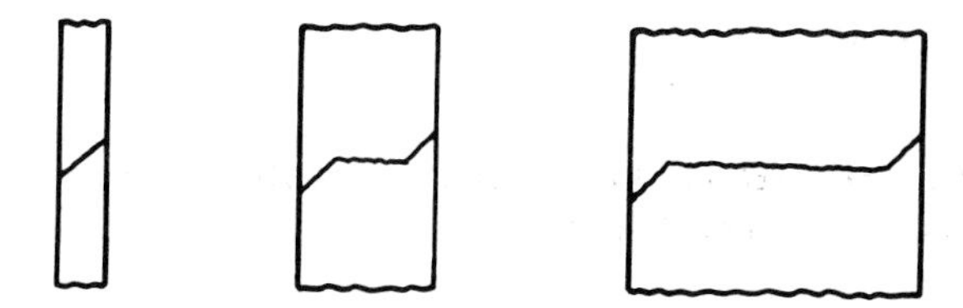

Figure 6.6 Effect of thickness on form of crack propagation

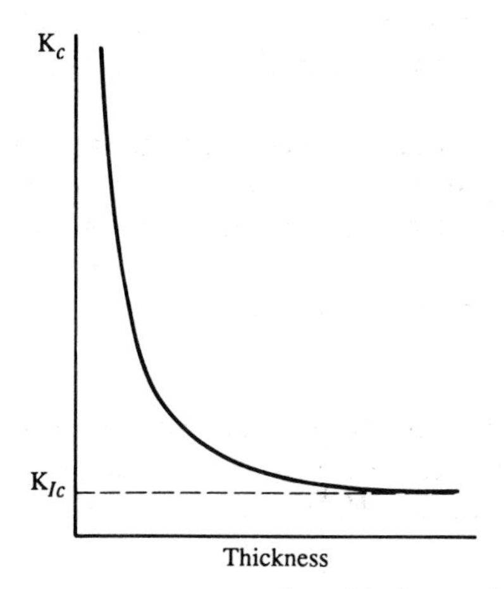

Figure 6.7 Effect of plate thickness on the critical stress intensity factor

The critical stress intensity factor K_c is a function of the material and plate thickness concerned. The thickness factor is because the form of crack propagation is influenced by the thickness of the plate. In thin plates failure is by shear on planes at 45° to the tensile forces across the crack (Figure 6.6). Thicker plates show a central flat fracture with 45° shear fractures at the sides. The thicker the plate the greater the amount of central flat fracture. The effect of this on the value of the critical stress intensity factor is shown in Figure 6.7. High values of K_c occur with thin sheets and K_c decreases as the thickness increases to become almost constant at large thicknesses. At such large thicknesses the portion of the fracture area which has sheared is very small, most of the fracture being flat and at right-angles to the tensile forces. This lower limiting value of the critical stress intensity factor is called the plane strain fracture toughness and is denoted by K_{Ic}. This factor is solely a property of the material. It is the value commonly used in design, for all but very thin sheets, since it is the lowest value of the critical stress intensity factor and hence is the safest value to use. The lower the value of K_c used the less tough the material is assumed to be. Table 6.1 gives some typical values of K_{Ic}.

A possible criterion used for the selection of a material could be the critical crack size. The critical crack size c is proportional to $(K_{Ic}/\sigma)^2$. Since materials

Table 6.1 *Typical values of K_{Ic} at 20°C*

Material	K_{Ic} (MN m$^{-3/2}$)
Metals	
Aluminium alloys	20 to 160
Steels	25 to 150
Titanium alloys	40 to 150
Polymers	
Cast acrylic sheet	1.6
General purpose polystyrene	1.0
Ceramics	
Alumina	4.9
Silicon carbide	4.0
Soda-line glass	0.7

are often used at a specific fraction of their yield stress a useful parameter for making comparisons between materials is the value of $(K_{Ic}/\sigma_y)^2$, where σ_y is the yield stress. Table 6.2 gives some typical values.

Table 6.2 *Typical values of $(K_{Ic}/\sigma_y)^2$*

Material	$(K_{Ic}/\sigma_y)^2$ (mm)
Aluminium alloys	1.1 to 13.4
Steels	0.3 to 16.4
Titanium alloys	2.1 to 44.5

If the critical crack size is to be the same for two materials then the value of $(K_{Ic}/\sigma)^2$ should be the same. Thus if, for example, we consider two aluminium alloys, having values of $(K_{Ic}/\sigma_y)^2$ of 2 and 10, then if f_1 and f_2 are the fractions of the yield stress at which the materials will be used and have the same critical crack length:

$$\left(\frac{2}{f_1}\right)^2 = \left(\frac{10}{f_2}\right)^2$$

$$\frac{f_1}{f_2} = \frac{2}{10}$$

Thus the alloy with the higher value of the parameter can be used at a higher fraction of its yield stress than the other material.

Factors affecting fracture toughness

Fracture toughness of a material, i.e. its resistance to crack propagation, is affected by a number of factors:

1 *Composition of the material* – different alloy systems have different fracture toughness. Thus, for example, aluminium alloys have lower values of plane strain toughness (K_{Ic}) than many steels. Within each alloy system there are, however, some alloying additions which markedly reduce toughness, e.g. phosphorus or sulphur in steels.
2 *Heat treatment* – heat treatment can markedly affect the fracture toughness of a material. Thus, for example, the toughness of a steel is markedly affected by changes in tempering temperature.
3 *Material thickness.*
4 *Service conditions* – service conditions such as temperature, corrosive environment, and fluctuating loads can all affect fracture toughness.

Fracture toughness tests

Fracture toughness test are generally concerned with the measurement of K_{Ic}. Various types of test pieces have been developed for which stress intensity factors, i.e. K values, can be calculated. For such test pieces

$$K = \sigma_n Y \sqrt{c}$$

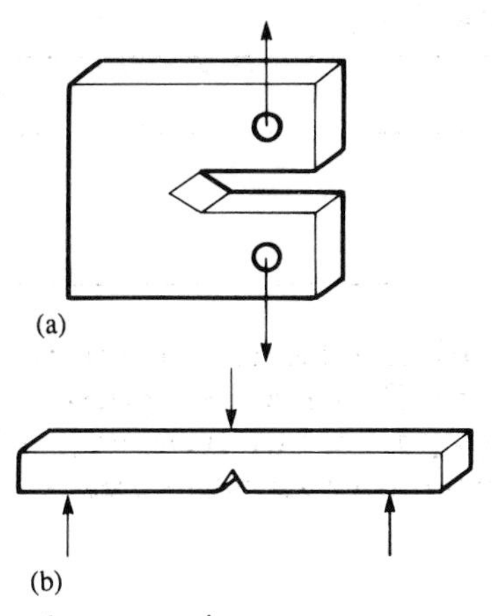

Figure 6.8 Fracture toughness test pieces

where σ_n is the nominal stress, Y is determined from the geometry of the test piece, and $2c$ is the crack length. Two commonly used forms of test piece are the so-called compact tension specimen (Figure 6.8a) for testing in tension and the single-edge notched type (Figure 6.8b) for testing by three-point bending. With all test pieces it is essential to use sharp initial notches. These are produced by initially machining a notch and then adopting a standardized procedure for producing a small fatigue crack at the tip of the machined notch. The test piece is then subjected to a standardized test procedure in which the applied force and the opening of the notch are monitored. A graph of applied force against crack opening displacement is a straight line until the point is reached when the crack begins to self-propagate. This condition is then determined and used to calculate K_{Ic}.

6.5 Failure with polymeric materials

Brittle failure with polymeric materials is a common form of failure with materials below their glass transition temperature. It arises when the molecules are unable to uncoil or slip past each other so that the applied load is essentially applied directly to the backbone of the molecular chains. Little extension is possible in this backbone and so failure occurs with comparatively little extension of the material occurring. The stress/strain graph of such a material is similar to that shown in Figure 6.9 and the resulting fracture surfaces show a mirror-like region, where the crack has grown slowly, surrounded by a region which is rough and coarse, where the crack has propagated fast.

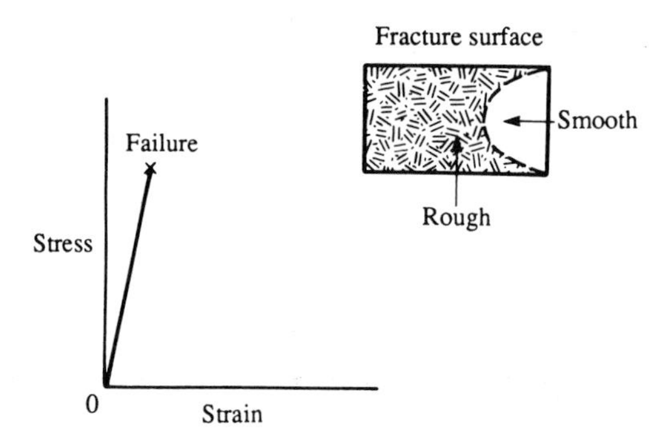

Figure 6.9 Fracture with a brittle polymeric material

Ductile failure with polymeric materials occurs in regions of the material where there is considerable, permanent, deformation. In some cases the failure is preceded by necking (see the section on orientation in Chapter 4). Prior to the material yielding and necking starting, the material is quite likely to begin to show a cloudy appearance. This is due to small voids being produced within the material. Figure 6.10 shows typical forms of stress/strain graphs and failure appearances for ductile materials.

Fracture of composites

Figure 6.11 shows some of the different modes of fracture that can occur with fibre-reinforced composites. The mode is determined by whether the fibres and the matrix are ductile or brittle in character. If the matrix is ductile

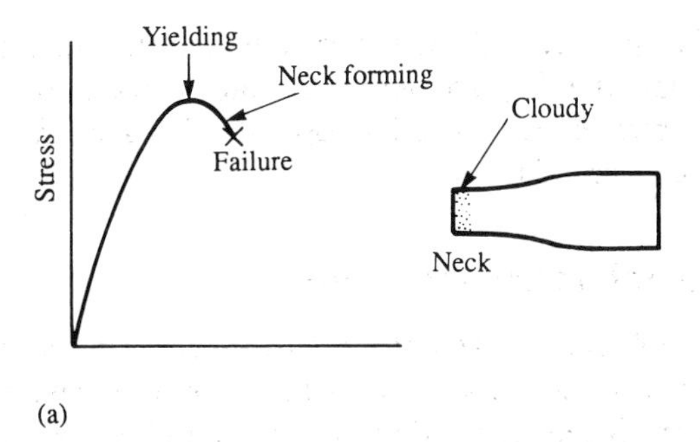

(a)

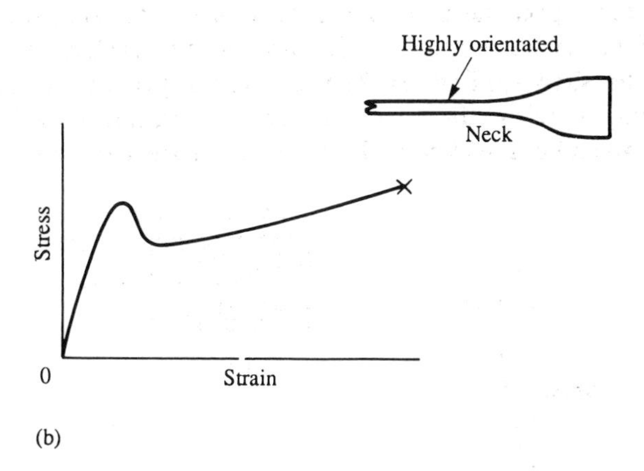

(b)

Figure 6.10 Fracture with ductile polymeric materials (a) Unplasticized PVC, (b) Polypropylene

compared with the fibres, failure occurs by the fracture of a fibre which then leads to fracture of further fibres and the consequential fracture of the intervening matrix. If the matrix is brittle compared with the fibres, failure occurs in the matrix and the result is a fractured surface with lengths of fibre sticking out from it, rather like bristles sticking out from a brush.

Problems

1 Distinguish between the forms of ductile and brittle fractures.
2 Given a fractured specimen what would you look for in order to determine whether the fracture was a ductile or a brittle fracture?
3 How does the presence of a notch or an abrupt change in section have an effect on the failure behaviour?
4 In 1944 in Cleveland, Ohio, a steel tank holding liquefied natural gas fractured and caused 128 deaths and considerable damage. The temper-

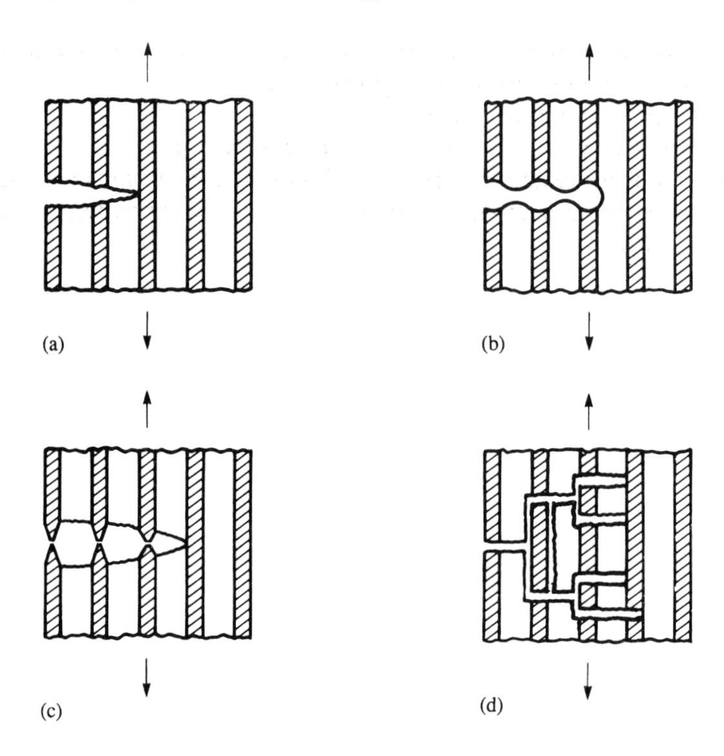

Figure 6.11 Fracture modes for composites (a) Brittle fibres and brittle matrix, (b) Brittle fibres and ductile matrix, (c) Ductile fibres and brittle matrix, (d) Debonding along fibres

ature to which the steel was exposed was very low, of the order of −160°C. What effect might this low temperature have had on the steel? Why would the choice of steel for such an application be important? What types of steel may be suitable?

5 Describe how, for a steel, the results given by an impact test such as the Charpy test, change with temperature when the steel shows a ductile-brittle transition.

6 Describe the nucleation and growth of a ductile fracture.

7 Explain brittle fracture in terms of elastic strain energy and the required surface energy for crack propagation.

8 Explain the significance of the Griffith equation

$$\sigma = \sqrt{\left(\frac{\gamma E}{c}\right)}$$

for crack propagation.

9 Explain what is meant by fraction toughness.

10 Explain the terms: stress intensity factor K, critical stress intensity factor K_c and plane strain fracture toughness K_{Ic}.
11 What factors can affect the values of K_{Ic}?
12 Describe the principles of fracture toughness testing.
13 Explain how the appearance of the fracture surfaces aids in the identification of fracture mode for (a) metals, (b) polymeric materials and, (c) composites.

7

Fatigue

7.1 Fatigue failure

In service many components undergo thousands, often millions, of changes of stress. Some are repeatedly stressed and unstressed, while some undergo alternating stresses of compression and tension. For others the stress may just fluctuate about some value. Many materials subject to such conditions fail, even though the maximum stress in any one stress change is less than the fracture stress as determined by a simple tensile test. Such a failure, as a result of repeated stressing, is called a *fatigue failure*.

The source of the alternating stresses can be due to the conditions of use of a component. Thus, in the case of an aircraft, the changes of pressure between the cabin and the outside of the aircraft every time it flies subject the cabin skin to repeated stressing. Components such as a crown wheel and pinion are subject to repeated stressing by the very way in which they are used, while others receive their stressing 'accidentally'. Vibration of the component can occur as a result of the transmission of vibration from some machine nearby. Turbine blades may vibrate in use in such a way that they fail by fatigue. It has been said that fatigue causes at least 80 per cent of the failures in modern engineering components.

A fatigue crack often starts at some point of stress concentration. This point of origin of the failure can be seen on the failed material as a smooth, flat, semicircular or elliptical region, often referred to as the nucleus. Surrounding the nucleus is a burnished zone with ribbed markings. This smooth zone is produced by the crack propagating relatively slowly through the material and the resulting fractured surfaces rubbing together during the alternating stressing of the component. When the component has become so weakened by the crack that it is no longer able to carry the load, the final, abrupt fracture occurs, which shows a typically crystalline appearance. Figure 7.1 shows the various stages in the growth of a fatigue crack failure.

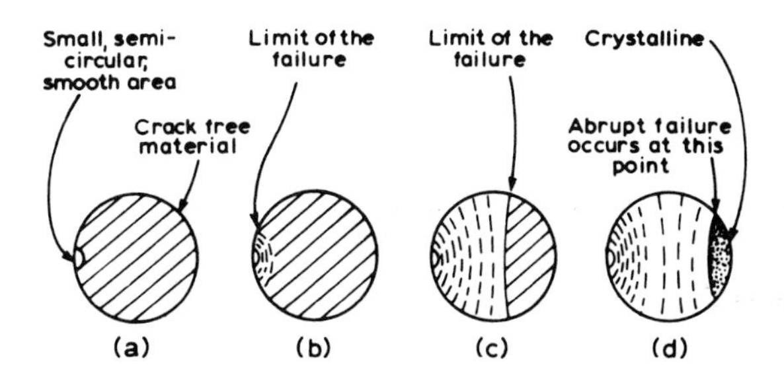

Figure 7.1 Fatigue failure with a metal (a) The nucleus, limit of the failure, (b) The crack grows slowly, (c) The crack continuing to grow slowly, (d) Complete failure

Stages in fatigue failure

A metal progressing to fatigue failure can be considered to pass through three stages.

Stage 1 – The nucleus and initial crack
This is the initiation zone of the fatigue crack. Under repeated stressing slip occurs as a result of the movement of dislocations. Slip may occur one way on one slip plane and the reverse way on an adjacent slip plane during the reverse stress cycle. The reverse direction of slip on the first slip plane is inhibited by local work hardening, i.e. an increase in dislocation density in this slip plane as a result of its plastic deformation. The result of such behaviour is a number of slip bands, essentially a series of grooves and extended tongues of metal from the metal surface (Figure 7.2). From this discontinuity the fatigue cracks

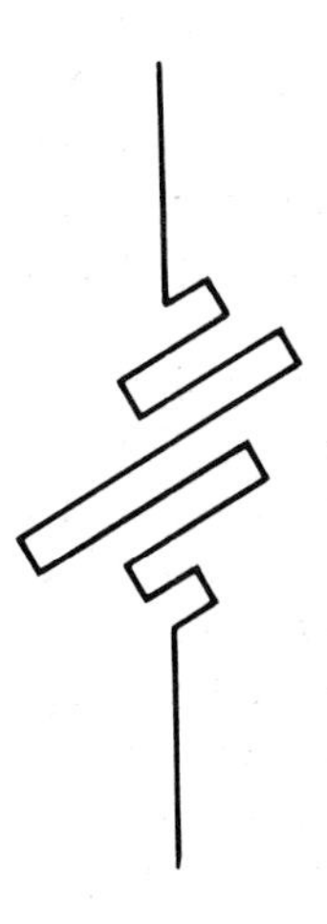

Figure 7.2 Slip bands at the nucleus of fatigue failure

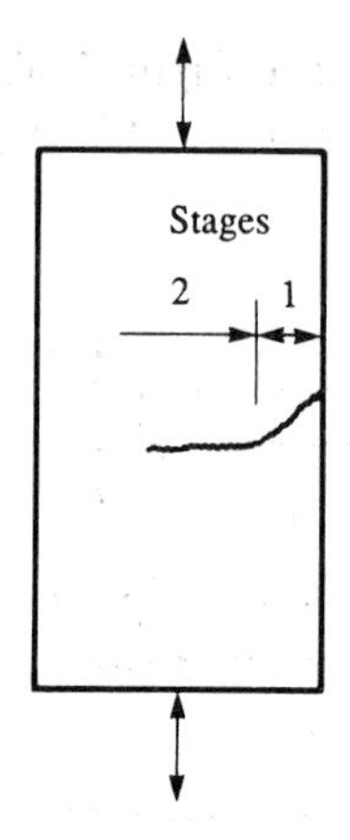

Figure 7.3 Stages 1 and 2 crack propagation

start. The initial crack growth is in a plane which is about 45° to the direction of the applied stress. This is a consequence of the slip occurring as a result of shear. After a few grains the direction of the crack changes to be at right-angles to the applied stress direction. This change in direction is the beginning of Stage 2.

Stage 2 – Crack propagation
The mode of failure of the material in Stage 1 is a shear mode, in Stage 2 the mode is a tensile mode with the crack being forced open (Figure 7.3). The Stage 2 crack surfaces show a series of fine striations. The striations are a series of ridges and result from the alternating stresses applied to the material, each stress cycle producing a striation. The striations are concentric on the nucleus and the spacing between successive striations increases as the crack propagates out from the nucleus.

Stage 3 – Failure
The Stage 2 crack continues growing until a point is reached when there is a sudden complete failure of the material. This final fracture zone is likely to show a crystalline surface appearance, quite different from the striations of Stage 2 (see Chapter 6 for possible mechanisms).

7.2 Fatigue tests

Fatigue tests can be carried out in a number of ways, the way used being the one needed to simulate the type of stress changes that will occur to the material of a component when in service. There are thus bending-stress machines which bend a test piece of the material alternately one way and then the other (Figure 7.4a), and torsional-fatigue machines which twist the test piece alternately one way and then the other (Figure 7.4b). Another type of machine can be used to produce alternating tension and compression by direct stressing (Figure 7.4c).

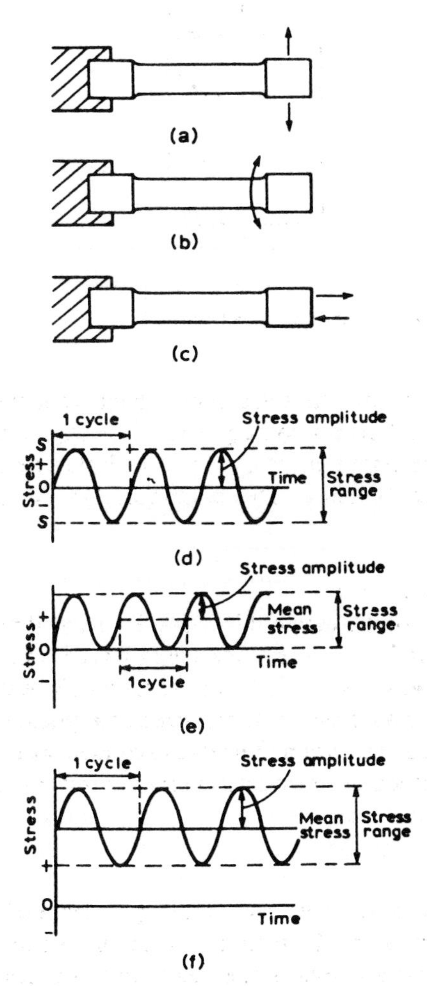

Figure 7.4 Fatigue testing (a) Bending, (b) Torsion, (c) Direct stress, (d) Alternating stress, (e) Repeated stress, (f) Fluctuating stress

The tests can be carried out with stresses which alternate about zero stress (Figure 7.4d), apply a repeated stress which varies from zero to some maximum stress (Figure 7.4e) or apply a stress which varies about some stress value and does not reach zero at all (Figure 7.4f).

In the case of the alternating stress (Figure 7.4d), the stress varies between $+S$ and $-S$. The tensile stress is denoted by a positive sign, the compressive stress by a negative sign; the stress range is thus $2S$. The mean stress is zero as the stress alternates equally about the zero stress. With the repeated stress (Figure 7.4e), the mean stress is half the stress range. With the fluctuating stress (Figure 7.4f) the mean stress is more than half the stress range.

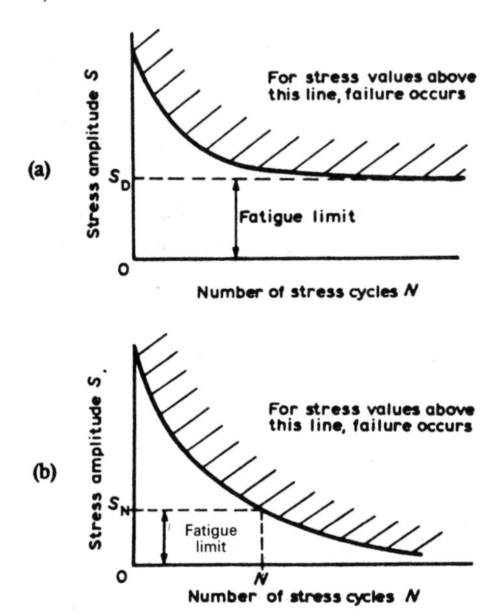

Figure 7.5 Typical *S/N* graphs for (a) A steel, (b) A non-ferrous alloy

During the fatigue tests, the machine is kept running, alternating the stress, until the specimen fails, the number of cycles of stressing up to failure being recorded by the machine. The test is repeated for the specimen subject to different stress ranges. Such tests enable graphs similar to those in Figure 7.5 to be plotted. The vertical axis is the *stress amplitude*, half the stress range. For a stress amplitude greater than the value given by the graph line, failure occurs for the number of cycles concerned. These graphs are known as *S/N graphs*, the S denoting the stress amplitude and the N the number of cycles.

For the *S/N* graph in Figure 7.5a there is a stress amplitude for which the material will endure an indefinite number of stress cycles. The maximum value, S_D, being called the *fatigue limit*. For any stress amplitude greater than the fatigue limit, failure will occur if the material undergoes a sufficient number of stress cycles. With the *S/N* graph shown in Figure 7.5b there is no stress amplitude at which failure cannot occur; for such materials a *fatigue limit* S_N is quoted for N cycles. The term *endurance limit* is sometimes used where the material will endure an infinite number of cycles.

The number of reversals that a specimen can sustain before failure occurs depends on the stress amplitude, the bigger the stress amplitude the smaller

Table 7.1 S/N *values for an aluminium alloy specimen*

Stress amplitude/MN m^{-2}	*Number of cycles before failure/$\times 10^6$*
185	1
155	5
145	10
120	50
115	100

the number of cycles of stress and reversals that can be sustained.

Some typical results for an aluminium alloy specimen are given in Table 7.1.

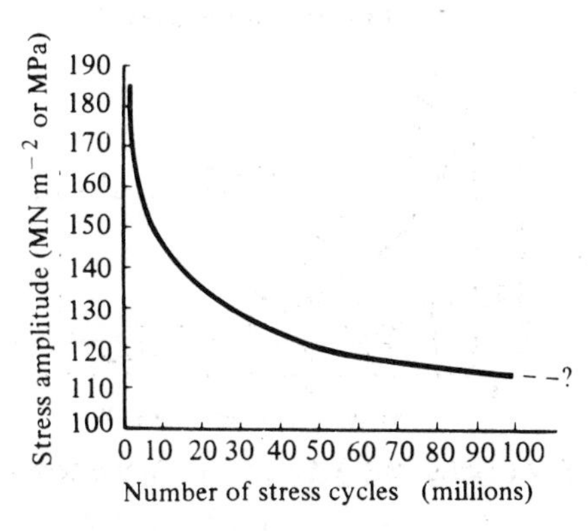

Figure 7.6 *S/N* graph for an aluminium alloy

With a stress amplitude of 185 MN m^{-2}, e.g. a stress alternating from +185 MN m^{-2} to −185 MN m^{-2}, one million cycles are needed before

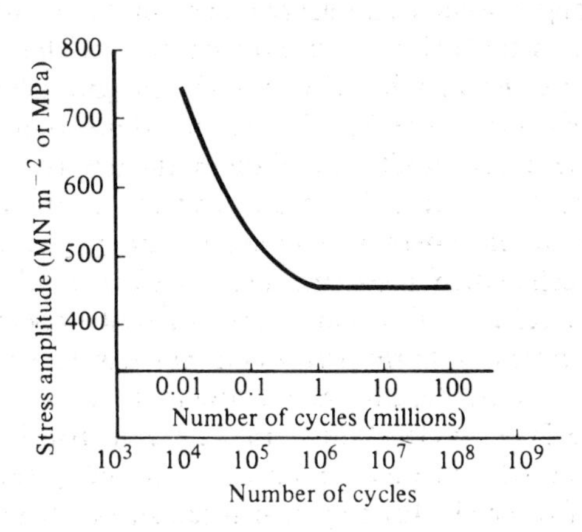

Figure 7.7 *S/N* graph for a steel. Note: number of cycles shown on logarithmic scale

failure occurs. With a smaller stress amplitude of 115 MN m^{-2} one hundred million cycles are needed before failure occurs. Figure 7.6 shows the *S/N* graph for the above data. Extrapolation of the graph seems to indicate that for a greater number of cycles, failure will occur at even smaller stress amplitudes. There seems to be no stress amplitude for which failure will not occur; the material has no fatigue limit. If a component made of that material had a service life of 100 million stress cycles then we could specify that during the lifetime, failure should not occur for stress amplitudes less than 115 MN m^{-2}. The endurance limit for 100 million cycles is thus 115 MN m^{-2}.

Table 7.2 gives some typical results for a steel, the fatigue tests being bending stress (Figure 7.7):

Table 7.2 *S/N values for a steel specimen*

Stress amplitude/MN m^{-2}	*Number of cycles before failure($\times 10^6$)*
750	0.01
550	0.1
450	1
450	10
450	100

With a stress amplitude of 750 MN m^{-2}, e.g. a stress alternating from +750 MN m^{-2} to −750 MN m^{-2}, 0.01 million cycles or ten thousand cycles are needed before failure occurs. For one million, ten million and one hundred million cycles the stress amplitude for failure is the same, 450 MN m^{-2}. For stress amplitudes below this value the material should not fail, however long the test continues. The fatigue limit is thus 450 MN m^{-2}.

The fatigue limit, or the endurance limit at about 500 million cycles, for metals tends to lie between about a third and a half of the static tensile strength. This applies to most steels, aluminium alloys, brass, nickel and magnesium alloys. For example, a steel with a tensile strength of 420 MN m^{-2} has a fatigue limit of 180 MN m^{-2}, just under half the tensile strength. If used in a situation where it were subject to alternating stresses, such a steel would need to be limited to stress amplitudes below 180 MN m^{-2} if it were not to fail at some time. A magnesium alloy with a tensile strength of 290 MN m^{-2} has an endurance limit of 120 MN m^{-2}, just under half the tensile strength. Such an alloy would need to be limited to stress amplitudes below 120 MN m^{-2} if it were to last to 500 million cycles.

It must be recognized that there is a relatively large scatter of results in any fatigue test. Thus an *S/N* graph is essentially drawn through data points which represent the mean value of the life at each stress range. This variation in life must be considered in interpreting *S/N* graphs.

Effect of mean stress

For any particular value of the mean stress it is possible to determine an *S/N* graph. When the mean stress is zero the fatigue limit that is given is that which occurs in the absence of any stress, however when the mean stress is tensile strength for the material, then the fatigue limit is zero since the material fails without any cycles being undertaken. Between these two limits of mean stress, increasing the mean stress decreases the fatigue limit. A number of empirical relationships have been devised to describe the relationship between fatigue limit and the mean stress.

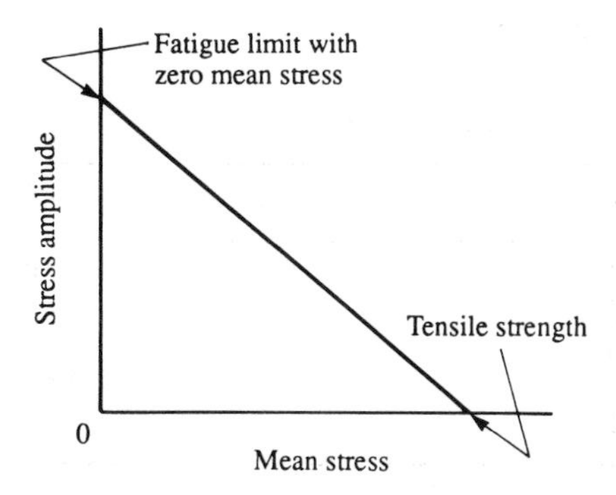

Figure 7.8 The Goodman relationship

Figure 7.8 shows the relationship proposed by *Goodman*. The graph between stress amplitude and mean stress is a straight line drawn between two points, the stress amplitude being the fatigue limit in the absence of mean stress at the zero value of mean stress and the stress amplitude being zero when the mean stress equals the tensile strength. If, for given conditions, the point representing the value of mean stress and stress amplitude lies below the

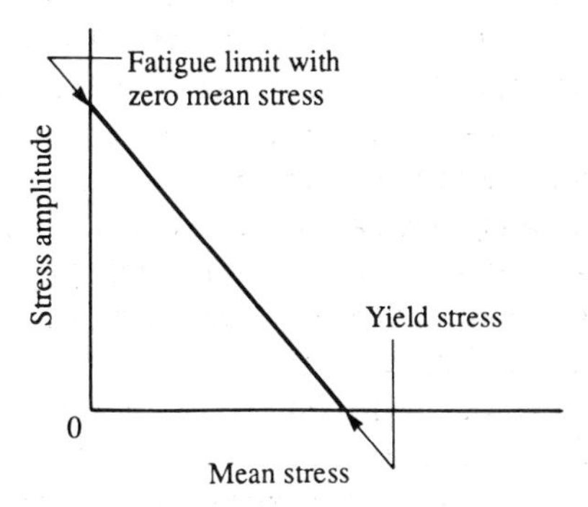

Figure 7.9 The Soderberg relationship

straight line in the graph then the Goodman relationship considers it will not fail by fatigue.

Another relationship is that put forward by *Soderberg*. He proposed that the graph of stress amplitude against mean stress should be drawn between the points of the stress amplitude being the fatigue limit in the absence of mean stress – at the zero value of mean stress and the stress amplitude being zero – when the mean stress equals the yield stress (Figure 7.9).

An important consequence of the above is that an *S/N* graph or values for fatigue or endurance limits should not be used without a consideration of the mean stress.

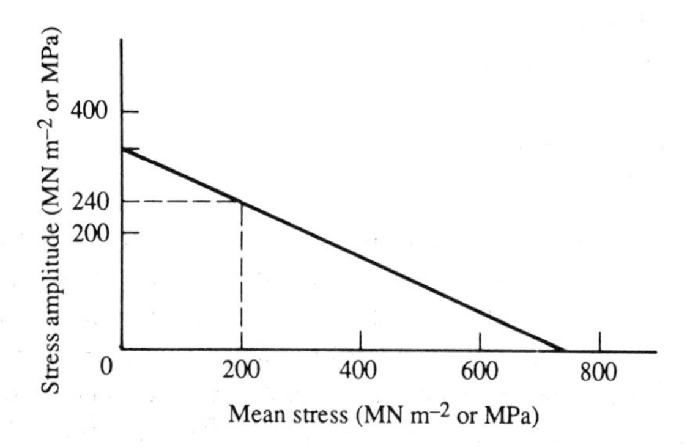

Figure 7.10 Goodman diagram for a 0.6 per cent carbon steel

As an illustration of the use of a Goodman diagram, consider a 0.6 per cent carbon steel with a fatigue limit of 320 MN m^{-2} and a tensile strength of 740 MN m^{-2}. Figure 7.10 shows the resulting Goodman diagram. Thus on the basis of this diagram we can predict that a stress amplitude of 240 MN m^{-2} will cause failure with a mean stress of 200 MN m^{-2} – any lower stress amplitude would be safe.

Cumulative damage

During service many components are subject to cyclic loading but not necessarily regularly repeated loading at the same stress amplitude. A simple relationship that is used to assess the effects of cycles at different stress amplitudes is that of Miner, the relationship being known as *Miner's law*. If the component is subject to n_1 cycles at a stress amplitude for which fatigue failure would occur at N_1 cycles, and n_2 cycles at a stress amplitude for which fatigue failure would occur at N_2 cycles, and n_3 cycles at a stress amplitude for which failure would occur at N_3 cycles, etc. then the component will fail if the sum of n/N ratios equals 1.

$$\frac{n_1}{N_1} + \frac{n_2}{N_2} + \frac{n}{N_3} + \text{etc.} = 1$$

The relationship must only be regarded as a useful approximation. The condition for failure is, for instance, affected by the sequence of the different load cycles.

7.3 Factors affecting the fatigue properties of metals

The main factors affecting the fatigue properties of a component are:

1 Stress concentrations caused by component design.
2 Corrosion.
3 Residual stresses.
4 Surface finish/treatment.
5 Temperature.
6 Microstructure of alloy.
7 Heat treatment.

Fatigue of a component depends on the stress amplitude attained, the bigger the stress amplitude the fewer the stress cycles needed for failure. Stress concentrations caused by sudden changes in cross-section, keyways, holes or sharp corners can thus more easily lead to a fatigue failure. The presence of a countersunk hole was considered in one case to have led to a stress concentration which could have led to a fatigue failure. Figure 7.11 shows the effect on the fatigue properties of a steel of a small hole acting as a stress raiser. With the hole, at every stress amplitude value less cycles are needed to reach failure. There is also a lower fatigue limit with the hole present, 700 MN m^{-2} instead of over 1000 MN m^{-2}.

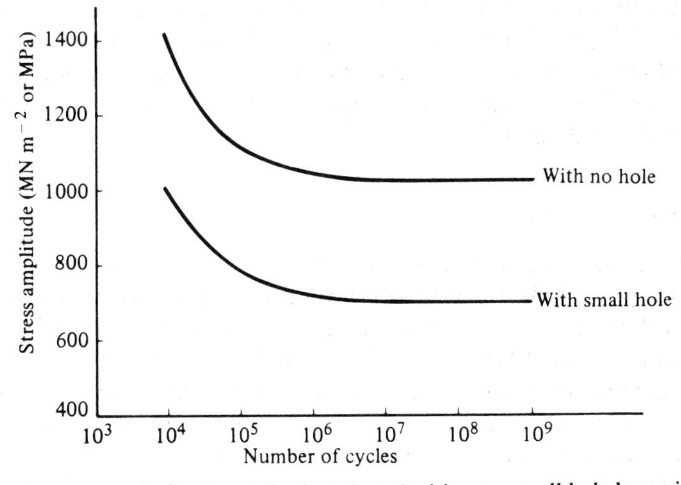

Figure 7.11 *S*/*N* graphs for a steel both with and without a small hole by acting as a stress raiser

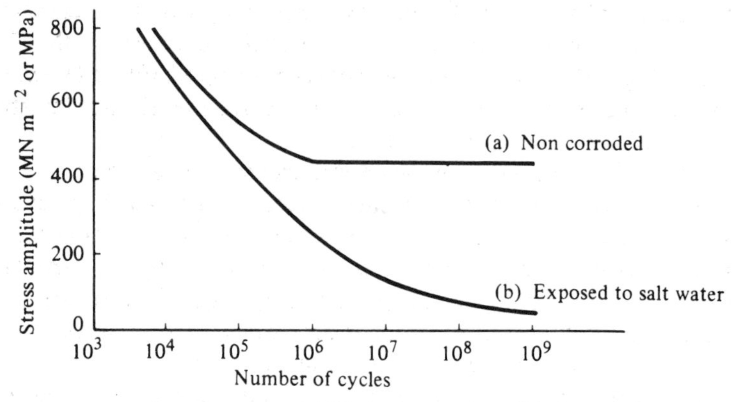

Figure 7.12 *S/N* graphs for a steel (a) With no corrosion, (b) Corroded to exposure to salt solution

Figure 7.12 shows the effect on the fatigue properties of a steel of exposure to salt solution. The effect of the corrosion resulting from the salt solution attack on the steel is to reduce the number of stress cycles needed to reach failure for every stress amplitude. The non-corroded steel has a fatigue limit of 450 MN^{-2}, the corroded steel has no fatigue limit. There is thus no stress amplitude below which failure will not occur. The steel can be protected against the corrosion by plating; for example, chromium or zinc plating of the steel can result in the same *S/N* graph as the non-corroded steel even though it is subject to a corrosive atmosphere.

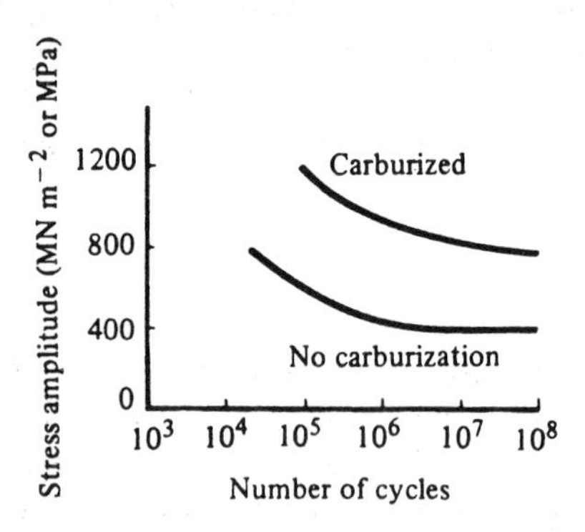

Figure 7.13 *S/N* graph for a steel, showing effect of carburization

Residual stresses can be produced by many fabrication and finishing processes. If the stresses produced are such that the surfaces have compressive residual stresses then the fatigue properties are improved, but if tensile residual stresses are produced at the surfaces then poorer fatigue properties result. The case-hardening of steels by carburizing results in compressive residual stresses at the surface, hence carburizing improves the fatigue properties. Figure 7.13 shows the effect of carburizing a hardened steel. Many machining processes result in the production of surface tensile residual stresses and so result in poorer fatigue properties.

The effect of surface finish on the fatigue properties of a component is very significant. Scratches, dents or even surface identification markings can act as stress raisers and so reduce the fatigue properties. Shot peening a surface produces surface compressive residual stresses and improves the fatigue performance. Some surface treatments e.g. conventional electroplating can, however, have a detrimental effect on the fatigue properties. This is because the surfaces end up with tensile residual stresses.

An increase in temperature can lead to a reduction in fatigue properties as a consequence of oxidation or corrosion of the metal surface increasing. For example, the nickel–chromium alloy Nimonic 90 undergoes surface degradation at temperatures around 700 to 800°C and there is a poorer fatigue performance as a result. In many instances an increase in temperature does result in a poorer fatigue performance.

The microstructure of an alloy is a factor in determining the fatigue properties. This is because the origins of fatigue failure are extremely localized, involving slip at crystal planes (see earlier in this chapter the section on stages in fatigue failure). Because of this, the composition of an alloy and its grain size can affect its fatigue properties. Inclusions, such as lead in steel, can act as nuclei for fatigue failure and so impair fatigue properties.

Heat treatment can change or produce residual stresses within a metal. As mentioned earlier, case hardening improves fatigue properties as a result of producing compressive residual stresses in surfaces. However, some heat treatments can reduce surface compressive stresses and so adversely affect fatigue properties. Some hardening and tempering treatments fall into this category.

Materials and fatigue resistance

Steels typically have a fatigue limit which is generally about 0.4 to 0.5 times the tensile strength of the material. Inclusions in steels can impair the fatigue properties, thus steels with lead or sulphur present to enhance machinability are to be avoided if good fatigue properties are required. The optimum structure for steels is tempered martensite for good fatigue resistance. Cast steels and cast irons tend to have relatively low endurance limits.

With steels there is generally a fatigue limit below which fatigue failure will not occur regardless of how many load cycles occur. However, with non-ferrous alloys this is generally not the case and a fatigue limit is quoted for a specific number of load cycles, usually 10^7 or 10^8 cycles. The fatigue limit for aluminium alloys is generally about 0.3 to 0.4 times the tensile strength of the material. Copper alloys tend to have fatigue limits about 0.4 to 0.5 times the tensile strength of the material.

Table 7.3 gives some typical values of tensile strength and fatigue limit, to about 10^7 or 10^8 cycles.

Table 7.3 *Tensile strength and fatigue limit data*

Material	*Tensile strength/ MN m^{-2}*	*Fatigue limit/ MN m^{-2}*
0.10% carbon steel, normalized	360	190
0.60% carbon steel, normalized	740	320
Alloy steel, 3.5% Ni, 1.55% Cr oil quenched and tempered 550°C	850	470
Stainless steel, 18% Cr, 8% Ni annealed	560	240
Grey cast iron	150 (min)	70
Aluminium, wrought alloy, 1.0% Mg, 0.27% Cu, 0.60% Si, 0.20% Cr, annealed	130	60
solution treated and aged	315	100
Aluminium, casting alloy, 5% Mg, as cast	170	50
Copper alloy, cupro-nickel, 30% Ni hard drawn	510	230

It should be realized that the above figures relate to the materials when used in perfect conditions. To illustrate this, consider the data in Table 7.4 for a steel with a tensile strength of about 800 MN m^{-2} and the effect on the fatigue limit of different conditions.

Table 7.4 *Effect of conditions on fatigue limit*

Condition	*Fatigue limit/MN m^{-2}*
Mirror polished surface, no flaws	470
Machined surface	420
Surface with a 0.1 mm notch	310
Hot worked surface	220
Under fresh water	170
Under sea water	120

7.4 The fatigue properties of plastics

Fatigue tests can be carried out on plastics in the same way as on metals. A factor not present with metals is that when a plastic is subject to an alternating stress it becomes significantly warmer. The faster the stress is alternated, i.e., the higher the frequency of the alternating stress, the greater the temperature rise. Under very high frequency alternating stresses the temperature rise may be large enough to melt the plastic. To avoid this, fatigue tests are normally carried out with lower-frequency alternating stresses than is usual with metals. The results of such tests, however, are not entirely valid if the alternating stresses experienced by the plastic component in service are higher than those used for the test.

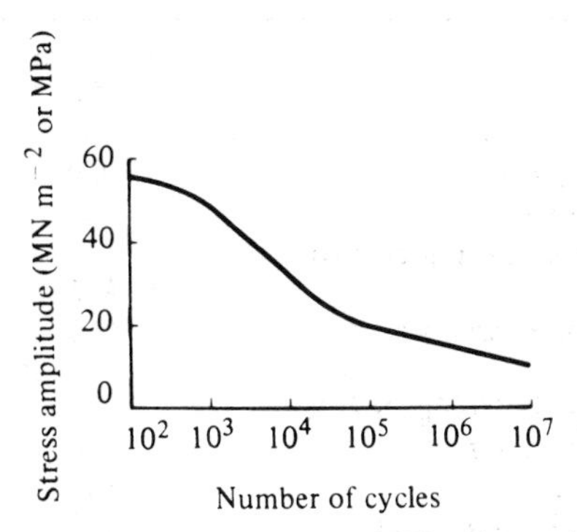

Figure 7.14 *S/N* graph for unplasticized PVC, alternating stress being a square waveform at frequency 0.5 Hz

Figure 7.14 shows an *S/N* graph for a plastic, unplasticized PVC. The alternating stresses were applied with a square waveform at a frequency of 0.5 Hz i.e. a change of stress every 2 s. The graph seems to indicate that there will be no stress amplitude for which failure will not occur; the material thus seems to have no fatigue limit.

Problems

1 Explain what is meant by fatigue failure.
2 List the types of test available for the determination of the fatigue properties of specimens.
3 Describe the various stages in the failure of a component by fatigue.
4 Explain the terms 'fatigue limit' and 'endurance limit'.
5 Figure 7.6 shows the *S/N* graph for an aluminium alloy.
   (a) For how many stress cycles could a stress amplitude of 140 MN m^{-2} be sustained before failure occurs?
   (b) What would be the maximum stress amplitude that should be applied if the component made of the material is to last for 50 million stress cycles?
   (c) The alloy has a tensile strength of 400 MN m^{-2} and a yield stress of 280 MN m^{-2}. What should be the limiting stress when such an alloy is used for static conditions? What should be the limiting stress when

the alloy is used for dynamic conditions where the number of cycles is not likely to exceed 10 million?

6 Explain an *S/N* graph and state the information that can be extracted from the graph.

7 What is the fatigue limit for the uncarburized steel giving the *S/N* graph in Figure 7.13

8 What is the endurance limit for the unplasticized PVC at 10^6 cycles that gave the *S/N* graph in Figure 7.14?

9 Plot the *S/N* graph for the nickel–chromium alloy Nimonic 90, which gave the following fatigue test results. Determine from the graph the fatigue limit.

Stress amplitude/MN mm^{-2}	*Number of cycles before failure*
750	10^5
480	10^6
350	10^7
320	10^8
320	10^9
320	10^{10}

10 Plot the *S/N* graph for the plastic (cast acrylic) which gave the following fatigue test results when tested with a square waveform at 0.5 Hz. Why specify this frequency? What is the endurance limit at 10^6 cycles?

Stress amplitude/MN mm^{-2}	*Number of cycles before failure*
70	10^2
62	10^3
58	10^4
55	10^5
41	10^6
31	10^7

11 Figure 7.15 shows the *S/N* graph for a nickel-based alloy Iconel 718.
(a) What is the fatigue limit?
(b) What is the significance of the constant stress amplitude part of the graph from 10^0 to 10^4 cycles?
(c) The graph is for the material at 600°C. The tensile strength at that temperature is 100 MN m^{-2}. What can be added to your answer to part (b)?

12 Figure 7.16 shows two *S/N* graphs, one for the material in an un-notched state, the other for the material with a notch. Which of the graphs would you expect to represent each condition? Give reasons for your answer.

13 List factors that contribute to the onset of fatigue failure and those which tend to resist fatigue.

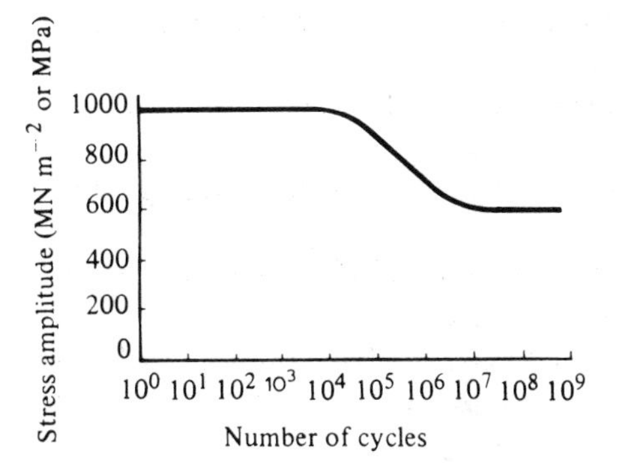

Figure 7.15 *S/N* graph for a nickel-based alloy, Iconel 718

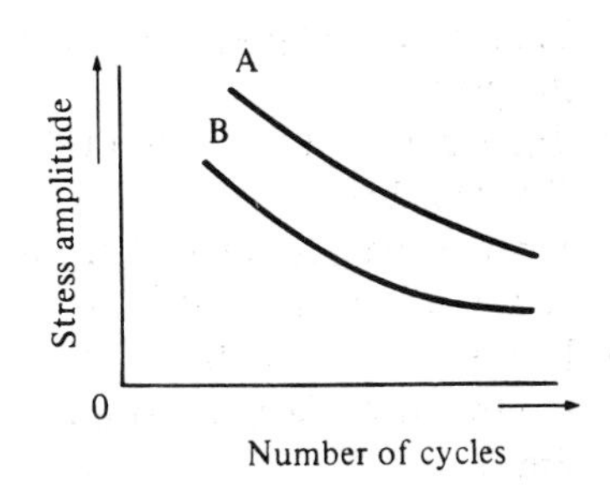

Figure 7.16 *S/N* graph

14 What is the effect on the *S/N* graph for a component of different mean stresses about which the alternating stresses occur?

15 Plot the Goodman diagram for a steel having a fatigue limit of 320 MN m^{-2} and a tensile strength of 740 MN m^{-2}. Predict, using the diagram, whether the following stress amplitudes are likely to lead to fatigue failure: (a) 100 MN m^{-2} at a mean stress of 200 MN m^{-2}, (b) 150 MN m^{-2} at a mean stress of 400 MN m^{-2}.

8

Creep

8.1 Short-term and long-term behaviour

There are many situations where a piece of material is exposed to a stress for a protracted period of time. The stress/strain data obtained from the conventional tensile test refer generally to a situation where the stresses are applied for quite short intervals of time and so the strain results refer only to the immediate values resulting from stresses. Suppose stress were applied to a piece of material and the stress remained acting on the material for a long time – what would be the result? If you tried such an experiment with a strip of lead you would find that the strain would increase with time – the material would increase in length with time even though the stress remained constant. This phenomenon is called *creep*, which can be defined as the continuing deformation of a material with the passage of time when the material is subject to a constant stress.

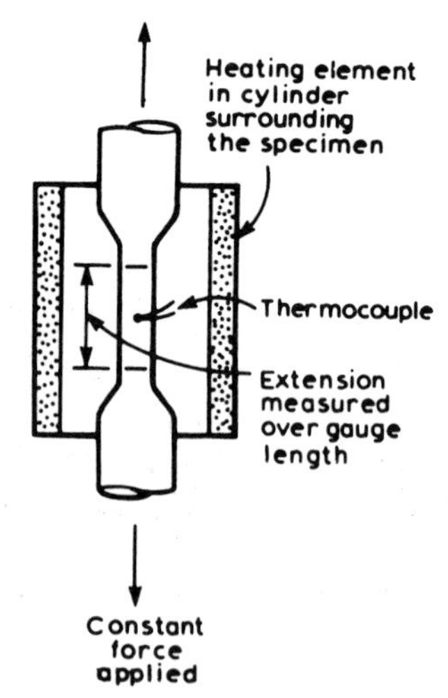

Figure 8.1 A creep test

For metals, other than the very soft metals like lead, creep effects are negligible at ordinary temperatures, but however become significant at high temperatures. For plastics, creep is often quite significant at ordinary temperatures and even more noticeable at higher temperatures.

Figure 8.1 shows the essential features of a creep test. A constant stress is applied to the specimen, sometimes by the simple method of suspending loads from it. Because creep tests with metals are usually performed at high temperatures a furnace surrounds the specimen, the temperature of the furnace being held constant by a thermostat. The temperature of the specimen is generally measured by a thermocouple attached to it.

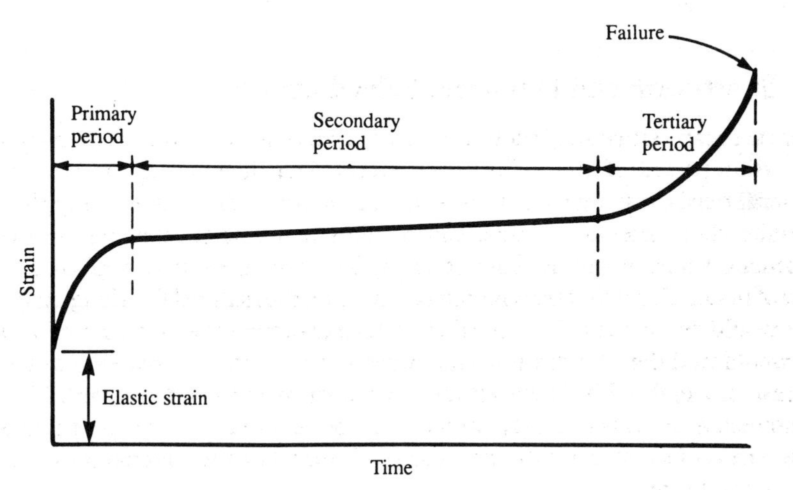

Figure 8.2 Typical creep curve for a metal

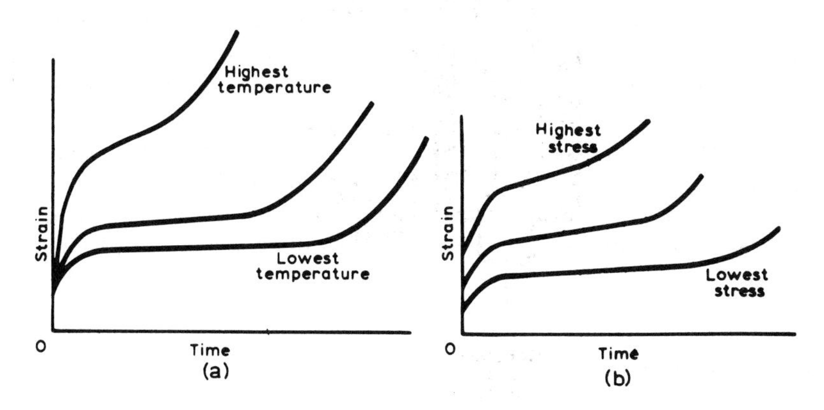

Figure 8.3 Creep behaviour for a material (a) At different temperatures but subject to constant stress, (b) At different stresses but subject to constant temperature

Figure 8.2 shows the general form of results from a creep test. Following an 'instantaneous', elastic strain region the curve generally has three parts. During the *primary creep* period the strain is changing but the rate at which it is changing with time decreases. During the *secondary creep* period the strain increases steadily with time at a constant rate. During the *tertiary creep* period the rate at which the strain is changing increases and eventually causes failure.

A family of graphs can be produced showing the creep for different initial stresses at a particular temperature and for different temperatures for a particular initial stress (Figure 8.3).

Stress to rupture

Because of creep an initial stress, which did not produce early failure, can result in a failure after some period of time. Such an initial stress is referred to as the *stress to rupture* in some particular time. Thus an acrylic plastic may have a rupture stress of 50 MN m^{-2} at room temperature for failure in one week, the value of the stress to rupture depending, for a particular material, on the temperature and the time.

Table 8.1 *Stress to rupture data for a 0.2 per cent carbon steel*

	Rupture stresses (MN m^{-2})	
Time (h)	*400°C*	*500°C*
1000	295	118
10 000	225	59
100 000	147	30

Table 8.1 shows the stress to rupture data that might be quoted for a 0.2 per cent carbon steel. The data means that at 400°C the carbon steel will rupture after 1000 hours if the stress is 295 MN m^{-2}. If the steel at 400°C is required to last for 10 000 hours then the stress must be below 147 MN m^{-2}. If, however, the temperature is 500°C then to last 100 000 hours the stress must be below 30 MN m^{-2}.

The stress to rupture a material in a particular time depends on the temperature. Figure 8.4 shows how the stress to rupture a Pireks 25/20 alloy in 10 000 hours depends on the temperature. At 800°C a stress of 65 MN m^{-2} will result in rupture, while at 1050°C a stress of only about 15 MN m^{-2} is needed.

The Larson–Miller parameter

In order to ascertain the stress to rupture for a material from data, account has to be taken of the stress, temperature and time. There are thus three variables

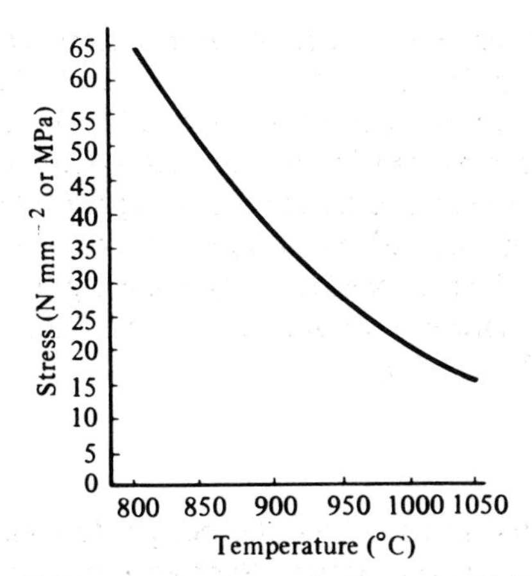

Figure 8.4 Data showing stress to rupture at 10 000 h for Pireks 25/20 alloy (Courtesy of Darwins Alloy Castings Ltd)

involved. Tables of data involving the three variables can be used, as indicated in the previous section of this chapter or, alternatively, the *Larson–Miller parameter* can be used. This is a relationship which, when used with a master graph for a material, enables the creep rupture time for any particular stress to be determined or the creep rupture stress for any particular time. The parameter P is a number obtained from the following relationship:

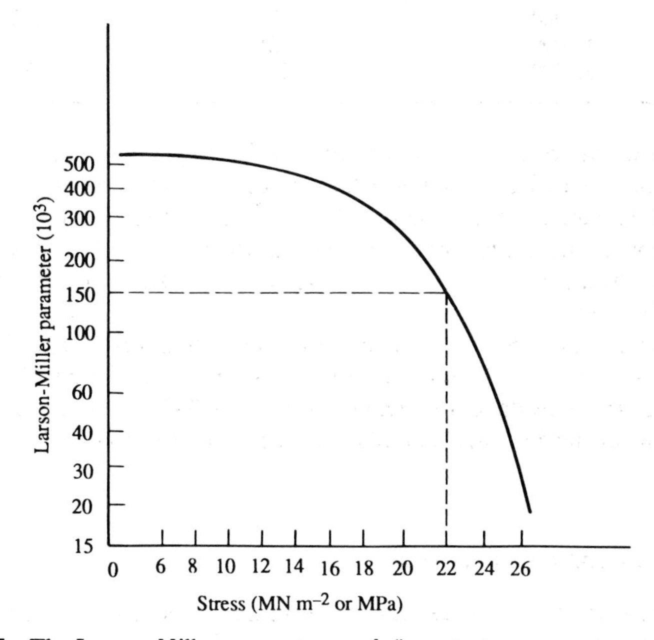

Figure 8.5 The Larson–Miller parameter graph for a steel

$P = T(C+\log_e t)$

where T is the temperature on the kelvin scale, C is a constant which is usually taken to have the value 20, and t is the time to rupture in hours.

To illustrate the use of the parameter, consider the master graph for a steel in Figure 8.5. To estimate the creep rupture time at 400°C (673 K) for a stress of 150 MN m^{-2} the graph is used to obtain the value of the parameter. This gives a value of 22×10^3. Hence the parameter equation becomes, when C is taken to have the value 20:

$$22 \times 10^3 = 673\,(20 + \log_e t)$$
$$\log_e t = \frac{22 \times 10^3}{673} - 20$$
$$t = e^{12.7}$$
$$= 3.24 \times 10^5 \text{ hours}$$

Design data

Creep can be an important consideration in the design of structures, particularly if they are to be used at elevated temperatures. One way of presenting data for design use is in terms of the design stress that can be permitted at any temperature if the creep is to be kept within specified limits within some time. Figure 8.6 shows the data for an alloy where the limit is 1 per cent creep in 10 000 hours. Thus for that alloy, a stress of about 59 MN m^{-2} will produce the 1 per cent creep in 10 000 hours at 800°C. Thus if the creep is not to exceed 1 per cent under these conditions 59 MN m^{-2} is the maximum permissible design stress. If the temperature is 1050°C the maximum permissible design stress for 10 000 hours is only about 10 MN m^{-2}.

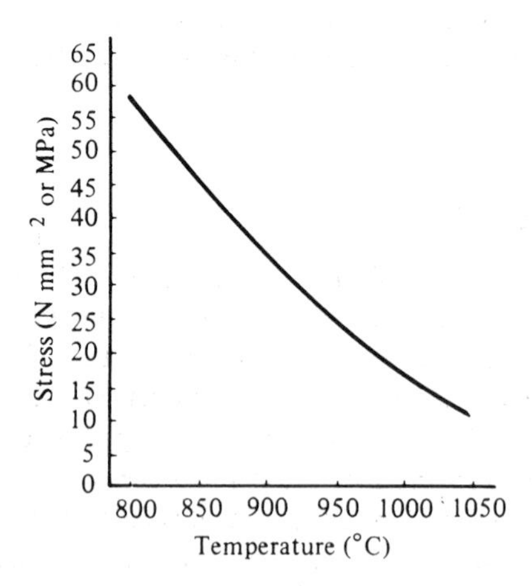

Figure 8.6 Stress to give 1 per cent creep in 10 000 h for Pireks 25/20 alloy (Courtesy of Darwins Alloy Castings Ltd)

8.2 Creep mechanisms in metals

Following the initial instantaneous strain, primary creep occurs as a result of movements of dislocations. Initially the dislocations can move easily and quickly, hence the rapid increase in strain with time. However, dislocation pile-up starts to occur at grain boundaries, i.e. work hardening occurs and this slows down the rate of creep, though some recovery does occur (see Chapter 2 for a discussion of these terms in relation to dislocation movements). The amount of recovery that occurs depends on the temperature – the higher the temperature the greater the amount of recovery. Therefore, at low temperatures when there is little recovery the work hardening quite rapidly reduces the rate of creep while at higher temperatures the work hardening effect is reduced by recovery and the creep rate is higher.

During the secondary creep state the rate of recovery is sufficiently fast to balance the rate of work hardening with the result that the material creeps at a steady rate. In addition to dislocation movements there is also, during this stage of creep, some movement of atoms along grain boundaries. This is referred to as grain boundary slide. This mechanism becomes more significant the higher the temperature. Thus, since fine grained materials contain more grain boundary regions per unit volume of a material than a coarse grained version of the same material, at high temperatures fine grained material creeps more than coarse grained. However, at low temperatures when there is little movement of atoms along grain boundaries, fine grained materials are more creep resistant than coarse grained ones since they inhibit the movement of dislocations.

With tertiary creep the rate of strain increases rapidly with time and finally results in fracture. This accelerating creep rate occurs when voids or microcracks occur at the grain boundaries. These are the result of vacancies migrating to such areas and the movement of atoms along grain boundaries. The voids grow and link up so that finally the material fails at the grain boundaries.

Creep-resistant materials

The movement of dislocations is an essential features of creep, hence creep can be reduced by reducing the movement. One mechanism is the use of alloying elements which give rise to dispersion hardening. Finely dispersed precipitates are effective barriers to the movement of dislocations. The Nimonic series of alloys, based on an 80/20 nickel–chromium alloy with the addition of small amounts of titanium, aluminium, carbon or other elements which form fine precipitates (see Table 8.2). The Nimonic alloys also have excellent resistance to corrosion at high temperatures and so this, combined with their high creep resistance, makes them useful for high temperature applications such as gas turbine blades.

Table 8.2 *Effect of alloy changes on creep resistance*

	Rupture stresses (MN m^{-2})					*Oxidation limit*
	Temp.	*700°C*		*800°C*		
Material and condition	*Time*	*1000 h*	*10 000 h*	*1000 h*	*10 000 h*	*(°C)*
80% Ni, 20% Cr ½ h 1050°C, air cool		100	59	31	18	900
80% Ni, 20% Cr, Co, Al, Ti 8 h 1080°C, air cool + 16 h 700°C, air cool		370	245	140	70	900

8.3 Factors affecting creep with metals

The main factors affecting creep behaviour with metals are:

1 *Temperature* – the higher the temperature the greater the creep at a particular stress (see Figure 8.3).

2 *Stress* – the higher the stress at a particular temperature the greater the creep (see Figure 8.3).

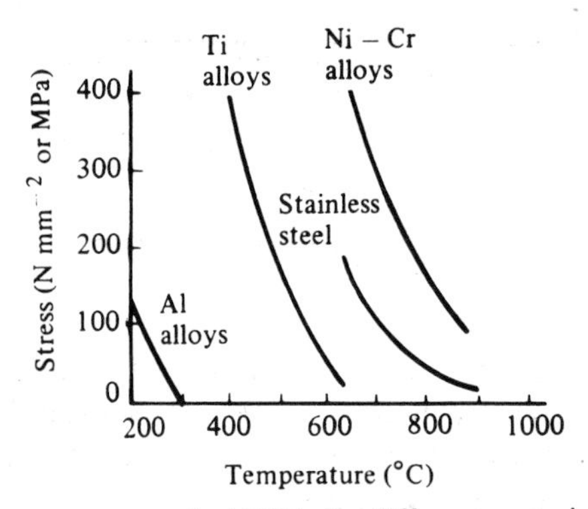

Figure 8.7 Stress to rupture in 1000 h for different materials

3 *Composition* – the creep behaviour of a metal depends on its composition. Figure 8.7 shows how the stress to rupture different materials in 1000 hours varies with temperature. Aluminium alloys fail at quite low stresses when the temperature rises over 200°C. Titanium alloys can be used at higher temperatures before the stress to rupture drops to very low values, while stainless steel is even better and nickel–chromium alloys offer yet better resistance to creep. As mentioned earlier in this chapter, the

addition of elements to an alloy in order to make it precipitation hardened increases its creep resistance.

4 *Grain size* – as mentioned earlier in this chapter grain size can have an effect on the creep behaviour of a material. At low temperatures fine grained material has the best creep properties while at high temperatures coarse grained material is best.

8.4 Factors affecting creep behaviour with plastics

While creep is significant mainly for metals at high temperatures, it can be significant with plastics at normal temperatures. The creep behaviour of a plastic depends on temperature and stress, just like metals. It also depends on the type of plastic involved – flexible plastics show more creep than stiff ones.

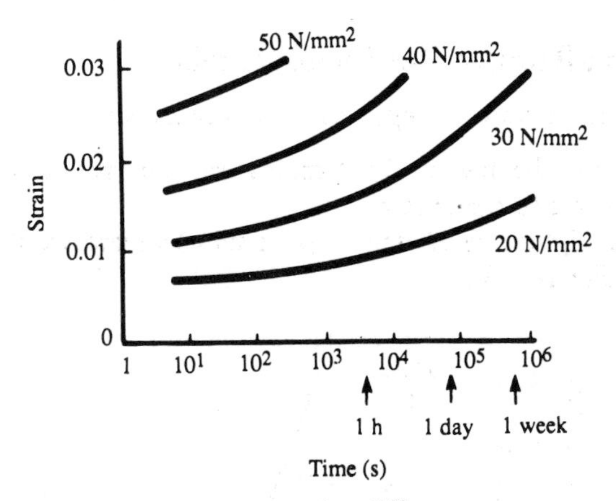

Figure 8.8 Creep behaviour of polyacetal at different stresses

Figure 8.8 shows how the strain on a sample of polyacetal at 20°C varies with time for different stresses. The higher the stress the greater the creep. As can be seen from the graph, the plastic creeps quite substantially in a period of just over a week, even at relatively low stresses.

There are three additioned ways by which creep data can be presented for plastics:

1 Isochronous stress/strain graphs
2 Creep modulus/time graphs
3 Isometric stress/time graphs.

Based on the Figure 8.9 graph of strain against time at different stresses a graph of stress against strain for different times can be produced. Thus for a time of 10^2 s, a vertical line drawn on Figure 8.8 enables the stresses needed for different strains after this time to be read from the graph (Figure 8.9a). The

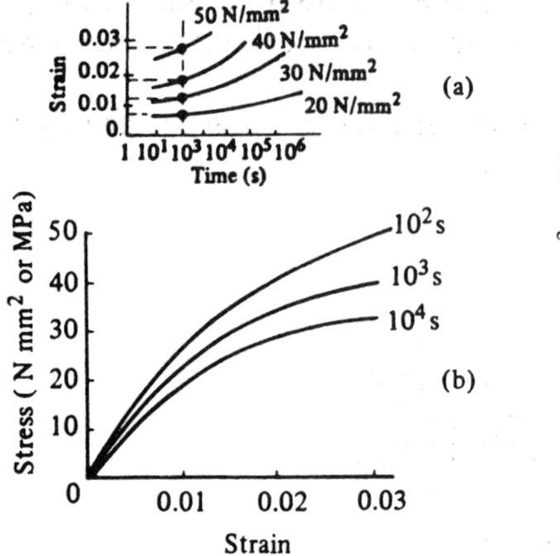

Figure 8.9 (a) Obtaining stress/strain data, (b) Isochronous stress/strain graph

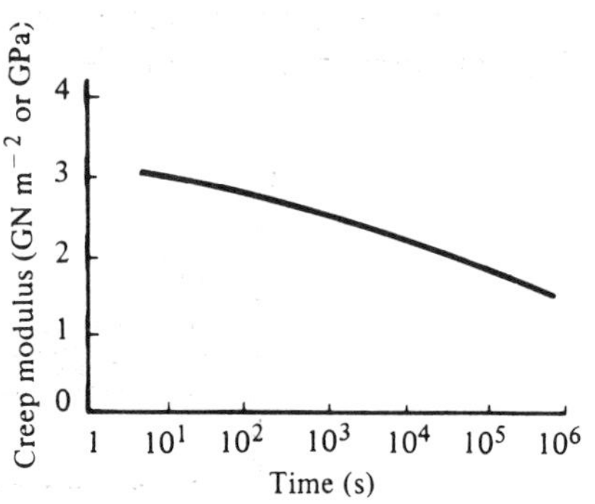

Figure 8.10 Variation of creep modulus with time for polyacetal at 0.5 per cent strain and 20°C

resulting stress/strain graph is shown in Figure 8.9b and is known as an *isochronous stress/strain* graph.

For a specific time the quantity obtained by dividing the stress by the strain for the isochronous stress/strain graph can be calculated and is known as the *creep modulus*. It is not the same as Young's modulus though it can be used to compare the stiffness of plastics. The creep modulus varies both with time and strain and Figure 8.10 shows how, at 0.5 per cent strain and 20°C it varies with time for the polyacetal described in Figures 8.8 and 8.9.

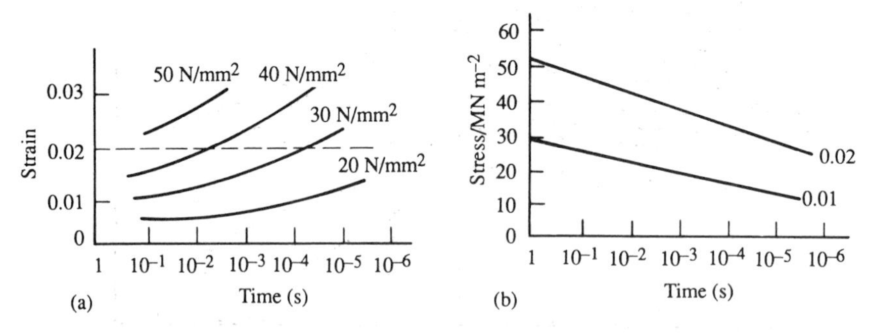

Figure 8.11 (a) Obtaining stress/time data, (b) Isometric stress/time graph

The isochronous stress/strain graph is based on data taken from the strain/time creep graph at constant time. The isometric stress/time graph is based on data taken from the strain/time graph at constant strain. Figure 8.11 shows how such a graph can be obtained from Figure 8.8. Thus for a strain of 0.02, a horizontal line drawn on Figure 8.8 enables the stresses at different times to be read from the graph.

Other creep data that can be given with specification for plastics is the way in which the stress to rupture varies with time.

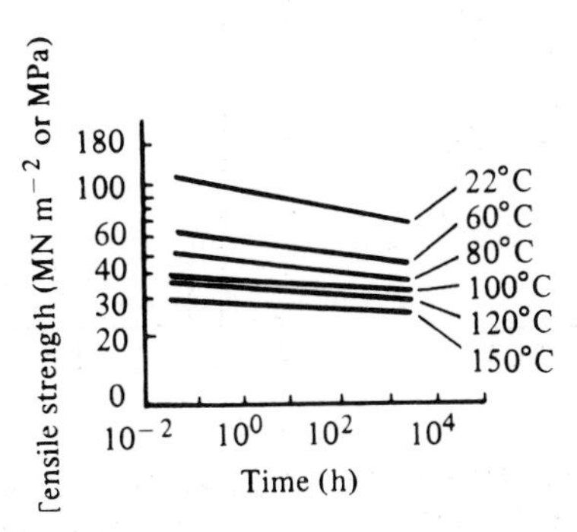

Figure 8.12 Creep rupture graph for Durethan BKV 30 (Courtesy Bayer UK Ltd)

Figure 8.12 shows how the stress to rupture a plastic, Durethanpolyamide, varies with time at different temperatures. The higher the temperature the lower the stress needed to rupture the material after any particular time.

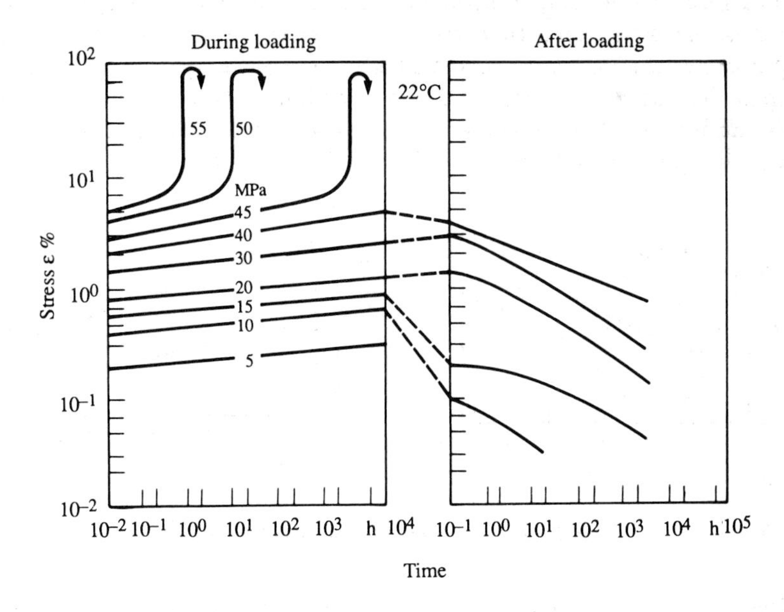

Figure 8.13 Creep and recovery properties of non-reinforced Makrolon 2800 (Courtesy Bayer UK Ltd)

Recovery from creep

At the end of a creep test on a metal when the load is removed, only that part of the deformation which was elastic is recoverable – the plastic deformation remains as a permanent set. Polymeric materials, however, behave differently. On removing the load the polymeric material can recover most, or even all, of the strain given sufficient time. Figure 8.13 shows the creep and recovery for Makrolon.

The time taken to recover depends on the initial strain and time for which the material was creeping under load. For this reason such data is often given in the form of a graph of fractional recovery of strain with reduced time.

$$\text{Fractional recovery of strain} = \frac{\text{Strain recovered}}{\text{Creep strain at time of load removal}}$$

$$\text{Reduced time} = \frac{\text{Recovery time}}{\text{Time under load before recovery starts}}$$

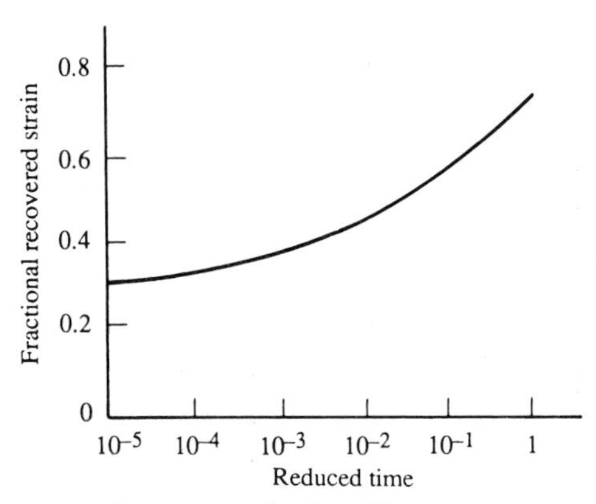

Figure 8.14 Recovery from creep of nylon 66

A fractional recovery of strain value of one means that the material has completely recovered and is back to its original size. A reduced time value of one means that the recovery time is the same as the creep time. Figure 8.14 shows the graph for nylon 66. Usually a number of lines are indicated on the graph for different creep conditions, e.g. low strain for a short time or high strain for a long time, or a band is shown within which all the data for the different conditions occurs.

Viscoelastic behaviour and models

The creep behaviour of polymeric materials shows that, even at room temperature, such materials cannot be regarded as elastic in the way that metals can. With metals the behaviour at room temperature is virtually independent of time thus when a load is applied the metal almost instantaneously becomes extended and maintains the same extension regardless of time. When the load is removed the metal almost instantaneously recovers its original dimensions (assuming it was not loaded to beyond its elastic limit). With polymeric materials there is a comparatively slow response to loading and to the release of loads. The response is, to some extent, like the flow of a very viscous liquid. With such a liquid the application of forces causes the liquid to flow and it will continue to flow as long as the forces are applied. Polymeric materials appear to behave somewhere between these two extremes of elastic and viscous behaviour, having some aspects of both. For this reason polymeric materials are referred to as showing *viscoelasticity*.

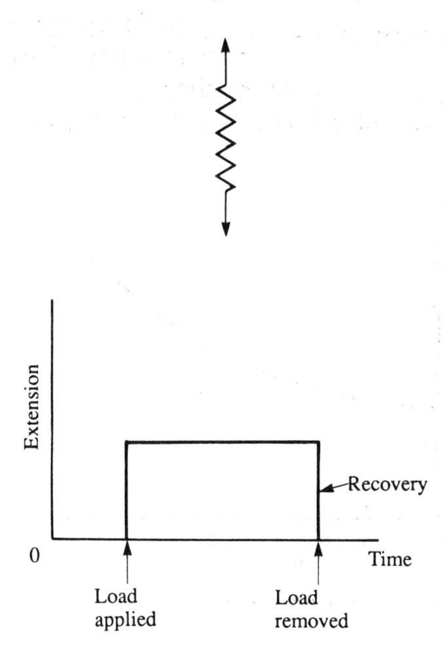

Figure 8.15 The extension–time relationship for a spring

An elastic material behaves rather like a spring. Figure 8.15 shows how the displacement of a spring will vary with time when loaded and unloaded. An ideal viscous fluid (see Chapter 1) has a displacement which is proportional to time.

$$\frac{F}{A} = \eta \frac{dv}{dx} = \eta \times \text{(rate of displacement)}$$

Hence the rate of displacement is a constant if F/A and η are constant – dv/dx is the velocity gradient when there is a viscous force acting over an area A with a fluid of viscosity η. This behaviour can be represented by a dashpot, i.e. a

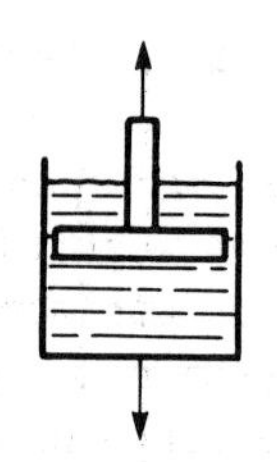

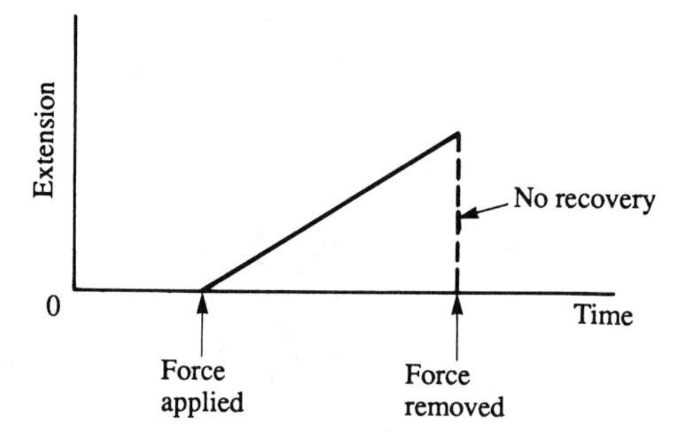

Figure 8.16 The extension–time relationship for a dashpot

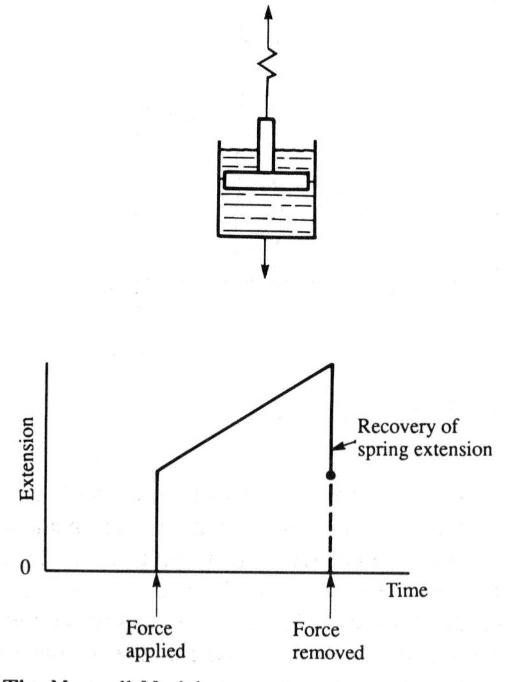

Figure 8.17 The Maxwell Model

loosely fitting piston able to move in a cylinder containing a viscous fluid (Figure 8.16).

The viscoelastic behaviour of polymeric materials can be simulated by a combination of springs and dashpots. Figure 8.17 shows the behaviour of a spring in series with a dashpot, the arrangement being known as the *Maxwell Model*. Because the two elements are in series, the extension at any instant will be the sum of the extensions due to each element. The result is a graph which represents how the strain will vary with time. This model does not, however, give a result which effectively replicates the actual behaviour of polymeric materials.

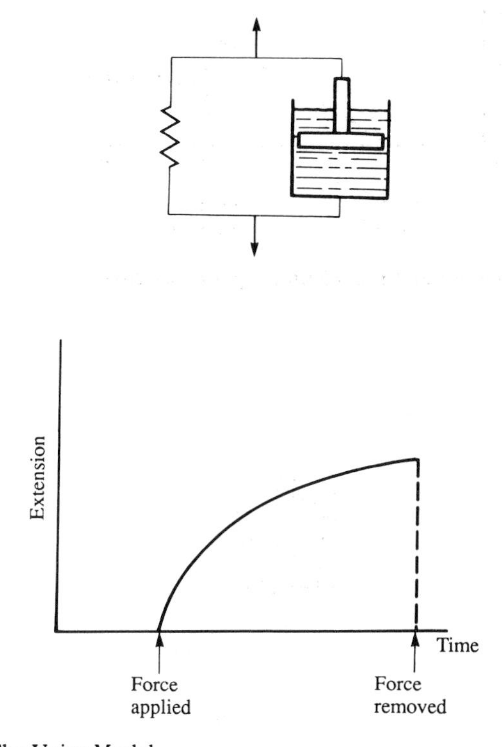

Figure 8.18 The Voigt Model

Another model used to simulate viscoelasticity is the *Voigt Model*. This consists of a spring in parallel with a dashpot and has the behaviour shown in Figure 8.18. Because the two elements are in parallel the extension at any instant of the spring must be equal to the extension of the dashpot. A consequence of this is that the model does not give any instantaneous extension but an extension which gradually increases with time and shows no recovery. This again does not replicate the behaviour of polymeric materials.

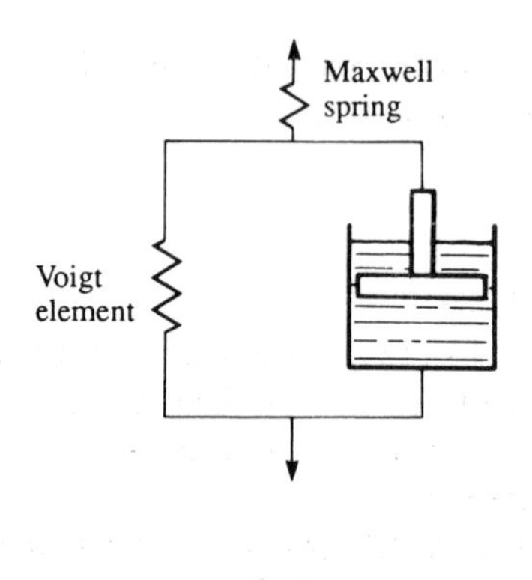

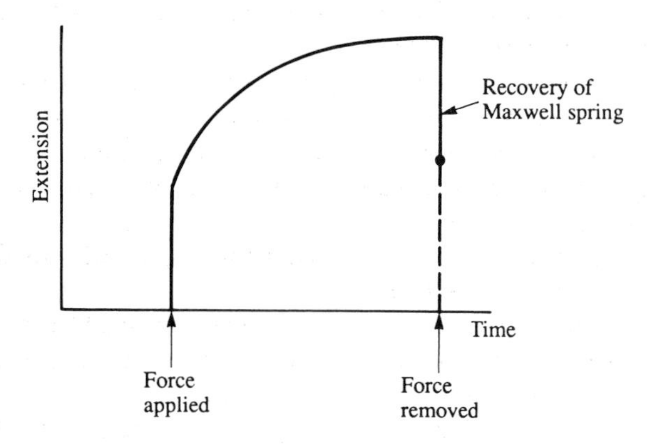

Figure 8.19 The combined Maxwell-Voigt Model

A more successful model is a combination of the Maxwell and Voigt Models, (Figure 8.19). This model shows an instantaneous extension, followed by a gradual creep with time, and a recovery of part of the extension when the load is removed.

Stress relaxation

Creep is defined as the continuing deformation of a material or change in strain when the material is subject to a constant stress. Stress relaxation is a related phenomenon, being the decay of stress with time when a material is subject to a constant strain. This means that with a bolted joint stress relaxation can cause leaks at the joint. With metals, stress relaxation, like creep, is mainly a concern at high temperatures. With plastics, however, stress relaxation can be significant at room temperature. This is particularly significant when the fit between a plastic component and a metal component depends on the stresses set up in the plastic component. Stress relaxation with time data is similar to the isometric stress with time graphs for a material.

Problems

1 Explain what is meant by 'creep'.
2 Describe the form of a typical strain/time graph resulting from a creep test and explain the significance of the slope of the graph at its various stages.
3 Describe the effect of (a) increased stress and (b) increased temperature on the creep behaviour of materials.
4 For the alloy described in Figure 8.6 estimate the stress that will result in a 1 per cent creep at a temperature of 900°C after 10 000 h.
5 For the alloy described in Figure 8.4, estimate the stress to rupture at 10 000 h for a temperature of 900°C.
6 Under what circumstances, would you consider that a metal like that described in Figures 8.6 and 8.4, would be necessary? What would you estimate the limiting temperature for use of such an alloy?
7 Explain how an isochronous stress/strain graph for a polymer can be obtained from creep test results.
8 Explain the significance of the graph shown in Figure 8.12 for the creep rupture behaviours of Durethan.
9 Durethan, as described by Figure 8.12 is used for car fan blades, fuse box covers, door handles and plastic seats. How would the behaviour of the material change when the temperature or stress rises?
10 Figure 8.20 shows how the strain changes with time for two different polymers when they are subjected to a constant stress. Describe how the materials will creep with time. Which material will creep the most?

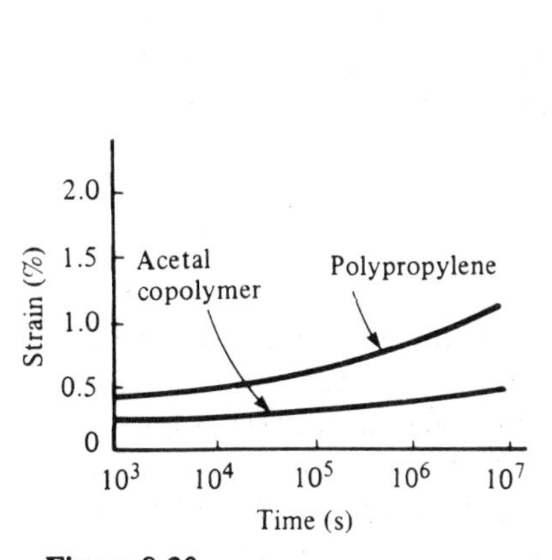

Figure 8.20

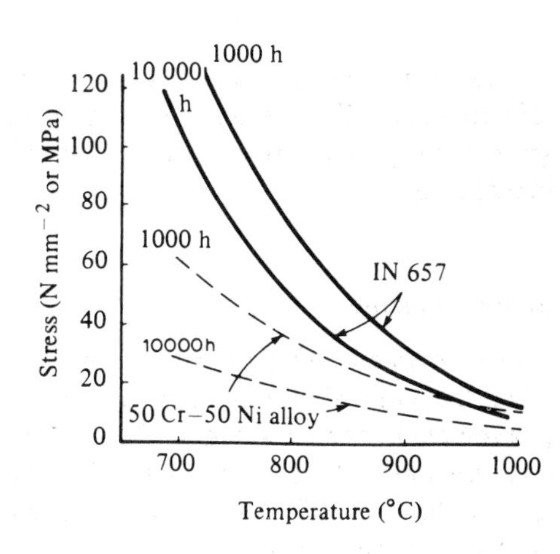

Figure 8.21 Creep rupture data from 'High chromium Cr–Ni alloys to resist residual fuel ash corrosion'. (Courtesy of Inco Europe Ltd)

11 Figure 8.21 shows the stress rupture properties of two alloys, one 50 per cent chromium and 50 per cent nickel, the other (IN 657) 48–52 per cent chromium, 1.4–1.7 per cent niobium, 0.1 per cent carbon, 0.16 per cent nitrogen, 0.2 per cent carbon + nitrogen, 0.5 per cent maximum silicon, 1.0 per cent iron, 0.3 per cent maximum manganese and the remainder nickel. The creep rupture data is presented for two different times, 100 h and 10 000 h.
   (a) What is the significance of the difference between the 1000 h and 10 000 h graphs?
   (b) What is the difference in behaviour of the 50 Cr–50 Ni alloy and the IN 657 alloy when temperatures increase?
   (c) The IN 657 alloy is said to show 'improved hot strength' when compared with the 50 Cr–50 Ni alloy. Explain this statement.

12 Explain what is meant by recovery, explaining the significance of the recovery data presented in Figure 8.13.

13 Describe the creep mechanisms that can occur with a metal in the primary, secondary and tertiary stages of creep.

9

Environmental stability of materials

9.1 Corrosion

The car owner can rightly be concerned about rust patches appearing on the bodywork of the car, as the rust not only makes the bodywork look shoddy but indicates a mechanical weakening of the material. The steel used for the car bodywork has thus changed with time due to an interaction between it and the environment. The possibility of such corrosion is therefore a factor that has to be taken into account when a material is selected for a particular purpose.

The term *corrosion* is used to describe an unintentional chemical or electrochemical reaction between a metal and its environment which results in the removal of the metal or its conversion into an oxide or some other compound.

The environments to which metals can be commonly exposed are the air, water and marine. In addition there are many more specialized environments such as steam, high temperatures, and chemicals.

Dry corrosion

The term *dry corrosion* is used to describe that corrosion which arises as a result of a reaction between a metal and an oxidizing gas, e.g. the oxygen in air.

Most metals react, at moderate temperatures, only slowly with the oxygen in the air. The result is the build up of a layer of oxide on the surface of the metal, which can insulate the metal from further reaction with the oxygen. Aluminium is an example of a metal that builds up a protective oxide layer which is a very effective barrier against further oxidation. If oxidation were the only reaction which tended to occur between a metal and its environment then probably corrosion would present few problems. The presence of moisture in the environment can very markedly affect corrosion, as can the presence of chemically active pollutants.

The reaction between a metal and oxygen can be summarized as:

Metal + oxygen → metal oxide

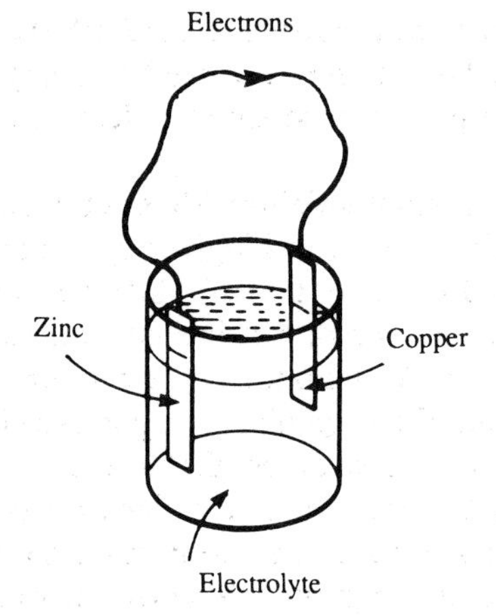

Figure 9.1 A simple cell

Galvanic corrosion

A simple electrical cell could be just a plate of copper and one of zinc (Figure 9.1). dipping into an *electrolyte*, a solution that conducts electricity. Such a cell gives a potential difference between the two metals – it can be measured with a voltmeter or used to light a lamp. Different pairs of metals give different potential differences. Thus a zinc–copper cell gives a potential difference of about 1.1 V, and an iron–copper cell about 0.8 V. A zinc–iron cell gives a potential difference of about 0.3 V; this value is however the difference between the potential differences of the zinc–copper and iron–copper cells. It is as though we had a cell made up with zinc–copper–iron.

By tabulating values of the potential differences between the various metals

Table 9.1 *Galvanic series*

Metal	*Potential difference/V*
Gold	+1.7
Silver	+0.80
Copper	+0.34
Hydrogen	0.00
Lead	−0.13
Tin	−0.14
Nickel	−0.25
Iron	−0.44
Zinc	−0.76
Aluminium	−1.67
Magnesium	−2.34
Sodium	−2.71

and a standard, a table can be produced from which the potential differences between any pair of metals can be forecast (see Table 9.1). The standard used is hydrogen and the table gives the potential differences relative to hydrogen for a number of metals.

The table gives, for a silver–aluminium cell, a potential difference of 2.47 V, that is $+0.80 - (-1.67)$ V. The metal highest up the table, as shown above, behaves as the negative electrode of the cell while the metal lowest in the table is the positive electrode. The term *cathode* is used for the negative electrode and *anode* for the positive electrode. Thus for the silver–aluminium cell the silver is the negative electrode and the aluminium the positive electrode.

With a copper–zinc cell it is found that the copper electrode remains unchanged after a period of cell use but the zinc electrode is badly corroded. With any cell the anode is corroded and the cathode not affected. The greater the cell potential difference the greater the corrosion of the anode.

The above table lists what is called the *electromotive series* or *galvanic series*. Tables of such series are available for metals in various environments. Such tables are of use in assessing the possibilities of corrosion when two different metals are in electrical contact, either directly or through a common electrolyte.

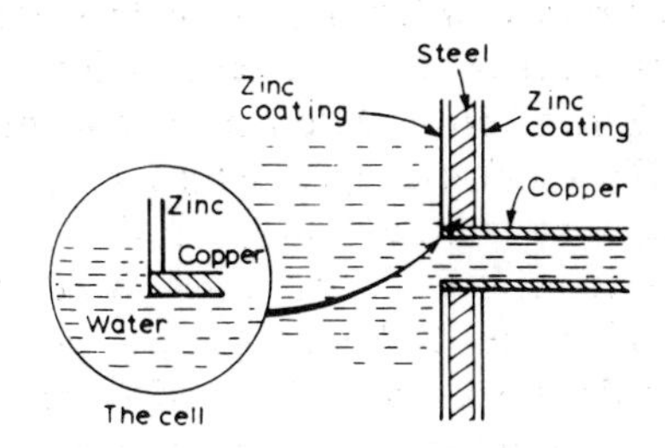

Figure 9.2 Copper–zinc cell for copper pipe connected to galvanized steel water tank

If a copper pipe is connected to a galvanized steel tank, perhaps the cold water storage tank in your home, a cell is created (Figure 9.2) and corrosion follows. Galvanized steel is zinc-coated steel; there is thus a copper–zinc cell. With such a cell the copper is the cathode and the zinc the anode; the zinc thus corrodes and so exposes the iron. Iron–copper is also a good cell with the iron as the anode. The result is corrosion of the iron and hence the overall result of connecting the copper pipe to the tank is likely to be a leaking tank.

Stainless steel and mild steel form a cell, the mild steel being the anode. Thus if a stainless steel trim, on say a car, is in electrical contact with mild steel bodywork, then the bodywork will corrode more rapidly than if no stainless steel trim were used. The electrical connection between the stainless steel and the mild steel may be through water gathering at the junction between the two. With oxygen-free clean water the cell potential difference may be only 0.15 V, but if the water were sea water containing oxygen the potential difference could become as high as 0.7 V. So the use of salt in Britain, to melt ice on the road, leads to greater corrosion of cars.

It is not only with two separate metals that cells can be produced and corrosion occur, as galvanic cells can be produced in a number of ways. An alloy or a metal containing impurities can give rise to galvanic cells within itself. For example, brass is an alloy of copper and zinc, a copper–zinc cell has the zinc as the anode and thus the zinc corrodes. The effect is called *dezincification*, one example of *demetallification*. After such corrosion the remaining metal is likely to be porous and lacking in mechanical strength.

A similar type of corrosion takes place for carbon steels in the pearlitic condition, the cementite in the steel acting as the cathode and the ferrite as the anode. The ferrite is thus corroded.

Cast iron is a mixture of iron and graphite, the graphite acting as the cathode and the iron as the anode. The iron is thus corroded the effect being known as *graphitization*.

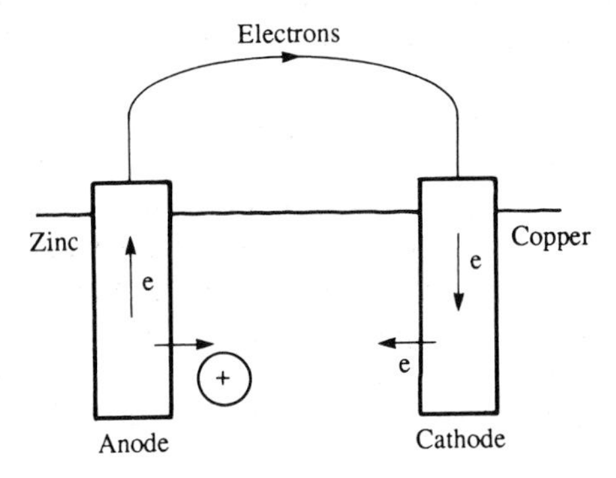

Figure 9.3 Cell action

Electrochemical corrosion

Galvanic corrosion is an *electrochemical reaction*. An electrochemical reaction involves the passage of electric currents through parts of the metal not being corroded. Thus in the case of the cell action described by Figure 9.3, the zinc undergoes a reaction in which zinc ions are produced and pass into solution while electrons remain behind in the metal. This anode reaction can be represented as:

Metal $\rightarrow$ positive metal ions + electrons

If the reaction is to continue then the electrons left behind in the metal as a result of the anode reaction must move away to a point where they can be 'used up'. In the case of the cell this is the copper. A cathode reaction occurs. This can take a number of forms. The electrons can combine with the positive metal ions released from the anode and in solution.

Electrons + positive ions → metal

If the solution is acidic the cathode reaction may involve the electrons combining with hydrogen ions that are in the solution to give hydrogen gas which then bubbles out of the solution at the cathode.

Electrons + hydrogen ions → hydrogen

Overall, at the anode electrons are left in the metal as metal ions move out of the metal while at the cathode electrons are absorbed by ions.

Concentration cells

Variations in concentration of the electrolyte in contact with a metal can lead to corrosion. That part of the metal in contact with the more concentrated electrolyte acts as a cathode while the part in contact with the more dilute electrolyte acts as an anode and so corrodes most. Such a cell is known as a *concentration cell.*

Another type of concentration cell is produced if there are variations in the amount of oxygen dissolved in the water in contact with a metal. That part of the metal in contact with the water having the greatest concentration of oxygen acts as a cathode while the part of the metal in contact with the water having the least concentration acts as an anode and so corrodes most. A drop of water on a steel surface is likely to have a higher concentration of dissolved oxygen near its surface where it is in contact with air than in the centre of the drop (Figure 9.4). The metal at the centre of the drop acts as an anode and so corrodes most.

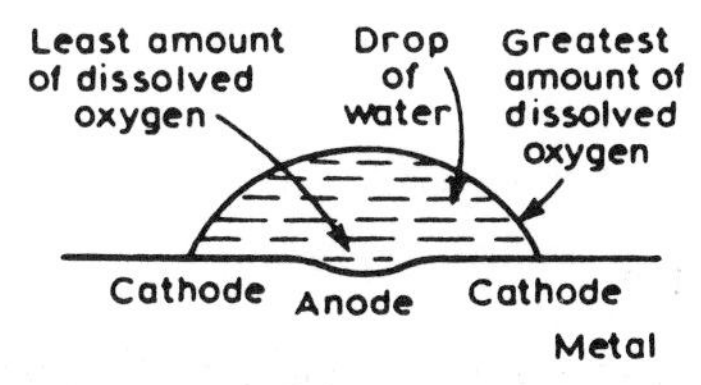

Figure 9.4 An example of a concentration cell

A similar situation arises with a piece of iron dipping into water. The water nearer the surface can have its oxygen replenished more readily than deeper levels of water. The result is that the iron nearer to the water surface becomes a cathode and that lower down an anode. The corrosion is thus most pronounced towards the lower end of the piece of iron (Figure 9.5).

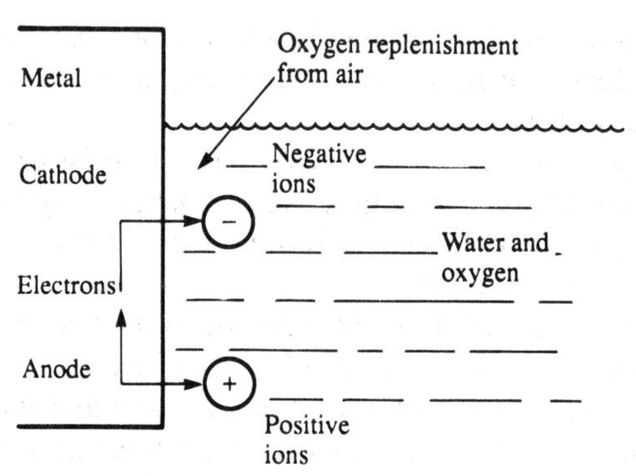

Figure 9.5 An example of a concentration cell

Rusting

Iron rusts when the environment contains both oxygen and moisture but not with oxygen alone or moisture alone. Iron nails kept in a container with dry air or in oxygen-free, e.g. boiled, water do not rust. But in moist air they rust readily.

The reaction can be considered to occur in a number of stages:

1 Atoms in the surface of the metal lose electrons and become positive ions.

Metal → positive metal ions + electrons

2 The liberated electrons move through the metal to where they can combine with water and oxygen molecules to give negative hydroxyl ions.

Water + oxygen + electrons → negative hydroxyl ions

3 The negative hydroxyl ions then combine with the positive metal ions to give the metal hydroxide.

Negative hydroxyl ions + positive metal irons → metal hydroxide

The overall result can be described as:

Metal + oxygen + water → metal hydroxide

This hydroxide is friable and powdery and so does not protect the metal surface from further attack. Indeed it can hold a layer of water close to the metal surface and accentuate the reaction in air. This hydroxide following reactions with oxygen gives a product we call rust.

Rusting is an example of a concentration cell, the process being therefore an electrochemical one.

Aqueous corrosion

Corrosion of a metal in contact with water depends on a number of factors.

1 *The metal* – gold does not corrode in water, iron does. The position of a metal in the electrochemical series is an indicator of the likelihood of that metal corroding in water.
2 *Metallurgical factors* – electrochemical corrosion requires an anode and a cathode to occur. These can be the result of chemical segregation occurring within a metal (as with dezincification with brass), inclusions, local differences in grain size, etc.
3 *Surface defects* – surface defects can allow water to come into contact with a metal as a result of removing a protective coating.
4 *Stress* – variations in stress within a metal or a component can lead to the production of cells and hence corrosion. A component which has part of it heavily cold-worked and part less-worked will contain internal stresses which can result in the heavily-worked part acting as an anode and the less-worked part as a cathode. Therefore the heavily cold-worked part corrodes most.
5 *The environment* – rusting of iron requires the presence of both water and oxygen, hence iron in oxygen-free water does not rust. Chemically-active pollutants in the environment can have a marked effect on corrosion, especially those that are soluble in water. Man-made pollutants such as the oxides of carbon and sulphur, produced in the combustion of fuels, dissolve in water to give acids which readily attack metals and many other materials. Marine environments also are particularly corrosive, due to the high concentrations of salt from the sea. The sodium chloride reacts with metals to produce chlorides of the metals which are soluble in water and thus cannot act as a protective layer on the surface of a metal as a non-soluble oxide may do. The salt may also destroy the protective oxide layer that has been acquired by a metal.
6 *Temperature* – increasing the temperature increases the rate at which dissolved gases diffuse through water and so increases the reaction rate. Other factors can also affect the corrosion rate, e.g. the development of protective scale when the temperature increases.
7 *Aeration* – the effect of aeration of the water can affect the corrosion rate, the effect, however, depending on the metal concerned. Aeration can, for instance, replace the oxygen used in the rusting of iron and so allow rusting to continue.
8 *Anodic reactions* – reactions may occur between the anodic region of a metal and the electrolyte which result in the development of surface films of oxides or absorbed gases which then form a barrier to the movement of metal ions into the electrolyte. Further corrosion is thus prevented. This action is said to lead to *passivity*.

9.2 Types of corrosion

There are many types of corrosion attack, all however being the result of chemical or electrochemical reactions. The following are some of the more common forms.

1 *Uniform corrosion* – this is where all the surface of a metal is corroded to the same extent. This could, for instance, be the result of a chemical reaction between the surface of a sheet of aluminium and oxygen resulting in the formation of a surface oxide layer.
2 *Galvanic corrosion* – galvanic corrosion is in fact the mechanism of many forms of corrosion. The term is, however, often used to just describe that form of corrosion occurring when two different metals which are in electrical contact are exposed to a corrosive environment. An electrical cell is produced which results in a corrosive attack on the metal constituting the anode.
3 *Pit and crevice corrosion* – pit corrosion can start at a surface discontinuity. Within the pit the ingress of oxygen is restricted and so a localized concentration cell can develop. A crevice is just another form of pit.
4 *Selective leaching* – this is where only certain parts of an alloy are attacked. One example of this is graphitization with cast iron where a galvanic cell is set up between the iron and graphite with the result that the iron leaches away. Similarly, dezincification occurs with brass as a result of the galvanic cell between the copper and zinc.
5 *Intergranular corrosion* – this is one form of selective leaching. Grain boundaries are generally more reactive to chemical attack than other parts of a metal.
6 *Stress corrosion* – this form of corrosion takes place when both stress and a corrosive environment are present, taking the form of a network of fine cracks propagating through the material. Fairly specific conditions are necessary for this to occur. Examples of this are brass exposed to ammonia and mild steel exposed to nitrates.
7 *Corrosion fatigue* – fatigue failure can be accelerated by corrosive environments, the rate at which cracks propagate being increased.
8 *Fretting corrosion* – this is a particular form of corrosion fatigue that can occur when closely-fitting metal surfaces are subjected to slight oscillatory slip in a corrosive environment. Due to the surfaces not being perfect, contacts between the surfaces occur at a few high points. The result is high stresses at these points. This leads to localized plastic flow and cold welding. The motion, however, ruptures these welds and loose metal particles are produced. This would occur regardless of environment. However, a corrosive environment such as air (oxygen) leads to these particles oxidizing. This makes them even more abrasive and increases the damage to the surfaces.
9 *Erosion corrosion* – the term erosion is used to describe the wear produced as a result of the relative motion past a surface of particles or bubbles in a liquid. This is essentially a mechanical process. The result of such

abrasive action can be to break down protective coatings on metals and lead to corrosion by concentration cells.

10 *Microbiological corrosion* – this is corrosion produced by the activity of microorganisms. For example, bacterial activity in sea water can lead to an increase in the level of hydrogen sulphide dissolved in the water. This can have a serious, corrosive effect on steel. The bacteria responsible for this thrive on a diet of sulphates.

9.3 Corrosion prevention

Methods of preventing corrosion, or reducing it, can be summarized as:

1 Selection of appropriate materials.
2 Selection of appropriate design.
3 Modification of the environment.
4 Use of protective coatings.

In selecting materials, care should be taken not to use two different materials in close proximity, particularly if they are far apart in the galvanic series, giving a high potential difference between them. The material which acts as the anode will be corroded in the appropriate environment. It is not desirable to connect copper pipes to steel water tanks. Steel pipes to a steel tank would be better.

However, there are situations where the introduction of a dissimilar metal can be used for protection of another metal. Pieces of magnesium or zinc placed close to buried iron pipes can protect the pipes in that the magnesium or zinc acts as the anode relative to the iron which becomes the cathode. The result is corrosion of the magnesium or zinc and not the pipe. Such a method is known as *galvanic protection*.

The steel hull of a ship can be protected below the water line by fixing pieces of magnesium or zinc to it. The steel then behaves as the cathode, the magnesium or zinc becoming the anode and so corroding. Another way of making a piece of metal act as a cathode and so not corrode, is to connect it to a source of e.m.f. in such a way that the externally applied potential difference makes the metal the cathode in an electrical circuit (Figure 9.6).

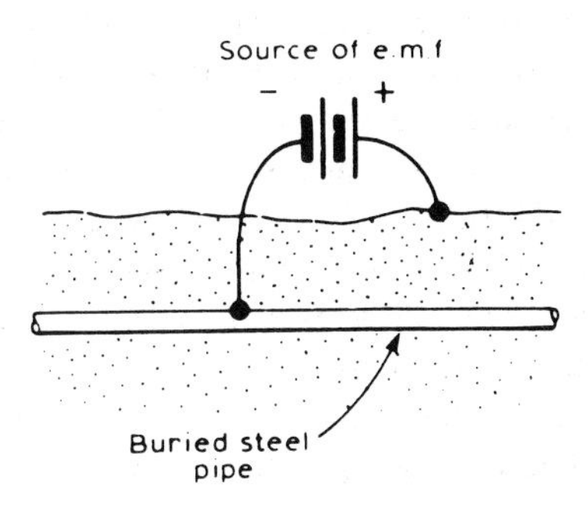

Figure 9.6 Corrosion prevention using an applied e.m.f.

Selection of an appropriate metal for a specific environment can do much to reduce corrosion. These are the properties of some metals in different environments.

1 *Copper* – when exposed to the atmosphere copper develops an adherent protective layer which then insulates it from further corrosion. Copper is used for water pipes, e.g. domestic water supply pipes, as it offers a high resistance to corrosion in such situations.
2 *Copper alloys* – demetallification can occur. Some of the alloying metals, however, can improve the corrosion resistance of the copper by improving the development of the adherent protective surface layer.
3 *Iron and steel* – this is corroded readily in many environments and exposure to sea water can result in graphitization. Stress corrosion can occur in certain environments.
4 *Alloy steels* – the addition of chromium to steel can improve considerably the corrosion resistance by modifying the surface protective film produced under oxidizing conditions. The addition of nickel to an iron–aluminium alloy can further improve corrosion resistance and such a material can be used in sea water with little corrosion resulting. Austenitic steels, however, are susceptible to stress corrosion in certain environments. Iron–nickel alloys have good corrosion resistance and such alloys with 20 per cent nickel and 2 to 3 per cent carbon are particularly good for marine environments.
5 *Zinc* – zinc can develop a durable oxide layer in the atmosphere and then becomes resistant to corrosion.
6 *Aluminium* – readily developing a durable protective surface film, aluminium is then resistant to corrosion. Aluminium alloys are subject to stress corrosion.

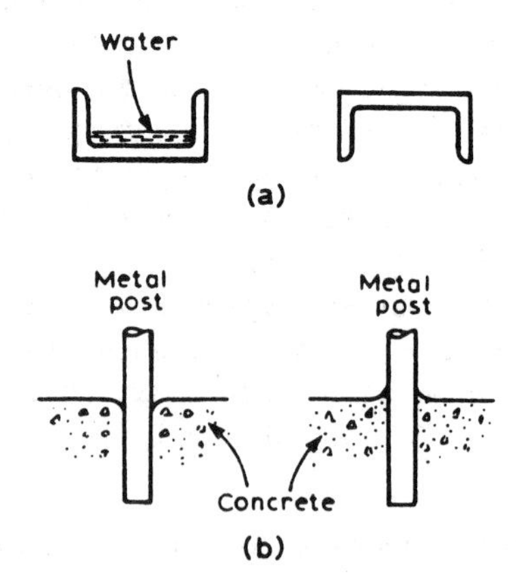

Figure 9.7 Ways of reducing corrosion by preventing water from collecting, (a) Simple inversion, (b) Using a fillet to eliminate crevice

The avoidance of potential crevices (Figure 9.7) which can hold water or some other electrolyte and so permit a cell to function, perhaps by bringing two dissimilar metals into electrical contact or by producing a concentration cell, should be avoided in design. Suitable design can also do much to reduce the incidence of stress corrosion.

Corrosion can be prevented or reduced by modification of the environment in which a metal is situated. Thus in the case of a packaged item an impervious packaging can be used so that water vapour cannot come into contact with the metal. Residual water vapour within the package can be removed by including a dessicant such as silica gel, within the package.

Where the environment adjacent to the metal is a liquid it is possible to add certain compounds to the liquid so that corrosion is inhibited, such additives being known as *inhibitors*. In the case of water-in-steel radiators or boilers, compounds which provide chromate or phosphate ions may be used as inhibitors, as they help to maintain the protective surface films on the steel.

Protective coatings

Coatings applied to the surfaces of metals can have one or more functions:

1 To protect against corrosion.
2 To protect against abrasion and wear.
3 To provide electrical or thermal insulation or conductivity.
4 To improve the appearance of the surface, perhaps by giving colour or a shiny appearance.

The protection functions are achieved by the coating either isolating the surface from the environment or by electrochemical action.

One way of isolating a metal surface from the environment is to cover the metal with a protective coating, which can be impervious to oxygen, water or other pollutants. Coatings of grease, with perhaps the inclusion of specific corrosion inhibitors, can be used to give a temporary protective coating. Plasticized bitumens or resins can be used to give harder but still temporary coatings, while organic polymers or rubber latex can be applied to give coatings which can be stripped off when required.

One of the most common coatings applied to surfaces in order to prevent corrosion is paint, different types having different resistances to corrosion environments. Thus some paints have a good resistance to acids while others are good for water. For paints to provide an efficient protective coating there needs to be:

1 An appropriate paint composition for the metal and environment concerned.
2 Proper surface preparation so that good adhesion occurs.
3 An adequate, uniform paint thickness.

Metallic coatings can be applied to metals to protect them by electrochemical action. The coating used can achieve corrosion resistance by being either a less noble or sacrificial metal or a more protective, more noble metal.

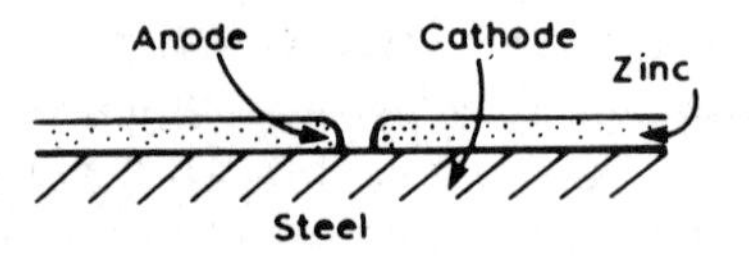

Figure 9.8 Galvanized steel

Zinc is less noble than steel and thus can be used to give sacrificial protection. Steel components can be protected by dipping them in molten zinc to form a thin surface coating of zinc on the steel surface. This method is known as *galvanizing*. The zinc acts as an anode with the steel being the cathode (Figure 9.8), and the zinc corrodes, rather than the steel, when the surface layer is broken. Small components can be coated with zinc by heating them in a closed container with zinc dust. This process is called *sherardizing*.

With noble metal coating of a metal surface, the protection occurs as a result of the coating being itself resistant to corrosion. Such a coating is, however, only completely protective if there is no exposed substrate; discontinuities in the coating can lead to rapid corrosion of the substrate. This is different to the sacrificial form of coating – breaks in such coats are not so detrimental since the corrosion is confined to the coating which is being sacrificed. Chromium is often applied to steel by electroplating to give a noble coating. This coating is often over a base coat of nickel, also electroplated. The nickel coat is also a noble coating with respect to the steel.

Another form of coating used with metals is a chemical conversion coating. Such a coating involves chemically producing changes in the surface layers of the metal. The application of solutions of chromates to some metals, e.g. aluminium, magnesium and copper, can result in the formation of a corrosion-resistant surface layer. The layer is composed of a mixture of oxides of chromium and the metal. Steel surfaces are often treated with phosphoric acid or solutions containing phosphate ions, the process being known as *phosphating*. The treatment results in the removal of surface rust and the formation of a protective phosphate coat. The coating, though offering some corrosion protection, is generally used as a precursor for other coatings, perhaps paint. Aluminium in the atmosphere generally has an oxide surface layer which offers some corrosion resistance and can be thickened by an electrolytic process. The treatment is known as *anodizing*.

9.4 The stability of polymers

Polymers have maximum service temperatures of 60° to 120°C, the value depending on the polymer concerned. While there is a quite significant deterioration of the tensile strength and hardness of thermoplastics approaching this temperature, thermosets are little affected. Thermoplastics, in particular, show creep and this is apparent at room temperature and increases quite markedly as the temperature is increased.

While some polymers are highly resistant to chemical attack, others are liable to stain, craze, soften, swell or dissolve completely. Thus, for instance, nylon shows little degradation with weak acids but is attacked by strong acids; it is resistant to alkalis and organic solvents. Polystyrene, however, though showing similar properties with acids and alkalis, is attacked by organic solvents. Polymers are generally resistant to water, hence their wide use for containers and pipes. However, there is generally a small amount of water absorption.

Polymers are generally affected by exposure to the atmosphere and sunlight. Ultra-violet light, present in sunlight, can cause a breakdown of the bonds in the polymer molecular chains and result in surface cracking. For this reason plastics often have a UV inhibitor mixed with the polymer when the material is compounded. Deterioration in colour and transparency can also occur. Most thermoplastics are affected by ultra-violet light, particularly cellulose derivatives and, to a limited extent, nylon and polyethylene.

Natural rubber is resistant to most acids and alkalis, however its resistance to petroleum product is poor. It also deteriorates rapidly in sunlight. Some synthetic rubbers such as neoprene, are more resistant to petroleum products and sunlight. Ozone can cause cracking of natural rubber, and many synthetic rubbers. Neoprene is one of the rubbers having a high resistance to ozone.

Polymeric materials can provide fire hazards in that they can help spread a fire and produce a dense, toxic smoke. Many polymers burn very easily and produce the toxic gas carbon monoxide; the nitrogen containing polymers, e.g. polyurethane, producing hydrogen cyanide. Polyurethane foams, used for cushioning in domestic furniture, burn very readily because of the high surface to volume ratio. They produce dense smoke and toxic gases and are particular hazardous in the home.

The stability of ceramics

Ceramics are relatively stable when exposed to the atmosphere, though the presence of sulphur dioxide in the atmosphere and its subsequent change to sulphuric acid can result in deterioration of ceramics. Thus building materials such as stone and brick can be severely damaged by exposure to industrial atmosphere in which sulphur dioxide is present.

Damage to ceramics may also result from the freezing of water which has become absorbed into pores of the material and from thermal shock. The low thermal conductivity of ceramics can result in large thermal gradients being set up and hence considerable stresses. This can lead to flaking of the surface. Ceramics used as furnace linings may well be affected by molten metals or slags. Furnace linings need to be chosen with care.

Problems

1. It has been observed that cars in a dry desert part of a country remain remarkably free of rust when compared with cars in a damp climate such as England. Offer an explanation for this.
2. Why does the de-aeration of water in a boiler reduce corrosion?
3. What criteria should be used if corrosion is to be kept to a low value when two dissimilar metals are joined together?
4. Aluminium pipes are to be used to carry water into a water tank. Possible materials for the tank are copper or galvanized steel. Which material would you advocate if corrosion is to be minimized?
5. It is found that for a junction between mild steel and copper in a sea water environment the mild steel corrodes rather than the copper. With a mild steel–aluminium junction in the same environment it is found that the aluminium corrodes more than the mild steel. Explain the above observations.
6. Pieces of magnesium placed close to buried iron pipes are used to reduce the corrosion of the iron. Explain.
7. Corrosion of a stainless steel flange is found to occur when a lead gasket is used. Explain.
8. Propose a method to give galvanic protection for a steel water storage tank.
9. Why should copper piping not be used to supply water to a galvanized steel water storage tank?
10. What are the main ways galvanic cells can be set up in metals?
11. Compare the use of zinc and tin as protective coatings for steel.
12. The following note appeared in the Products and Techniques section of *Engineering*. Explain the purpose and mode of action of the anodes referred to in the note.

    '*Corrosion anodes*
    Solid anodes for the cathodic protection of water tanks are now available in a variety of sizes and three types – freestanding, suspended and weld attached. No prior preparation is required before installation, and protection begins at once. Being non-toxic, the Metalife anodes can be used in tanks for drinking water.
    Belonza Molecular Metalife Ltd, Harrogate.'
13. What further information do you think you should need to assess the merits of the anodes referred to in the previous question? Do you think it would matter what type of material the water tank was made of?
14. In an article on underwater equipment design the author states that good design required the avoidance of crevices and high stress concentrations and also that the number of different metals should be kept to a minimum. Give explanations for these criteria.

PART THREE

Metals

10

Forming processes with metals

10.1 The main processes

The range of forming processes can be divided broadly into four categories:

1 Casting, shaping a material by pouring the liquid material into a mould.
2 Manipulative processes, shaping a material by plastic deformation processes.
3 Powder techniques, producing a shape by compacting a powder.
4 Cutting and grinding, producing a shape by metal removal.

The choice of forming process will depend on a number of factors:

1 The quantity of items required.
2 The dimensional accuracy required.
3 The surface finish required.
4 The size of the items, both overall size and section thicknesses.
5 The requirement for holes, inserts and undercuts.

Another factor to be considered is whether a joining process would be more economic or more suitable for the form of the item concerned. Linked with the choice of process is the choice of material, one cannot be considered without the other.

10.2 Casting

Most metal products have at some stage in their manufacture been cast. *Casting* is the shaping of an object by pouring the liquid metal into a mould and then allowing it to solidify. The resulting shape may be that of the final manufactured object, or one that requires some machining, or an ingot which is then further processed by manipulative processes.

The mould used to form the shape into which the liquid metal is poured has to be designed in such a way that, however complicated the shape, the liquid metal flows easily and quickly to all parts. This has implications for the finished casting in that sharp corners and re-entrant sections have to be avoided and gradually tapered changes in sections used. Account has also to be taken of the fact that the dimensions of the finished casting will be less than those of the mould due to shrinkage occurring when the metal cools from the liquid state to room temperature. Moulds are generally made in two or more parts, which are clamped together while the liquid metal is poured into them, then separated, when the metal has solidified, to enable the finished casting to be extracted. Complex castings can be achieved by the use of moulds having a number of parts. Hollow castings or holes or cavities can be achieved by incorporating separate loose pieces inside the mould, known as cores.

There are a number of casting methods possible and the factors determining the choice of a particular method are:

1 Size of casting required.
2 The number of castings required.
3 The cost per casting.
4 Complexity of casting.
5 The mechanical properties required for the casting.
6 The surface finish required.
7 Dimensional accuracy required.
8 The metal to be used.

Sand casting involves the making of a mould using a mixture of sand with clay, for the traditional moulding material. This is packed around a pattern of the casting, generally of a hard wood and larger than the required casting to allow for shrinkage. The mould is made in two or more parts so that the pattern can be extracted after the sand has been packed round it (Figure 10.1). Sand casting can be used for a wide range of casting sizes and for small- or large-number production. It is the cheapest process for small-number production and a reasonably priced process for large-number production. Complex

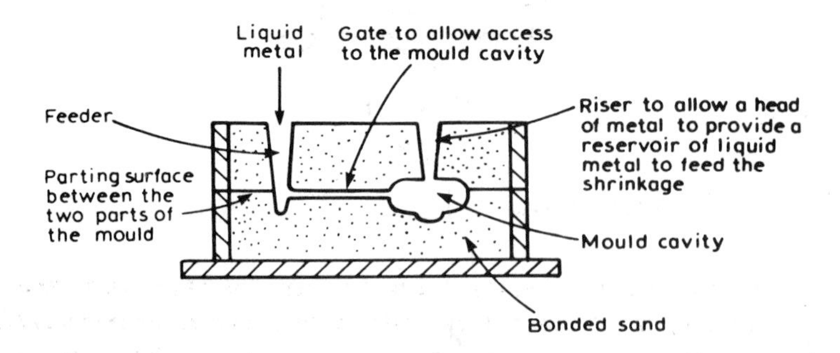

Figure 10.1 Sand casting

castings can be produced by this method. The mechanical properties, surface finish and dimensional accuracy of the casting are however limited. A wide range of alloys can be cast by this process.

Die casting involves the use of a metal mould. Two types of die casting are used. *Gravity die casting* is similar to sand casting in that the metal mould has the liquid metal poured into it in a similar way to that adopted with a sand casting. The head of liquid metal in the feeder forces the metal into the various parts of the mould. With *pressure die casting* the liquid metal is injected into the mould under pressure. This has the advantage that the metal can be forced into all parts of the mould cavity and thus very complex shapes with high dimensional accuracy can be produced. There are limitations to the size of the casting that can be produced by die casting, that for pressure die casting being smaller than that for gravity die casting. The cost of the mould is high and thus the process is relatively uneconomic for small-number production. Large-number production is necessary to spread the cost of the mould. These initial high costs may, however, be more than compensated for with large-number production by the reduction or complete elimination of machining or finishing costs. The mechanical properties, surface finish and dimensional accuracy of the casting are very good. The metals that can be used for this process are, however, restricted to the lower melting point metals and alloys, e.g. aluminium, copper, magnesium and zinc and their alloys.

Another method which is used to force the liquid metal into the various parts of the mould is known as *centrifugal casting*. The mould is rotated (Figure 10.2) and the forces resulting from this rotation force the metal against the sides of the mould. This method is used for simple geometrical shapes, e.g. large-diameter pipes. The method is not suitable for complex castings.

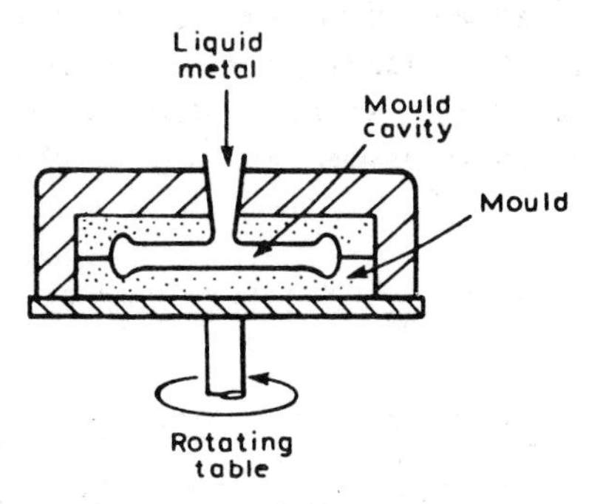

Figure 10.2 Centrifugal casting

Investment or *lost wax casting* is a process that can be used for metals that have to withstand very high temperatures, and so have high melting points, and for which high dimensional accuracy is required; areo engine blades are a typical product. The process is not restricted to high-melting-point metals but can also be used with low-melting-point metals. It is, however, the only casting method that can be used for the high-melting-point metals. Such metals cause rapid die failures when used with die casting. Modern investment casting uses metal moulds to produce wax patterns. The wax patterns are then coated with a ceramic paste. When this coated wax pattern is heated the ceramic hardens and the wax melts to give a ceramic mould. The liquid metal is then injected into this ceramic mould by pressure or the centrifugal process. After the mould has cooled the ceramic is broken away to release the casting. The size of castings that can be produced in this way is limited and it is an expensive process for large-number production. It is, however, relatively cheap for small-number production, particularly where high dimensional accuracy and good surface finish are required. It is suitable for complex castings.

Casting and grain structure

Casting involves the shaping of a product by the pouring of liquid metal into a mould. The grain structure within the product is determined by the rate of cooling. Thus the metal in contact with the moulds cools faster than that in the centre of the casting. This gives rise to small crystals, termed *chill crystals*, near the surfaces. These are smaller because the metal has cooled too rapidly for the crystals to grow to any size. The cooling rate nearer the centre is, however, much less and so some chill crystals can develop in an inward direction. This results in large elongated crystals perpendicular to the mould walls called *columnar crystals*. In the centre of the mould the cooling rate is the lowest. While growth of the columnar crystals is taking place small crystals are growing in this central region. These grow in the liquid metal which is constantly on the move due to convection currents. The final result is a central region of medium-sized, almost spherical, crystals called *equiaxed crystals* (Figure 2.1, page 28 shows all these types of crystals in a casting section).

In general a casting structure having entirely small equiaxed crystals is preferred. This type of structure can be promoted by a more rapid rate of cooling for the casting. Castings in which the mould is made of sand tend to have a slow cooling rate as sand has a low thermal conductivity. Thus sand castings tend to have large columnar grains and hence relatively low strength. Die casting involving metal moulds has a much faster rate of cooling and so gives castings having a bigger zone of equiaxed crystals. As these are smaller than columnar crystals the casting has better properties. Table 10.1 shows the types of differences that can occur with aluminium casting alloys.

Castings do not show directionality of properties, the properties being the same in all directions. They do, however, have the problems produced by working from a liquid metal of blowholes and other voids occurring during solidification.

Table 10.1 *Effect of casting process on properties*

Material	*Tensile strength (MPa)* *Sand cast*	*Die cast*	*Percentage elongation* *Sand cast*	*Die cast*
5% Si, 3% Cu	140	150	2	2
12% Si	160	185	5	7

10.3 Manipulative processes

Manipulative processes involve the shaping of a material by plastic deformation processes. Where the deformation is carried out at a temperature in excess of the recrystallization temperature of the metal, the process is said to involve *hot working*. Plastic deformation at temperatures below the recrystallization temperature is called *cold working*. The main hot working processes are rolling, forging and extrusion. Cold working processes are cold rolling, drawing, pressing, spinning and impact extrusion.

An increase in temperature at which a metal is worked means less energy is required to work the metal, i.e. the metal is more malleable. High temperatures can mean, however, surface scaling or damage occurring. The initial cast metal has coarse grains; hot working breaks the grains down to give a finer structure and thus better mechanical properties.

In addition to the alloying elements present in a metal there are impurities derived from the fluxes and slags used in the melting operation. With the cast

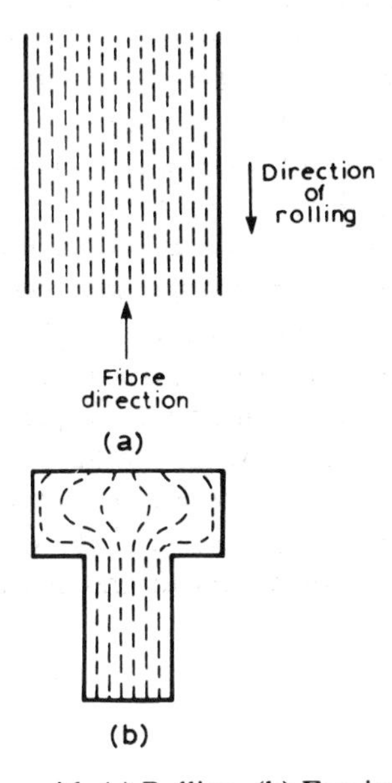

Figure 10.3 Fibre directions with (a) Rolling, (b) Forging

ingot these impurities are reasonably randomly distributed but with hot working they tend to become oriented as *fibres* in the direction of the working. Thus with rolled products the fibre lines tend to be in a direction parallel to the direction of rolling (Figure 10.3a). With a forging the work pattern, shown in Figure 10.3 b is more complex and so the fibre direction is more complex. The effect of the fibres having a specific direction is to give a corresponding *directionality* of mechanical properties. Thus although hot working improves the mechanical properties it does lead to the properties varying in different directions. The fibres can act as lines along which cracks can be propagated, so the design of a product should, as far as is possible, be such as to have the fibre direction, parallel to the tensile stresses and not at right angles to them.

Cold working involves the use of a greater amount of energy than a hot working process to obtain a particular amount of deformation. During cold working the crystal structure becomes broken up and distorted, leading to an increase in mechanical strength and hardness. Unlike hot working, cold working can give a clean, smooth surface finish.

Cold working

The term cold working is applied to any process which results in plastic deformation at a temperature which does not alter the structural changes produced by the working. Table 10.2 shows some of the changes that take place when a sheet of annealed aluminium is rolled and its thickness reduced.

Table 10.2 *Effect of work hardening on properties*

Reduction in sheet thickness %	*Tensile strength (MVm^{-2} or MPa)*	*Elongation %*	*Hardness HV*
0	92	40	20
15	107	15	28
30	125	8	33
40	140	5	38
60	155	3	43

(Based on a table in John, V. B., *Introduction to Engineering Materials*, Macmillan)

As the amount of plastic deformation is increased so the tensile strength increases, the hardness increases and the elongation decreases. The material is becoming harder as a result of the cold working, hence the term sometimes applied to cold working of *work hardening*. The more the material is worked the harder it becomes. Also, as the percentage elongation results above indicate the more a material is worked the more brittle it becomes. A stage can, however, be reached when the strength and hardness are a maximum and the elongation a minimum and further plastic deformation is not possible, the material is too brittle. With the rolled aluminium sheet referred to in the above table, this condition has been reached with about a 60 per cent reduction in sheet thickness. The material is then said to be *fully work hardened.*

The structure of cold worked metals

When stress is applied to a metal grain, deformation starts along the slip planes most suitably orientated. The effect of this is to cause the grains to become elongated and distorted. Figure 10.4 shows the results of heavy rolling of a tin bronze ingot (4 per cent tin). The grains have become elongated into fibre-like structures, which has the effect of giving the material different mechanical properties in different directions; a greater strength along the grain than at right angles to the grain. This effect can be used to advantage by the designer.

Figure 10.4 The grains in a heavily rolled tin bronze ingot (From Rollason, E. C., *Metallurgy for Engineers*, Edward Arnold)

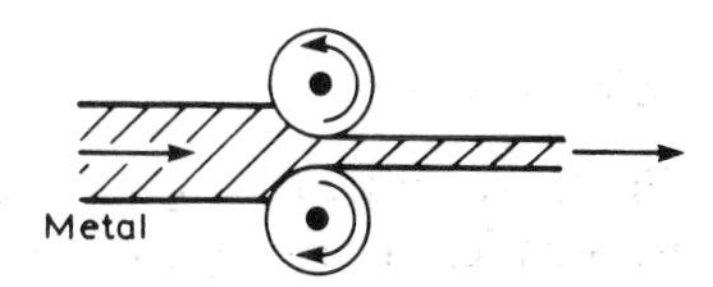

Figure 10.5 Basic principle of rolling

Cold working processes

Cold rolling is the shaping of metal by passing it at normal temperature, between rollers (Figure 10.5). Sheet and strip metal are often cold rolled as a cleaner, smoother finish to the metal surfaces is produced than if hot working is used. The process also gives a harder product. the aluminium foil used for wrapping sweets, such as chocolate, is an example of a cold rolled product. Cold rolling requires more energy than hot rolling.

Drawing involves the pulling of metal through a die (Figure 10.6). Wire manufacture can involve a number of drawing stages in order that the initial material can be brought down to the required size. As cold working hardens a metal there may have to be annealing operations between the various drawing stages to soften the material for further drawing to take place.

With *deep drawing*, sheet metal is pushed through an aperture by a punch (Figure 10.7). The more ductile materials such as aluminium, brass and mild steel are used and the products are deep cup-shaped articles such as cartridge cases.

With deep drawing the sheet metal is not clamped round the edges and so is

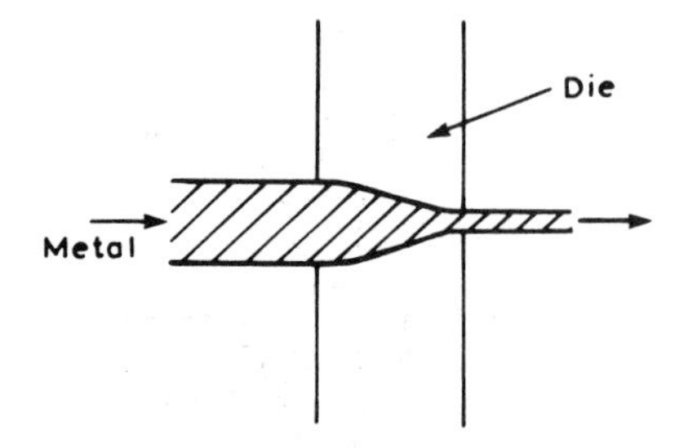

Figure 10.6 Drawing a wire

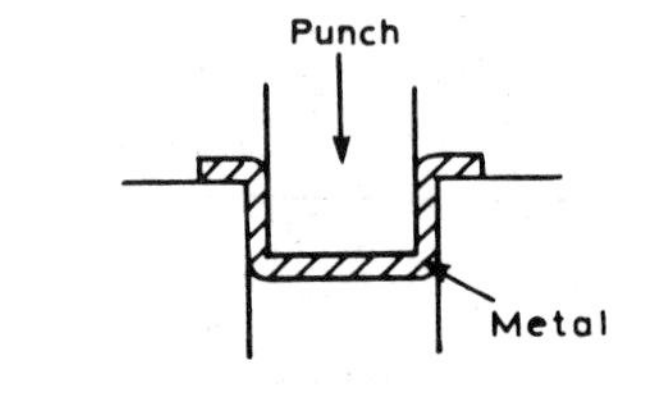

Figure 10.7 Deep drawing

drawn into the die by the pressure from the punch. If the material is clamped round the edges then the process is known as *pressing* (Figure 10.8). Car body panels, kitchen pans and other cooking utensils are typical examples of the products obtained by pressing. Ductile materials are used.

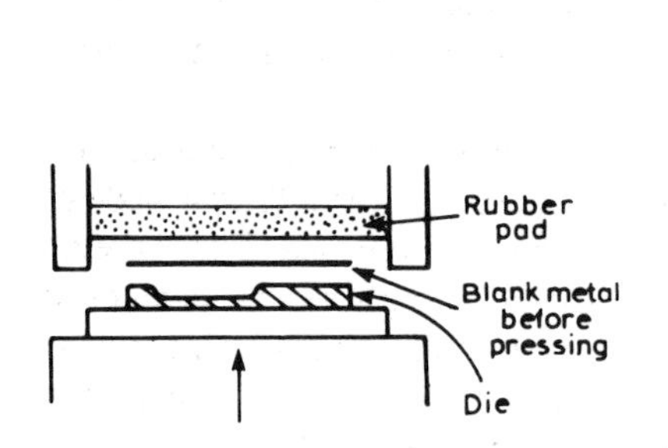

Figure 10.8 Pressing

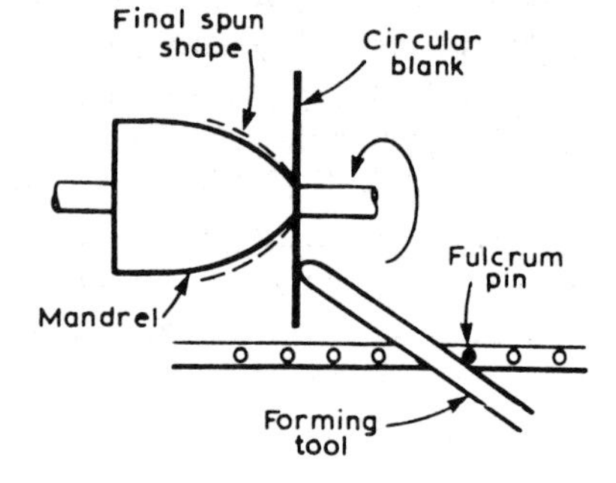

Figure 10.9 Spinning with a hand-held tool

Spinning is a process that can be used for the production of circular section objects. A circular blank of metal is rotated in a lathe type of machine and then pressure applied to deflect the blank into the required shape (Figure 10.9). Aluminium, brass and mild steel are examples of materials used for forming by this process. Spinning is an economic method for producing products required in small numbers and can be used also for large products where other forms of forming would be too expensive.

Explosive forming is used mainly for the forming of sheets of relatively large surface area on a comparatively small number production basis. An explosive charge is detonated under water and the resulting pressure wave used to press a metal blank against a die (Figure 10.10). Communication reflectors and contoured panels are examples of the products of this process.

Impact extrusion is the process used for the production of rigid or collapsible tubes or cans, e.g. zinc dry battery cases and toothpaste tubes, in softer

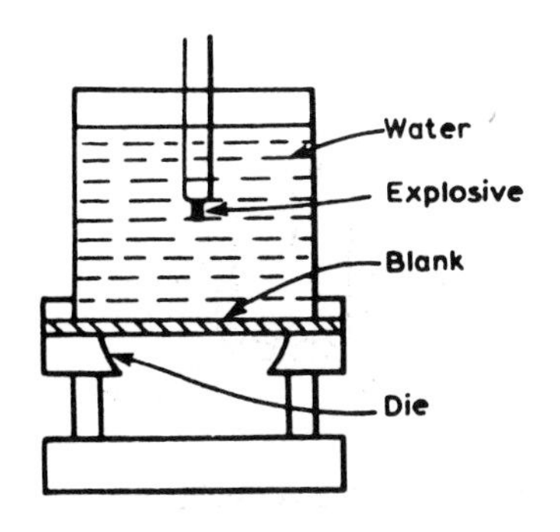

Figure 10.10 Explosive forming

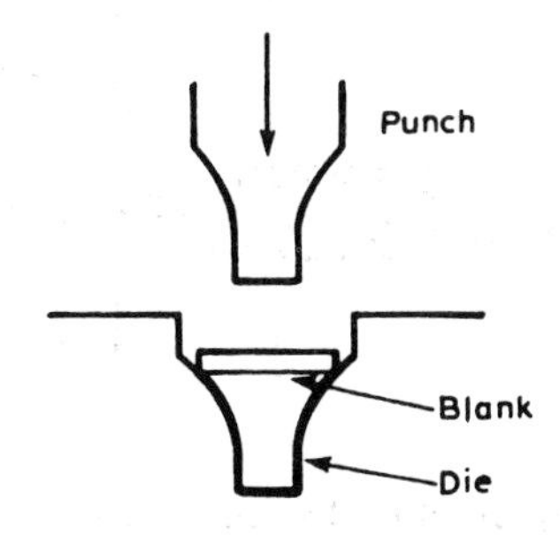

Figure 10.11 Impact extrusion

materials such as zinc, lead and aluminium. Figure 10.11 shows the essential features of the process, a punch forcing a blank to flow into the die. The punch descends very rapidly and hence the term 'impact' for this form of extrusion.

The effect of heat on cold-worked metals

Cold-worked metals generally have deformed grains and have often become rather brittle due to the working. In this process of deforming the grains, internal stresses build up.

When a cold-worked metal is heated to temperatures up to about 0.3 T_m where T_m is the melting point of the metal concerned on the Kelvin scale of temperature, then the internal stresses start to become relieved. There are no changes in grain structure during this but just some slight rearrangement of atoms in order that the stresses become relieved. This process is known as *recovery*. Copper has a melting point of 1083°C, or 1356 K. Hence stress relief with copper requires heating to above about 407 K, i.e. 134°C.

If the heating is contained to a temperature of about 0.3 to 0.5 T_m there is a very large change in hardness. The strength and also the structure of the metal change. Table 10.3 shows how the hardness of copper changes, the copper having been subject to a 30 per cent cold working.

Table 10.3 *Effect of temperature on hardness*

Temperature /°C	/K		*Hardness HV*
Initially			86
150	423	(0.3 T_m)	85
200	473		80
250	523	(0.4 T_m)	74
300	573		61
350	623	(0.5 T_m)	46
450	723		24
600	873	(0.6 T_m)	15

Figure 10.12 shows the results of Table 10.3 graphically. Between 0.3 T_m and 0.5 T_m there is a very large change in hardness. The strength also decreases while the elongation increases. What is happening is that the metal is recrystallizing.

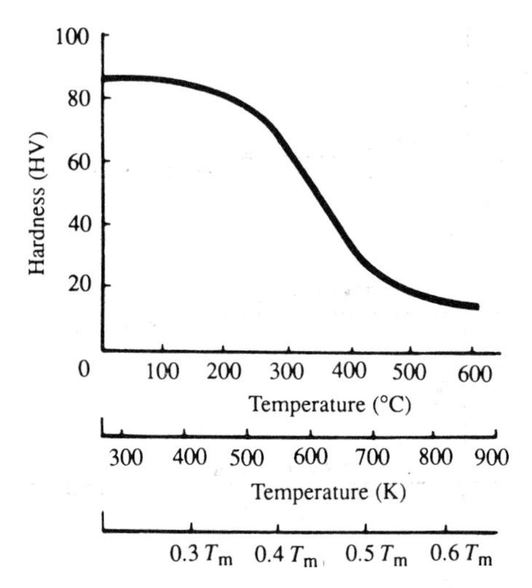

Figure 10.12 The effect of heat treatment on cold-worked copper

With *recrystallization* crystals begin to grow from nuclei in the most heavily deformed parts of the metal. The temperature at which recrystallization just starts is called the *recrystallization temperature*. This is, for pure metals, about 0.3 to 0.5 T_m.

Table 10.4 *Recrystallization temperatures*

Material	*Melting point* /°C	/K	*Recrystallization temperature* /°C	/K	
Aluminium	660	933	150	423	.05 T_m
Copper	1083	1356	200	473	0.3 T_m
Iron	1535	1808	450	723	0.4 T_m
Nickel	1452	1725	620	893	0.5 T_m

As the temperature is increased from the recrystallization temperature so the crystals grow until they have completely replaced the original distorted cold worked structure. Figure 10.13 illustrates this sequence and its relationship to the changes in physical properties.

The sequence of events that occurs when a cold-worked metal is heated can be broken down into three phases:

1 *Recovery* – the only significant change during this phase is the relief of internal stresses.

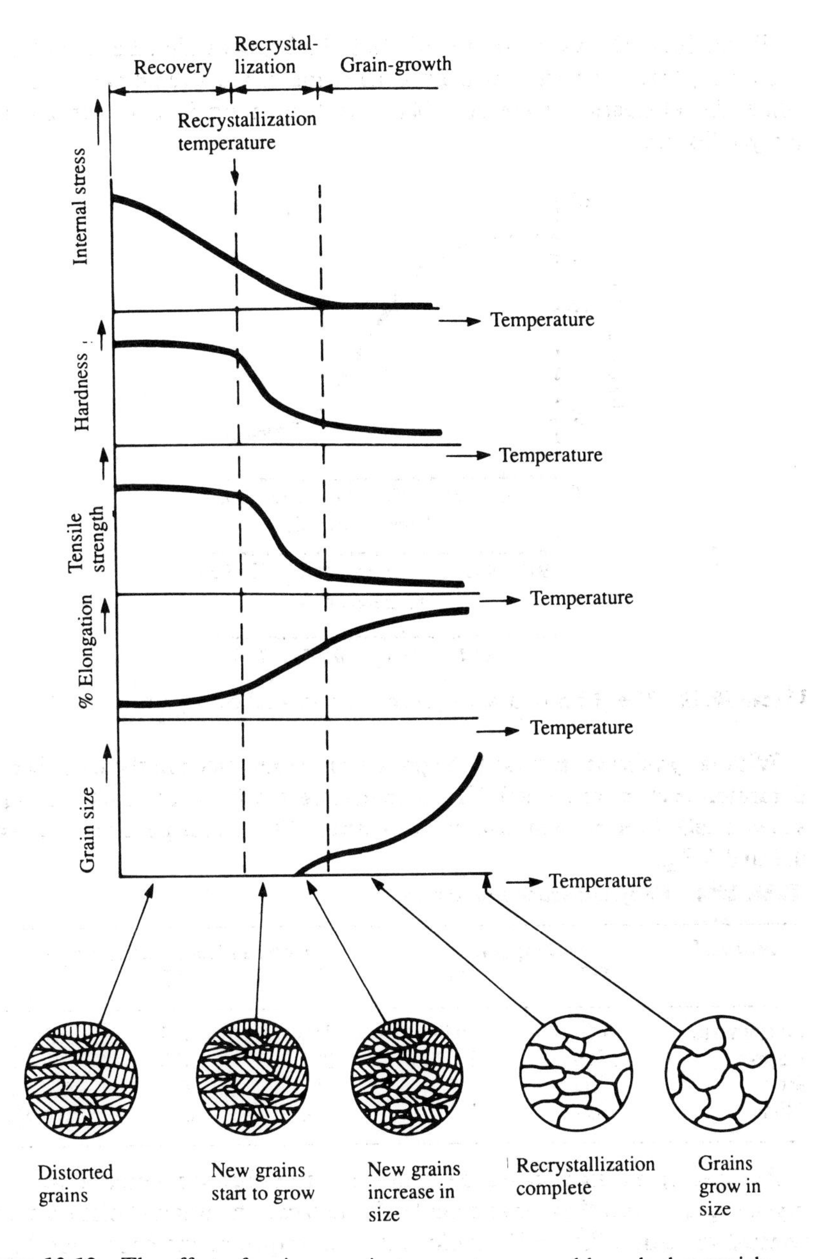

Figure 10.13 The effect of an increase in temperature on cold-worked materials

2 *Recrystallization* – the hardness, tensile strength and percentage elongation all change noticeably during this phase.
3 *Grain growth* – the hardness, tensile strength and percentage elongation

change little during this phase. The only change is that the grains grow and the material becomes large-grained.

During the grain-growth phase the newly-formed grains grow by absorbing other neighbouring grains. The amount of grain growth depends on the temperature and the time for which the material is at that temperature (Figure 10.14).

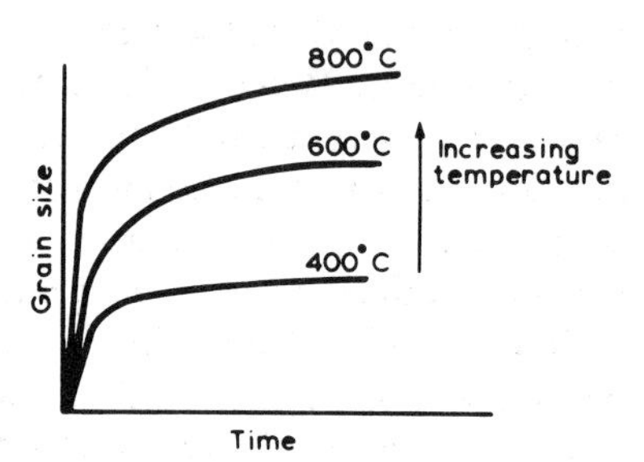

Figure 10.14 The effect of time and temperature on grain growth

The term *annealing* is used for the heating process used to change the properties of a material. Thus, in the case of aluminium that has been cold worked and has become too brittle to work further, heating to above the recrystallization temperature of 150°C enables new grains to grow and the material to become more ductile. The aluminium can then be worked further. This sequence of events, cold working followed by annealing and then further cold working, is used in many manufacturing processes.

Factors affecting recrystallization

1 A minimum amount of deformation is necessary before recrystallization can occur. The permanent deformation necessary depends on the metal concerned.
2 The greater the amount of cold work the lower the crystallization temperature for a particular metal.
3 Alloying increases the recrystallization temperature.
4 No recrystallization takes place below the recrystallization temperature. The higher the temperature above the recrystallization temperature the shorter the time needed at that temperature for a given crystal condition to be attained.
5 The resulting grain size depends on the annealing temperature. The higher the temperature the larger the grain size.
6 The amount of cold work prior to the annealing affects the size of the grains. The greater the amount of cold work the smaller the resulting grain size. The greater the amount of cold work the more centres are produced for crystal growth.

Hot working

Hot working processes
The main hot working processes are rolling, forging and extrusion. *Rolling* is the shaping of metal by the passing of the hot metal between rollers. *Forging* is the shaping of metal by a succession of hammer blows or application of pressure. *Extrusion* shapes metal by a hot ingot being forced, under pressure, through a die.

Rolling is a continuous process in which the metal is passed through the gap between a pair of rotating rollers. When cylindrical rollers are used, the product is in the form of a bar or sheet, but profiled rollers can be used to produce contoured surfaces, e.g. structural beam sections. A wide variety of cross-sections can be produced. These are often used as the basis for structures, e.g. window frames, the rolled product being only trimmed to size and perhaps joined with other rolled shapes. In some instances the product may be further shaped by cold working to give the final product.

With forging, the metal is squeezed between a pair of dies. *Heavy smith's forging* or *open die forging* involves the metal being hammered by a vertically moving tool against a stationary tool (Figure 10.15). This type of forging is like that once carried out by every village blacksmith, only now the ingot being hammered is likely to be considerably larger. The process is used nowadays, mainly for an initial rough shaping of an ingot before shaping with another process. *Closed die forging* involves the hot metal being squeezed between two shaped dies which effectively form a complete mould (Figure 10.16). The metal flows under the pressure into the die cavity. In order to completely fill the die cavity a small excess of metal is allowed and this is squeezed outwards to form a flash which is later trimmed away. *Drop forging* is one form of closed die forging and uses the impact of a hammer to cause the metal billet to flow and fill the die cavity.

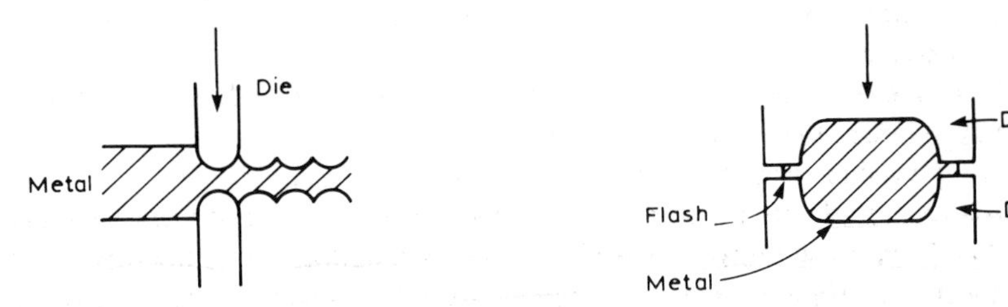

Figure 10.15 Open die forging

Figure 10.16 Closed die forging

Closed die forging can be used to produce large numbers of components with high dimensional accuracy and better mechanical properties than would be produced by casting or machining. This is because the fibre direction can be arranged to give the greatest strength.

With hot extrusion the hot metal is forced, under pressure, to flow through a die, i.e. a shaped orifice (Figure 10.17). It is rather like squeezing toothpaste out of its tube. Quite complex sections can be extruded.

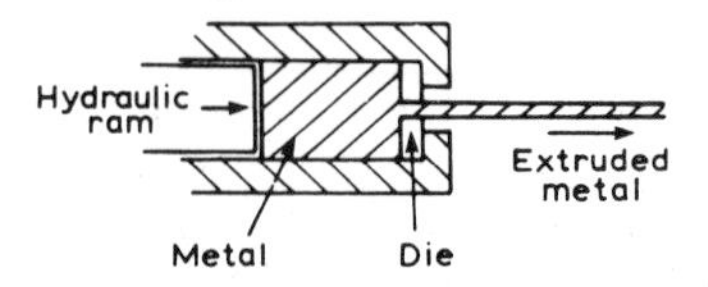

Figure 10.17 Extrusion

10.4 Powder techniques

Shaped metal components can be produced from a metal powder. The process, called *sintering*, involves compacting the power in a die, then heating to a temperature high enough to knit together the particles in the powder. Sintering of tungsten takes place at about 1600°C, considerably below the melting point of 3410°C. The sintering temperature for iron is about 1100°C.

Sintering is a useful method for the production of components from brittle materials like tungsten or composite materials. Cobalt-bonded tungsten carbide tools are produced in this way. The method is useful also for high melting point materials for which the forming processes involving melting become expensive. The degree of porosity of the metal product can be controlled during the process, which is thus useful for the production of porous bronze bearings. The bearings are soaked in oil before use and can then continue in service for a considerable period of time.

10.5 Machining

Machining is the removal of material from the workpiece, the block of material being machined, by the action of a tool. The tool moves relative to the workpiece and detaches thin layers of the unwanted material, known as 'chips'. Figure 10.18 shows some of the main machining methods.

Machining is generally a secondary process, following a primary process such as casting or forging, and is used to produce the final shape to the required accuracy and surface finish. Machining results invariably in waste material being produced and thus any costing of a machining process has to allow for this. To keep the waste to a minimum the primary process should give a product as near the final required dimensions as possible, bearing in mind any need to remove material to give a good surface finish.

Machining involves using a tool that is of a harder material than that of the workpiece. Tools may be made of high speed steels, metal carbides or ceramics. Metal carbide and ceramic tools are made by sintering appropriate

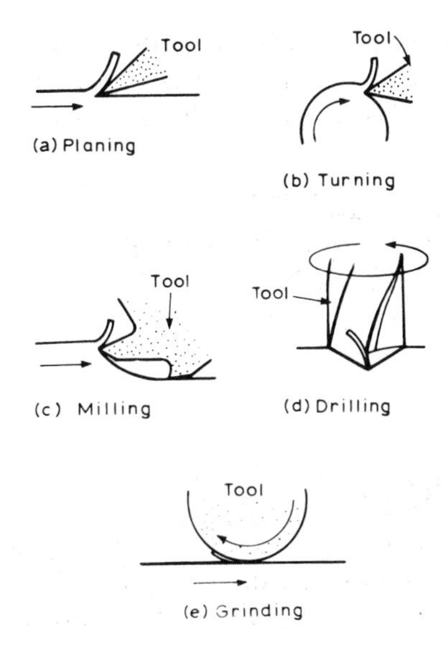

Figure 10.18 Machining methods (a) Planing, (b) Turning, (c) Milling, (d) Drilling, (e) Grinding

mixtures of powders (see previous section on powder techniques).

In machining, the cutting tool causes the workpiece material at the cutting edge to become highly stressed and subject to plastic deformation. The more ductile the material the greater the amount of plastic deformation and the more the material of the workpiece spreads along the tool face. The more this happens the greater the force needed to machine the material and so the greater the expenditure of energy in the machining process. A ductile material on machining gives rise to a continuous chip while a more brittle material leads to small discontinuous chips being produced. Less energy is needed for machining in this case.

The term *machinability* is used to describe the ease of machining. A material with good machinability will produce small chips, need low cutting forces and energy expenditure, be capable of being machined quickly and give a long tool life. Ductile and soft materials have poor machinability. A relative measure of machinability is given by a *machinability index*. It is only a rough guide to machinability, but the higher the index the better the machinability. These are some typical values:

Stainless steel	45
Wrought iron	50
Copper, ¼ hard rolled	60
Aluminium bronze	60
Cast steel	70

Free-cutting mild steel	100
Free-cutting α or β brass	200 to 400

The machinability of a metal can be improved if its ductility is decreased. Work hardening can make an improvement. Some multi-phase alloys have good machinability as the insoluble phase can provide discontinuities within the material and assist in breaking up a continuous chip to give small chips. The addition of a small amount of lead to a mild steel improves its machinability. For optimum machinability, the workpiece material must have low ductility and also low hardness.

Problems

1 Distinguish between cold and hot working processes.
2 Describe the effect on the mechanical properties of a metal of cold working.
3 Explain what is meant by the recrystallization temperature.
4 What is meant by 'work hardening'?
5 What are the properties of a fully work-hardened material in comparison with its properties before any working occurs?
6 How does cold working change the structure of a metal?
7 Describe how the mechanical properties of a cold-worked material change as its temperature is raised from room temperature to about 0.6 T_m, where T_m is the melting point temperature in degrees Kelvin.
8 What factors affect the recrystallization temperature of a metal?
9 How is the recrystallization temperature of a pure metal related to its melting point?
10 Zinc has a melting point of 419°C. Estimate the recrystallization temperature for zinc.
11 Magnesium has a melting point of 651°C. What order of temperatures would be required to (a) stress relieve, (b) anneal a cold-worked piece of magnesium?
12 What is the effect on the grain size of a metal after annealing of the amount of cold work the material had originally been subject to?
13 How does the temperature at which hot working is carried out determine the grain size and so the mechanical properties?
14 Describe the grain structure of a typical casting.
15 Why are the properties of a material dependent on whether it is sand cast or die cast?
16 What is meant by directionality in wrought products?
17 Why are the mechanical properties of a rolled metal different in the direction of rolling from those at right angles to this direction?
18 How does a cold-rolled product differ from a hot-rolled product?
19 Brasses have recrystallization temperatures of the order of 400°C. Roughly what temperature should be used for hot extrusion of brass?

20 A brass, 65 per cent copper and 35 per cent zinc, has a recrystallization temperature of 300°C after having been cold worked so that the cross-sectional area has been reduced by 40 per cent.

(a) How will further cold working change the structure and the properties of the brass?
(b) To what temperature should the brass be heated to give stress relief?
(c) To what temperature should the brass be heated to anneal it and give a relatively small grain size?
(d) How would the grain size, and the mechanical properties change if the annealing temperature used for (c) was exceeded by 100°C?

21 Give examples of the types of product obtained by: (a) sand casting, (b) die casting, (c) centrifugal casting, (d) investment casting.
22 Explain the meaning of the term 'directionality' and describe its effect on the properties of a product produced by hot working.
23 What shaping processes might be used to produce a can, perhaps a Coca Cola can?
24 Describe the different cold working processes and the types of products that they can produce.
25 Explain how the machining of, say, steel can be improved by heat treatment or alloying additions.
26 What type of materials are suitable for: (a) pressing, (b) impact extrusion?
27 Compare the different methods of producing castings from the point of view of economic large number production.

11

Ferrous alloys

11.1 Iron alloys

Pure iron is a relatively soft material and is hardly of any commercial use in that state. Alloys of iron with carbon are, however, very widely used and are classified according to their carbon content:

Material	*Percentage carbon*
Wrought iron	0 to 0.05
Steel	0.05 to 2
Cast iron	2 to 4.3

The percentage of carbon alloyed with iron has a profound effect on the properties of the alloy. The term *carbon steel* is used for those steels in which essentially just iron and carbon are present. The term *alloy steel* is used where other elements are included. The term *ferrous alloy* is used for all iron alloys.

Crystal structures of iron

Pure iron at room temperature exists as a body-centred cubic structure, this being known as *ferrite* or alpha iron. This form continues to exist up to 912°C. At this temperature the structure changes to a face-centred cubic one, known as *austenite* or gamma iron. At 1394°C this form changes to a body-centred cubic structure known as *delta ferrite*. At 1538°C the iron melts.

The body-centred cubic structure of ferrite is formed by iron atoms of diameter 0.256 nm. Between these atoms the structure leaves voids which can accommodate atoms in one type of void up to 0.070 nm in diameter and in the other 0.038 nm in diameter. Carbon atoms have a diameter of 0.154 nm. To accommodate carbon atoms in the ferrite requires a severe distortion of the lattice. Because of this, carbon has only a limited solubility in ferrite.

The face-centred cubic structure of austenite has voids which can accommodate atoms in one type of void up to 0.104 nm in diameter and in the other 0.056 nm in diameter. Carbon atoms with their diameter of 0.154 nm can thus be accommodated in one of the voids with some slight distortion of the lattice. Carbon thus has a higher solubility in austenite than in ferrite. Thus when iron containing carbon is cooled from the austenitic state to the ferrite state, the reduction in the solubility of carbon in iron means that some carbon must come out of solution. This is achieved by the formation of a compound between iron and carbon called *cementite*. This has one carbon atom for every three iron atoms, hence it is often referred to as Fe_3C (iron carbide).

11.2 Plain carbon steels

Figure 11.1 shows the iron–carbon system thermal equilibrium diagram. The α iron will accept up to about 0.2 per cent of carbon in solid solution. The γ iron will accept up to 2.0 per cent of carbon in solid solution. With these amounts of carbon in solution the α iron still retains its body-centred cubic structure, and the name ferrite, and the γ iron its face-centred structure, and the name austenite. The solubility of carbon in iron, in both the austenite and ferrite forms, varies with temperature. With slow cooling, carbon in excess of

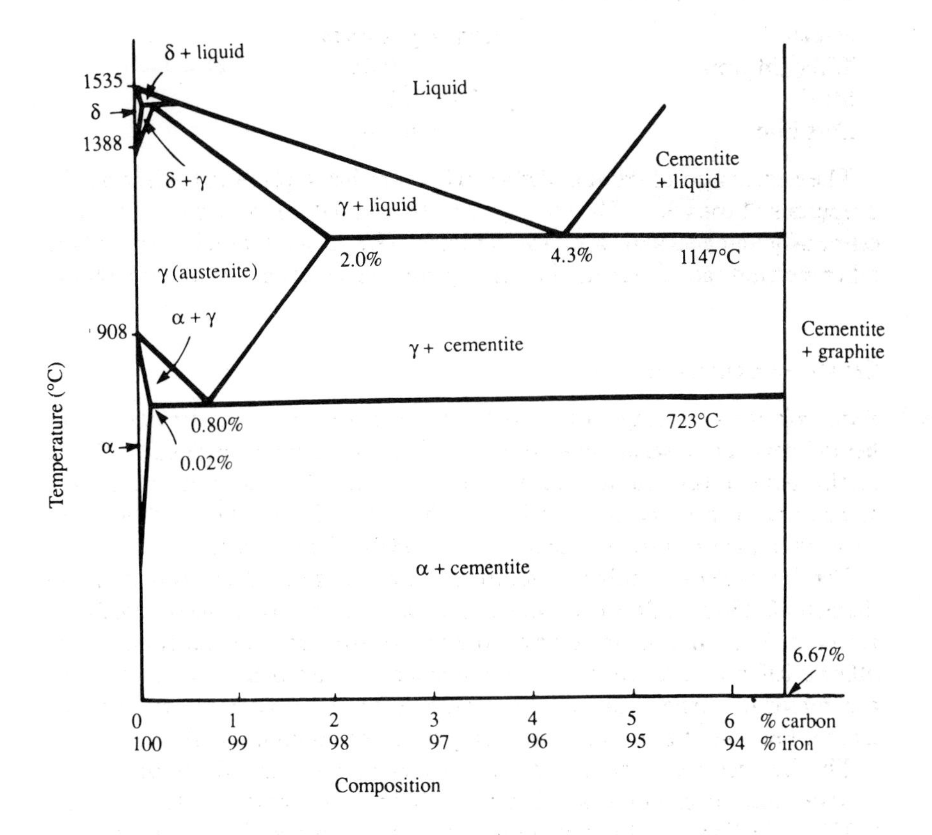

Figure 11.1 The iron–carbon system

that which the α or γ solid solutions can hold at a particular temperature will precipitate. The precipitate is not, however, carbon but as iron carbide (Fe_3C), a compound formed between the iron and the carbon. This iron carbide is known as *cementite*. Cementite is hard and brittle.

Consider the cooling from the liquid of an alloy with 0.80 per cent carbon (Figure 11.2). For temperatures above 723°C the solid formed is γ iron, i.e. austenite. This is a solid solution of carbon in iron. At 723°C there is a sudden change to give a laminated structure of ferrite plus cementite. This structure is called *pearlite* (Figure 11.3). This change at 723°C is rather like the change that occurs at a eutectic, but there the change is from a liquid to a solid, here the change is from one solid structure to another. This type of change is said to give a *eutectoid*. The eutectoid structure has the composition of 0.8 per cent carbon – 99.2 per cent iron in this case.

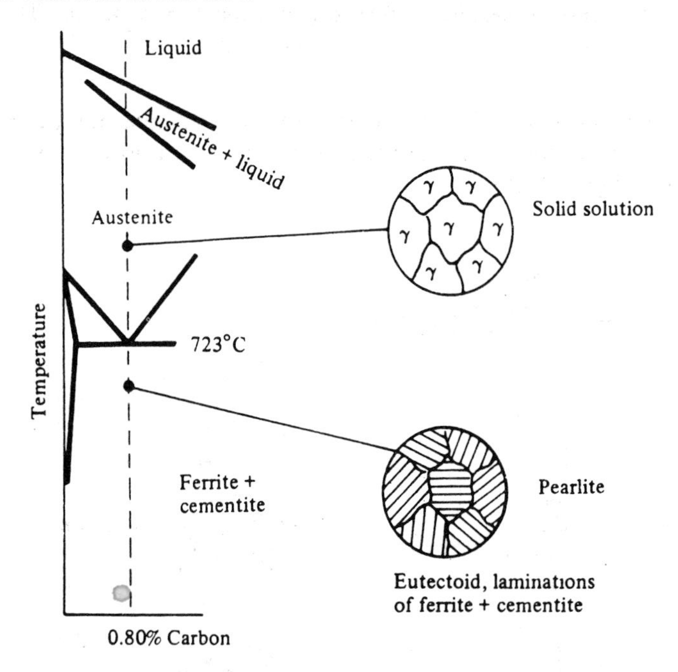

Figure 11.2 Slow cooling of 0.80 per cent carbon steel

Steels containing less than 0.80 per cent carbon are called hypoeutectoid steels, those with between 0.80 per cent and 2.0 per cent carbon are called hypereutectoid steels.

Figure 11.4 shows the cooling of a 0.4 per cent carbon steel, a hypoeutectoid steel, from the austenite phase to room temperature. When the alloy is cooled below temperature T_1, crystals of ferrite start to grow in the austenite. The ferrite tends to grow at the grain boundaries of the austenite crystals. At 723°C

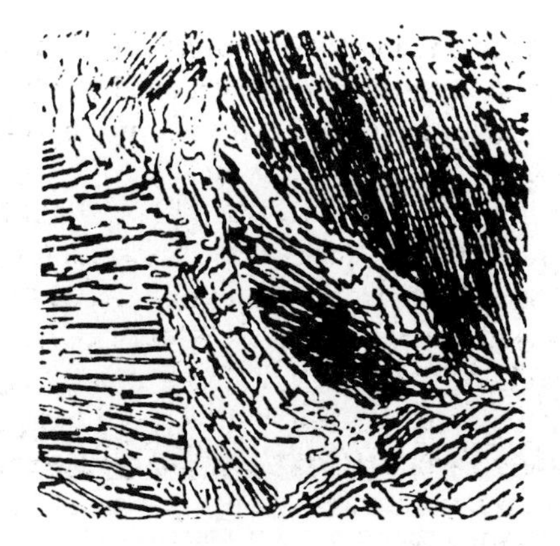

Figure 11.3 Lamellar pearlite × 600 magnification (From Monks, H. A. and Rochester, D. C., *Technician Structure and Properties of Metals*, Cassell)

the remaining austenite changes to the eutectoid structure, i.e. pearlite. The result can be a network of ferrite along the grain boundaries surrounding areas of pearlite (Figure 11.5).

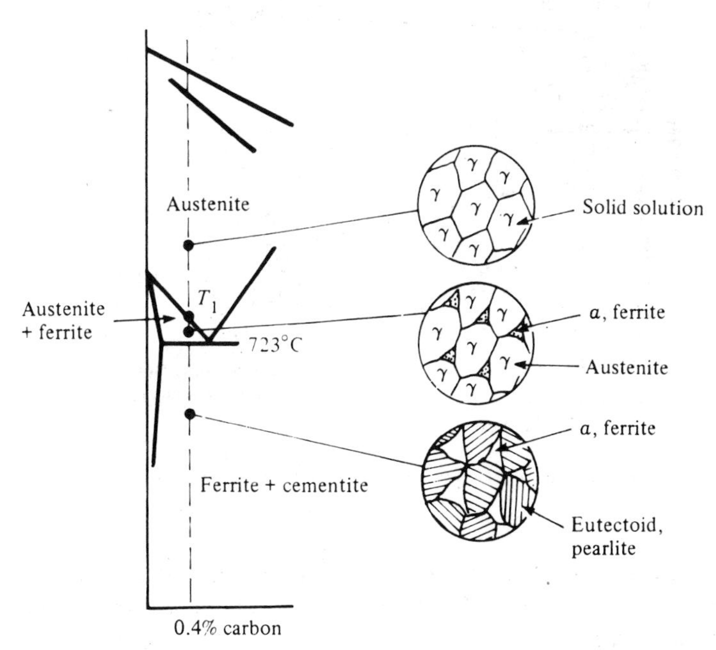

Figure 11.4 Slow cooling of a 0.4 per cent carbon steel

Figure 11.6 shows the cooling of a 1.2 per cent carbon steel, a hyper-eutectoid steel, from the austenite phase to room temperature. When the alloy

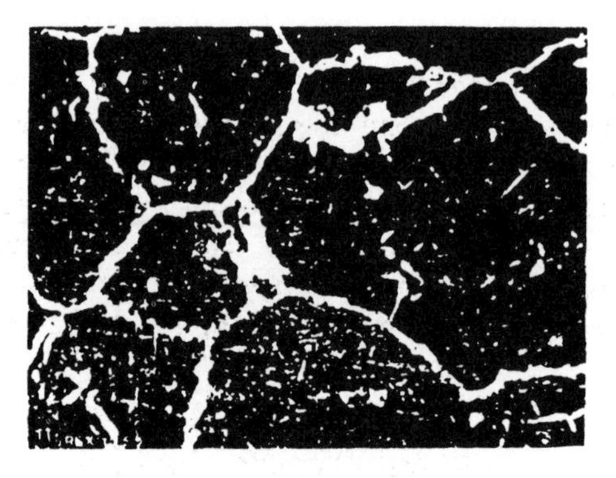

Figure 11.5 A 0.5 per cent carbon steel, slow cooled. Shows network of ferrite (From Rollason, E. C., *Metallurgy for Engineers*, Edward Arnold)

is cooled below the temperature T_1, cementite starts to grow in the austenite at the grain boundaries of the austenite crystals. At 723°C the remaining austenite changes to the eutectoid structure, i.e. pearlite. The result is a network of cementite along the grain boundaries surrounding areas of pearlite.

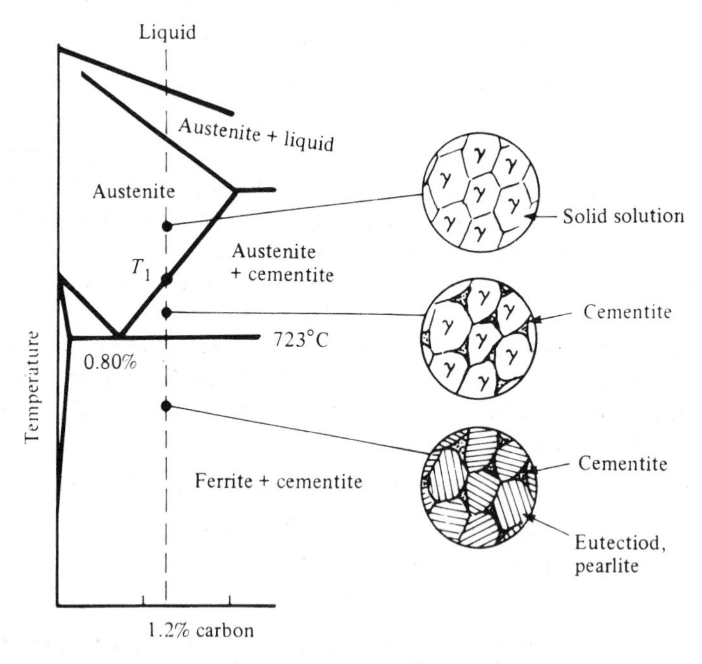

Figure 11.6 Slow cooling of a 1.2 per cent carbon steel

Thus *hypo*eutectoid carbon steels consist of a *ferrite* network enclosing pearlite and *hyper*eutectoid carbon steels consist of a *cementite* network enclosing pearlite.

Cementite

Cementite is the compound iron carbide with the composition of Fe_3C, i.e. one carbon atom to every three carbon atoms. By weight 6.67 per cent of cementite is carbon. Cementite has negligible solubility for iron or carbon thus when the carbon content reaches 6.67 per cent the thermal equilibrium diagram shows a single vertical line. Beyond 6.67 per cent carbon all that is present is cementite and surplus carbon in the form of graphite. Below 6.67 per cent there is insufficient carbon for all the iron to form cementite. Cementite is a hard and brittle material.

Pearlite

Pearlite is made up of alternate lamellae of ferrite and cementite. Pearlite is not a phase but a mixture of two phases, with the two being present in a definite ratio. This ratio can be calculated using the lever rule (see Chapter 3) to a slowly cooled 0.80 per cent eutectoid steel at a temperature just below the eutectoid temperature of 723°C. Using the thermal equilibrium diagram in Figure 11.1, then:

$$\text{Mass of ferrite} \times (0.80 - 0.02) = \text{mass of cementite} \times (6.67 - 0.80)$$

$$\frac{\text{mass of ferrite}}{\text{mass of cementite}} = \frac{6.67 - 0.80}{0.80 - 0.02}$$
$$= 7.52$$

Pearlite thus has 7.52 as much ferrite as cementite, i.e. 88 per cent of the pearlite is ferrite and 12 per cent cementite. This is the composition at almost 723°C and since the solubilities change very little with temperature this will be the composition at room temperature. Since the densities of ferrite and cementite are almost the same, this mass ratio means that the volume of pearlite that is ferrite is about 7.5 times that of the cementite.

The proportion of a hypoeutectoid or hypereutectoid steel that is pearlite can also be determined using the lever rule. Thus for 0.40 per cent carbon steel at a temperature just above 723°C (as described by the thermal equilibrium diagram shown in Figure 11.4), then:

$$\text{Mass of ferrite} \times (0.40 - 0.02) = \text{mass of austenite} \times (0.80 - 0.40)$$

$$\frac{\text{mass of ferrite}}{\text{mass of austenite}} = \frac{0.80 - 0.40}{0.40 - 0.02}$$
$$= 1.05$$

Thus at this temperature about 51 per cent is ferrite and about 49 per cent austenite. This austenite forms pearlite at 723°C and thus about 49 per cent will become pearlite. There will be little further change on cooling to room temperature and thus the composition of a 0.40 per cent hypoeutectoid steel will be 51 per cent ferrite and 49 per cent pearlite.

For a 1.2 per cent hypereutectoid steel (as in Figure 11.6) at just above 723°C:

$$\text{Mass of cementite} \times (1.20 - 0.80) = \text{mass of austenite} \times (6.67 - 1.2)$$

$$\frac{\text{mass of cementite}}{\text{mass of austenite}} = \frac{6.67 - 1.2}{1.2 - 0.80}$$
$$= 13.7$$

Thus at this temperature about 93 per cent is cementite and 7 per cent austenite. The austenite forms pearlite at 723°C and thus about 7 per cent will become pearlite. There will be little further change on cooling to room temperature and thus the composition of a 1.2 per cent hypereutectoid steel will be 93 per cent cementite and 7 per cent pearlite.

The effect of carbon content

Ferrite is a comparatively soft and ductile material. Pearlite is a harder and much less ductile material. Thus the relative amounts of these two substances in a carbon steel will have a significant effect on the properties of that steel. Figure 11.7 shows how the percentages of ferrite and pearlite change with

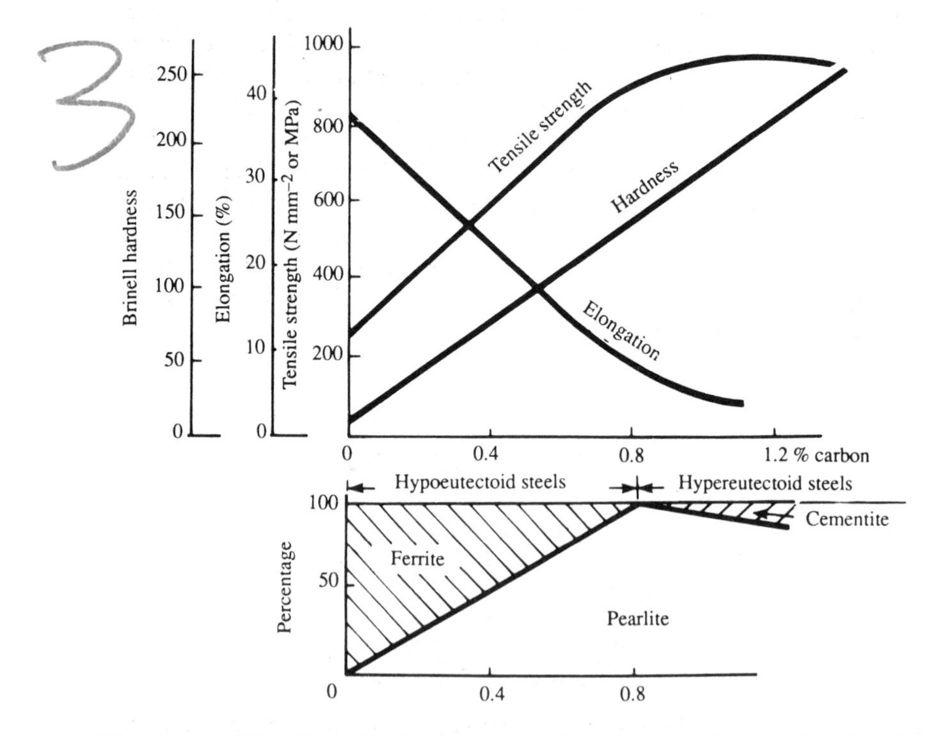

Figure 11.7 The effect of carbon content on the structure and properties of steels

percentage carbon and also how the mechanical properties are related to these changes. The data refers only to steels cooled slowly from the austenitic state.

Up to the euctectoid composition carbon steel, i.e. for hypoeutectoid steels, the decreasing percentage of ferrite and the increasing percentage of pearlite results in an increase in tensile strength and hardness. The ductility decreases, the elongation at fracture being a measure of this. For hypereutectoid steels, increasing the amount of carbon decreases the percentage of pearlite and increases the percentage of cementite. This increases the hardness but has little effect on the tensile strength. The ductility also changes little.

Heat treatment

Heat treatment can be used to change the microstructure and hence the properties of carbon steels. *Annealing* involves heating the steel to a temperature at which it is completely austenite, then cooling very slowly so that the structure has time to convert to ferrite and pearlite, also for any stresses to be eliminated and for the grain size to be reduced. The steel is then in its softest condition. *Normalizing* is essentially the same as annealing but with the cooling from the austenitic state being in air rather than in the furnace, hence a more rapid cooling. This results in a slightly harder material.

For steels with more than about 0.3 per cent carbon, rapid cooling, i.e. quenching, from the austenitic state can produce structural changes which result in the formation of *martensite*. This gives rise to a much harder material.

Martensite is produced when an iron–carbon alloy is cooled rapidly from the austenitic state. The rate of cooling has to be sufficiently rapid not to allow time for carbon atoms to diffuse out of the face-centred cubic form of austenite, and produce the body-centred form of ferrite, with its lower solubility for carbon. Below about 0.3 per cent carbon the rate of cooling needed to prevent the diffusion of the relatively small number of carbon atoms is too high to be achievable by quenching the steel in cold water. Such steels are thus not subject to this form of heat treatment.

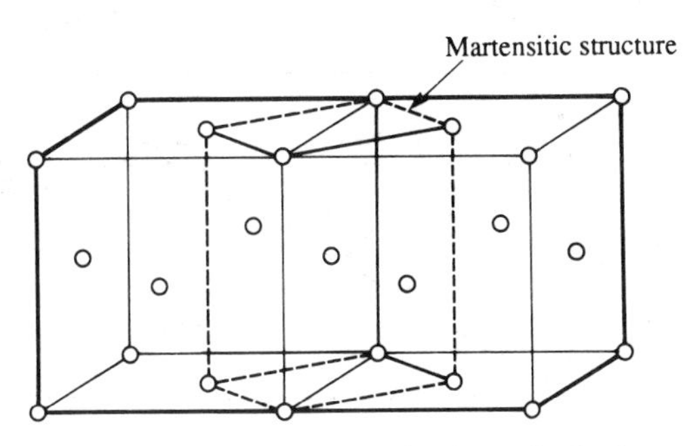

Figure 11.8 Two unit cells of austenite, showing how the body-centred tetragonal unit cell of martensite evolves

The transformation from austenite to martensite depends on the rate of cooling. The minimum cooling rate that will do this is called the *critical cooling rate*. Cooling rates faster than this give a completely martensitic structure but rates slower than this will not. The value of the critical cooling rate depends on the percentage of carbon present, the smaller the percentage of carbon the higher the rate.

Figure 11.8 shows two unit cells of austenite and how the body-centred unit cell of martensite evolves. For simplicity the figure only shows the position of the iron atoms, no carbon atoms being shown. The solubility of the carbon atoms in the body-centred structure is considerably exceeded and thus the structure has to distort from a cubic to a tetragonal one (Figure 11.9), a consequence of this being displacements of the metal surface where martensite grows. The greater the carbon content of the martensite the greater the distortion.

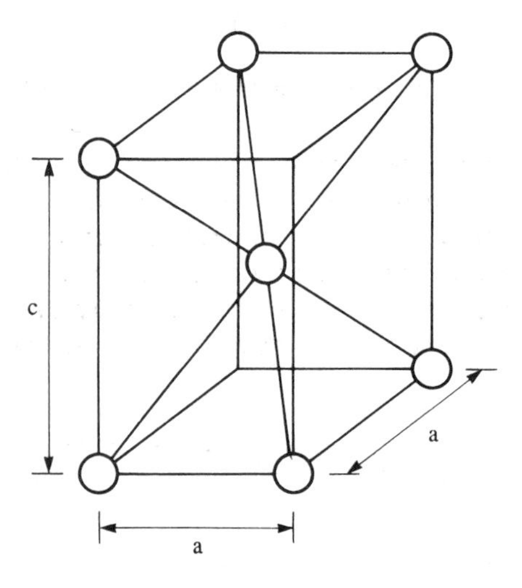

Figure 11.9 Body-centred tetragonal unit cell of martensite, $c > a$

As the temperature falls from the austenitic state the transformation of austenite starts at a temperature referred to as the *martensite start temperature* (M_s). The transformation of all the austenite to martensite is complete by a temperature referred to as the *martensite finish temperature* (M_f). The values of M_s and M_f depend on the composition of the alloy. Thus for plain carbon steels the relationship is as shown in Figure 11.10. This means that for a plain carbon steel with 0.4 per cent carbon, M_s is about 940°C and M_f about 720°C, and that by cooling to room temperature it is possible to change all the austenite to martensite. However, for a plain carbon steel with 1.0 per cent carbon, M_f is below room temperature and so all the austenite cannot be converted to martensite by cooling to room temperature. M_f is below room temperature for plain carbon steels with more than about 0.7 per cent carbon, hence all such steels with these amounts of carbon will contain both martensite and austenite.

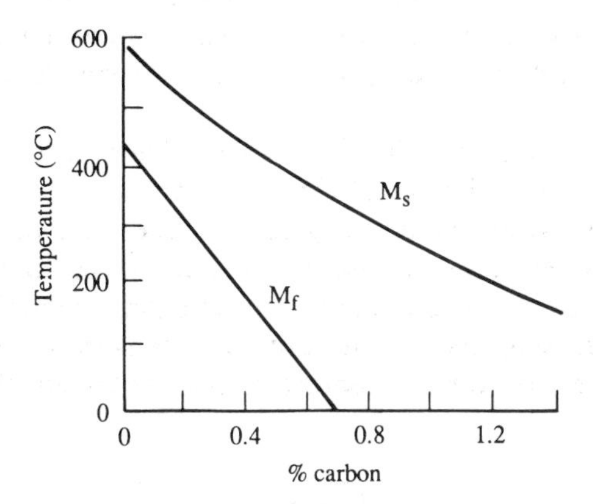

Figure 11.10 Effect on M_s and M_f for plain carbon steels of carbon content

The hardness of martensite increases with an increase in the carbon content. This means that plain carbon steels cooled at room temperature from the austenitic state will show an increase in hardness up to a carbon content of about 0.7 per cent; while above that the increasing hardness of the martensite is balanced by an increasing proportion of austenite, a relatively soft material. Figure 11.11 shows how the hardness thus depends on carbon content for such steels.

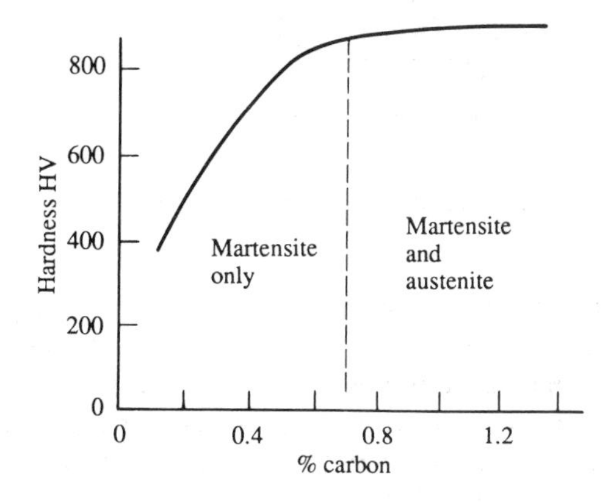

Figure 11.11 Hardness of water-quenched plain carbon steels

The rates of cooling at the centres of different diameter bars quenched to the same temperature will depend on their size – the bigger the diameter the slower will be the rate of cooling at the centre of the bar. There will, therefore, be a variation in properties across the section which will depend on the diameter. For this reason the mechanical properties of steels are quoted for different size bars. The term *limiting ruling section* is used, this being the maximum diameter of round bar at the centre of which the specified properties may be obtained.

Martensite has a low ductility because in a plain carbon steel the carbon atoms are locked in interstatial voids which are too small for them without lattice strain occurring. The result is that a large number of dislocations are produced and this high dislocation density hinders the movement of dislocations through the lattice. Some ductility can be restored to a martensitic structure by *tempering*. This process involves heating the material, perhaps a few hundred degrees Celsius, and then air cooling it. This allows carbon atoms to diffuse out of the martensite and so reduces its hardness.

Properties of carbon steels

Table 11.1 shows typical mechanical properties of a range of carbon steels after different heat treatments.

Table 11.1 *Properties of plain carbon steels*

Percentage carbon	*Condition*	*Tensile strength (MPa)*	*Elongation (%)*	*Hardness HB*
0.2	Annealed	400	37	115
	Normalized	450	36	130
0.4	Annealed	520	30	150
	Normalized	600	28	170
	Quench, temper 200°C	910	16	510
	Quench, temper 430°C	850	21	350
0.6	Annealed	635	23	180
	Normalized	790	18	230
	Quench, temper 200°C	1120	13	320
	Quench, temper 430°C	1090	14	310
0.8	Annealed	620	25	170
	Normalized	1030	11	290
	Quench, temper 200°C	1330	12	390
	Quench, temper 430°C	1300	13	375
1.0	Annealed	660	13	190
	Normalized	1030	10	290
	Quench, temper 200°C	1300	10	400
	Quench, temper 430°C	1200	12	360

Carbon steels are grouped according to their carbon content. *Mild steel* is a group of steels having between 0.10 per cent and 0.25 per cent carbon, *medium-carbon steel* has between 0.20 per cent and 0.50 per cent carbon, *high-carbon steel* has more than 0.5 per cent carbon. In addition such steels contain manganese.

Mild steel is a general purpose steel and is used where hardness and tensile strength are not the most important requirements. Typical applications are sections for use as joists in buildings, bodywork for cars and ships, screws, nails, wire. Medium-carbon steel is used for agricultural tools, fasteners, dynamo and motor shafts, crankshafts, connecting rods, gears. High-carbon steel is used for withstanding wear, where hardness is a more necessary requirement than ductility. It is used for machine tools, saws, hammers, cold chisels, punches, axes, dies, taps, drills, razors. The main use of high-carbon steel is thus as a tool steel.

Limitations of carbon steels

Plain carbon steels have limits on the engineering applications because:

1 A high-strength steel can only be obtained by increasing the carbon content to such a level that the material becomes brittle. High strength cannot be obtained with good ductility and toughness.
2 Hardness requires water quenching. The severity of this rapid rate of cooling often leads to distortion and cracking of the steel.
3 Large sections cannot be hardened uniformly. The hardness depends on the rate of cooling and this will vary across a large section.
4 Plain carbon steels have poor resistance to corrosion and oxidation at high temperatures.

11.3 Alloy steels

The term *alloy steel* is used to describe those steels to which one or more alloying elements, in addition to carbon, have been deliberately added in order to modify the properties of the steel. Plain carbon steels are, however, not just iron and carbon and invariably include some manganese and silicon. These are added during the steel-making in order to overcome the effects of the two main impurities – sulphur and oxygen – that are invariably present.

The term *low alloy* is used for alloy steels when the alloying additions are less than 2 per cent, *medium alloy* when the additions are between 2 and 10 per cent, and *high alloy* when they are above 10 per cent. In all cases the amount of carbon present is less than 1 per cent. Common elements that are added are aluminium, chromium, cobalt, copper, lead, manganese, molybdenum, nickel, phosphorus, silicon, sulphur, titanium, tungsten and vanadium.

There are a number of ways in which the alloying elements can have an effect on the properties of the steel. The main effects are to:

1 *Solution harden the steel* – most of the elements used in alloy steels form substitutional solid solutions with the iron. This increases the tensile and impact strengths (Figure 11.12).

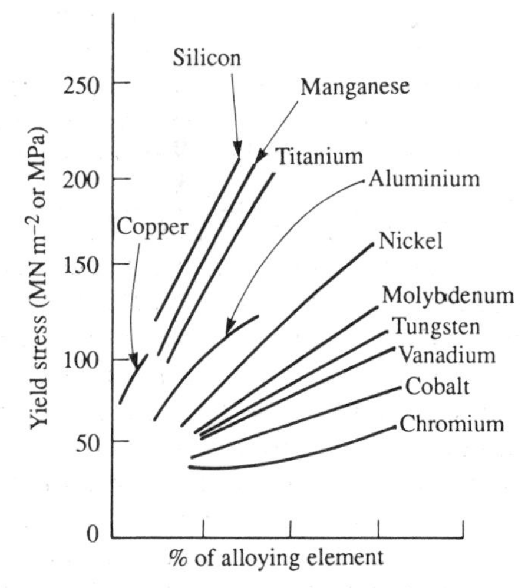

Figure 11.12 Effect of alloying element on yield strength as a result of solid solution hardening

2 *Form carbides* – they may form stable, hard carbides which may, if in an appropriate form such as fine particles, increase the strength and hardness. Manganese, chromium and tungsten have this effect.
3 *Form graphite* – they could cause the breakdown of cementite and lead to the presence of graphite in the alloy. Silicon and nickel have this effect. The result is a decrease in strength and hardness.
4 *Stabilize austenite or ferrite* – they may lower the temperature at which austenite is formed on heating the steel. Manganese, nickel, copper and cobalt have this effect. The lowering of this temperature means a reduction in the temperature to which the steel has to be heated for hardening by quenching. If a sufficiently high percentage of one of these elements is added to the steel the transformation temperature to austenite may be decreased to such an extent that the austenite is retained at room temperature. With manganese about 11 to 14 per cent produces what is known as an *austenitic steel*. Such steels have relatively good hardness combined with ductility and so are tough. They may increase the temperature at which austenite is formed on heating the steel. Chromium, molybdenum, tungsten, vanadium, silicon and aluminium have this effect. This raises the temperature to which the steel has to be heated for hardening. If a sufficiently high percentage of one of these elements is added to the steel, the transformation from ferrite to austenite may not take place before the steel reaches its melting point temperature. Such steels are known as *ferritic steels*, 12 to 25 per cent of chromium with a 0.1 per cent carbon steel gives such a type of steel. Such a steel cannot be hardened by quenching.

5 *Change the critical cooling rate* – the *critical cooling rate* is the minimum rate of cooling that has to be used if all the austenite is to be changed into martensite. With the critical cooling rate, or a higher rate of cooling, the steel has maximum hardness. With slower rates of cooling a less hard structure is produced. Most alloying elements reduce the critical cooling rate. The effect of this is to make air or oil quenching possible, rather than water quenching. It also increases the hardenability of steels.
6 *Improve corrosion resistance* – they can improve the corrosion resistance. Some elements promote the production of adherent oxide layers on the surfaces of the steel and so improve its corrosion resistance. Chromium is particularly useful in this respect. If it is present in a steel in excess of 12 per cent the steel is known as *stainless steel* because of its corrosion resistance. Copper is also used to improve corrosion resistance.
7 *Change grain growth* – they can influence grain growth. Some elements accelerate grain growth while others decrease grain growth. The faster grain growth leads to large grain structures and consequently to a degree of brittleness. The slower grain growth leads to smaller grain size and so to an improvement in properties. Chromium accelerates grain growth and thus care is needed in heat treatment of chromium steels to avoid excessive grain growth. Nickel and vanadium decrease grain growth.
8 *Improve machinability* – they can affect the machinability of the steel. Sulphur and lead are elements that are used to improve the chip formation properties of steels.

Table 11.2 indicates the main effects and functions of the various alloying elements. An alloying element generally affects the properties of the alloy in more than one way, some of the ways not always being beneficial. Alloying elements can thus be chosen to counteract the effects of impurities, counteract the effects of other alloying elements and improve the properties.

Low alloy structural steels

Mild steel, with tensile strengths up to about 450 MN m^{-2} (MPa), or the high-carbon steel, with a tensile strength of the order of 600 MN m^{-2}, has been widely used for structural work, axles to bridges. However, the higher the percentage of carbon in the steel the more brittle the material becomes (see Figure 11.7). There is thus a problem of combining strength with toughness. This can be overcome by the use of a low alloy steel.

The properties required of a constructional steel are:

1 High yield strength, this representing the maximum stress to which the material could be exposed in use.
2 Tough. Not brittle.
3 Good weldability.
4 Good corrosion resistance.
5 Low cost.

Table 11.2 *Effects and functions of alloying elements*

Element	*Main effects*	*Main functions*
Aluminium	Ferrite stabilizer	Aids nitriding
Chromium	Carbide former Ferrite stabilizer Forms surface oxide layers	Improves corrosion resistance Increases hardenability Improves high temperature properties Improves abrasion and wear resistance
Cobalt	Austenite stabilizer	Improves strength at high temperatures
Copper	Austenite stabilizer	Improves corrosion resistance
Lead	Improves chip formation	Improves machinability
Manganese	Solid solution hardening Carbide former Austenite stabilizer	Increases hardenability Combines with the sulphur in steel to reduce brittleness
Molybdenum	Ferrite stabilizer Carbide former Inhibits grain growth	Improves hardenability Restricts grain growth and so improves strength and toughness Increases hot strength and hardness Improves abrasion resistance with high carbon
Nickel	Austenite stabilizer Solid solution hardening Graphite former Inhibits grain growth	Improves strength and toughness
Phosphorus	Solid solution hardening Improves chip formation	Improves machinability
Silicon	Ferrite stabilizer Solid solution hardening	Deoxidation of liquid steel Improves fluidity in casting Strengthens low alloy steel
Sulphur	Improves chip formation	Improves machinability
Titanium	Carbide former Solid solution hardening	Forms compounds with carbon to improve chromium steels
Tungsten	Carbide former Ferrite stabilizer	Hot hardness and strength Hard, abrasion-resistant carbides in tool steels
Vanadium	Ferrite stabilizer Inhibits grain growth Carbide former	Restricts grain coarsening Increases hardenability Improves hot hardness

The carbon content of the alloy has to be kept low if brittle martensite is not to be formed in the heat-affected zone when welds are made. The carbon is thus kept to below about 0.4 per cent. To maximize the yield strength, alloying elements can be added to give solution hardening of the ferrite and to reduce grain size. Most of the common alloying elements increase the yield strength by solution hardening, unfortunately, however, most of them increase the temperature at which steel changes from being ductile to brittle. This is not desirable so the choice is restricted to those few elements which will give solution hardening without inceasing the transition temperature. For this reason, nickel, manganese and titanium are the only options. Nickel also inhibits grain growth. Unfortunately, nickel promotes the breakdown of cementite to give graphite. To counteract this, chromium might be added since it promotes the formation of carbides. A nickel–chromium steel can show brittleness after tempering, but this effect can be reduced by the addition of a small percentage of molybdenum, about 0.3 per cent.

Thus, a structural steel might consist of less than 0.4 per cent carbon, a small percentage of nickel, perhaps 1 per cent or less manganese, and possibly a small amount of chromium.

In addition to solution hardening and reducing grain size, alloys can also have their yield strength increased by the distribution of a hard carbide phase as a fine precipitate through the alloy. All these are used with high-strength, low alloy (HSLA) steels. Manganese or nickel are used for solution hardening; aluminium, vanadium, niobium or nickel for reduction of grain size; and niobium, titanium or vanadium to produce a fine carbide precipitate. Niobium is the best element to add since it both restricts grain size and produces a precipitate. Thus a typical HSLA steel contains 0.12 per cent carbon, 1.5 per cent manganese, 0.3 per cent silicon, 0.3 per cent niobium, 0.08 per cent vanadium and 0.005 per cent sulphur. After rolling and ageing such a steel can have a yield stress of the order of 600 MN m^{-2} (MPa) and a tensile strength of 700 MN m^{-2} (MPa).

Medium alloy nickel–chromium steels

Medium alloy steels based on the alloying elements of nickel and chromium are very widely used. The effect of adding nickel and chromium is to change the critical cooling rate so that martensite is formed at slower rates of cooling. This means that larger sections do not run the risk of cracking through having to be cooled quickly to obtain the required hardness. A slower rate of cooling also means that the differences in the rates of cooling of the core of a thick section compared with its surfaces are less and there is less variation of hardness with depth. A slower rate of cooling also means less chance of distortion occurring during cooling. The rate of cooling can be reduced to such an extent with some alloys that air cooling is possible and a fully martensitic steel is still produced.

Also included with the nickel and chromium is manganese. This is there to combine with the impurity sulphur in the steel and reduce brittleness that would otherwise occur.

Nickel–chromium steels tend to contain about 0.1 to 0.55 per cent carbon, 1.0 to 4.75 per cent nickel, 0.45 to 1.75 per cent chromium and 0.3 to 0.8 per cent manganese. Figure 11.13 shows the general form of the properties of such steels after quenching and then tempering to different temperatures. It should be notes that a characteristic of nickle–chromium steels is that the impact strength shows a marked reduction if tempering at about 300°C occurs. Because of this, tempering should not be used at such temperatures. This also has the effect that the impact strength resulting from slow cooling from a higher tempering temperature is much lower than that when rapid cooling is used. This effect is referred to as *temper brittleness*. It can be overcome by adding a small amount of molybdenum, e.g. 0.3 per cent, to the alloy.

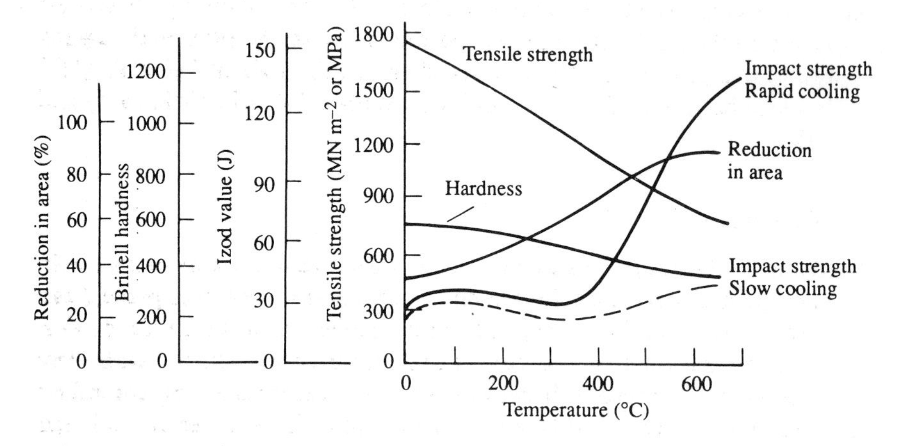

Figure 11.13 Effect of tempering temperature on the properties of a typical nickel–chromium steel

Chromium, molybdenum, tungsten and vanadium all produce hard carbides which, when a steel is tempered, produce a precipitation hardening effect, hence the addition of vanadium to nickel–chromium steels. A typical nickel–chrome–vanadium steel has 0.3 per cent carbon, 1.8 per cent nickel, 0.87 per cent chromium, 0.75 per cent manganese, 0.4 per cent molybdenum and 0.1 per cent vanadium.

Maraging steels

Maraging steels are high strength, high alloy steels which can be precipitation-hardened. The term maraging derives from the transformations which occur after the steel is cooled from the austenitic state, the production of *mar*tensite followed by *ageing* (mar-ageing), i.e. precipitation hardening.

The alloys have a high nickel content (18 to 22 per cent) and a carbon content less than 0.3 per cent. Other elements such as cobalt, titanium and molybdenum are present. These elements form intermetallic compounds with nickel. A typical maraging steel might contain 18 per cent nickel, 7 or 8 per cent cobalt, 5 per cent molybdenum and about 0.5 per cent titanium. The carbon content is kept low since otherwise the high nickel content could lead to the formation of graphite in the structure and a consequential drop in strength and hardness.

The steel is heated to about 830°C and then air cooled. This results in a martensitic structure being formed. Because of the low amounts of carbon in the steel this martensitic structure is of low strength and is easily machined or worked. Following machining and working the steel is then precipitation-hardened by heating it to about 500°C for two or three hours. During this time precipitates of intermetallic compounds form. The effect of this ageing process is to increase the tensile strength and hardness. Typically, prior to the ageing process, the material might have a tensile strength of about 700 MPa (MN m^{-2}) and a hardness of 300 HV and afterwards 1700 MPa (MN m^{-2}) and 550 HV.

Stainless steels

The addition of small percentages of chromium to a plain carbon steel results in the formation of hard carbides which, if dispersed as fine precipitates, can increase strength and hardness, and cause a reduction in the critical cooling rate. This means that a fully martensitic structure can be achieved with a less rapid cooling to improve hardenability. The percentages of chromium involved in the above would be less than 2 per cent. If, however, much larger percentages are used the effect is to give the steel exceptionally good corrosion resistance (Figure 11.14). The corrosion rate reaches a very low value when the chromium content is about 12 per cent.

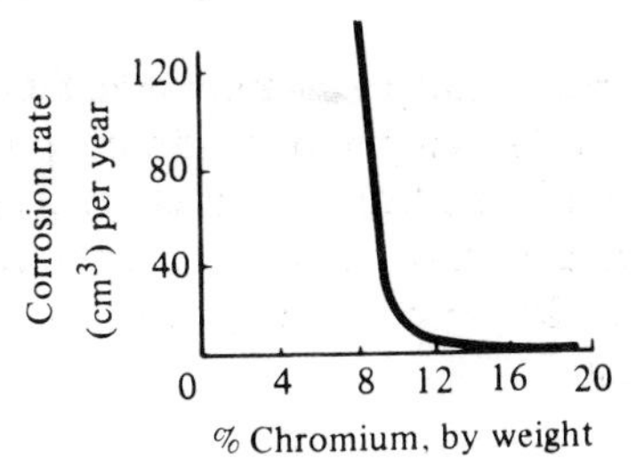

Figure 11.14 The effect of per centage of chromium on corrosion rate

There are several types of stainless steel: ferritic, martensitic and austenitic. *Ferritic steels* contain between 12 and 25 per cent chromium and less than 0.1 per cent carbon. Figure 11.15a shows the form of the thermal equilibrium diagram for iron–chromium alloys with this low percentage of carbon. Between about 12 and 25 per cent chromium, on cooling from the liquid there is only one change and that is to ferrite. Because austenite cannot be formed with such a steel, hardening by quenching to give martensite cannot occur. Martensite can only be produced by quenching from the austenite state. Thus, such steels cannot be hardened by heat treatment. They can, however, be hardened by work hardening. Since they work harden slowly they are suitable for forming by deep drawing, spinning, etc. Ferrite steels are used for mouldings and trim for car bodies; furniture; gas and electric stoves and other domestic appliances; spoons and forks; car silencers; nuts, bolts, screws etc. In general the applications are where good corrosion resistance is required without the need for high tensile strength.

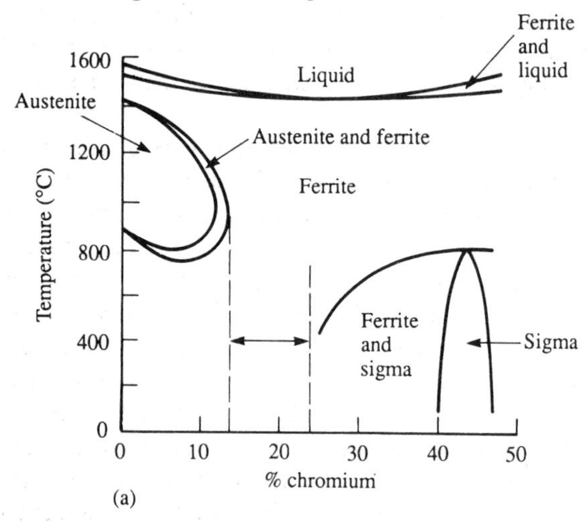

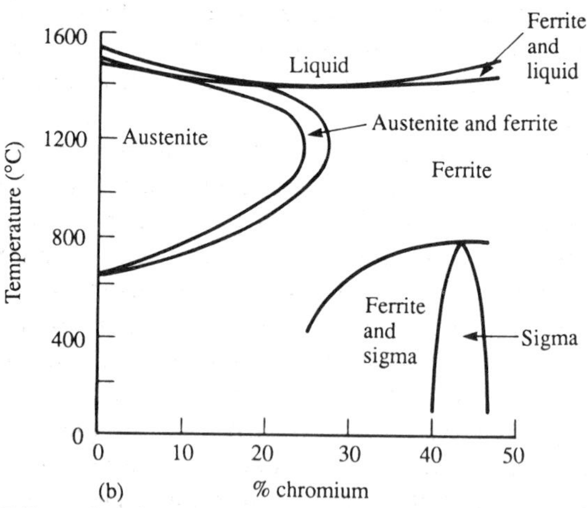

Figure 11.15 Effect of carbon content on thermal equilibrium diagrams for iron–chromium alloys (a) 0.1 per cent carbon, (b) 1 per cent carbon

Martensitic steels contain between 12 and 18 per cent chromium and between about 0.1 and 1.2 per cent carbon, the higher chromium content being with the higher carbon content. Increasing the carbon content increases the stability of the austenite phase, thus the loop on the thermal equilibrium diagram for the austenite phase increases as the carbon content increases (Figure 11.15), the loop being often referred to as the gamma loop. With virtually zero carbon the austenite phase can exist only out to a chromium content of about 12 per cent, with 0.35 per cent carbon austenite can exist out to about 15 per cent chromium. With higher percentages of carbon it extends even further. Thus, a suitable combination of carbon and chromium content means that a steel will have an austenitic phase. The existence of such a phase means that the steel can be quenched to give martensite and so harden by heat treatment. The result is steel with much higher tensile strength and hardness than is possible with ferritic steel.

Martensitic steels are subdivided into three groups: stainless irons, stainless steels and high-chromium stainless steels. *Stainless irons* contain about 0.1 per cent carbon and 12 to 13 per cent chromium. Quenching from about 900 to 1000°C gives a fully martensitic structure. Tempering at about 650 to 750°C gives a structure consisting of ferrite with fine carbides dispersed through it. Such a material can be easily machined. *Stainless steels* contain about 0.25 to 0.30 per cent carbon with 11 to 13 per cent chromium. This form of steel is often referred to as cutlery stainless steel since it is widely used for cutlery. Quenching such a steel from about 850°C results in a martensitic structure containing carbides and consequently with high hardness. *High-chromium stainless steels* contain 16 to 18 per cent chromium, about 2 per cent nickel and 0.5 to 0.15 per cent carbon. Without the nickel such a low percentage of carbon with the high chromium content would lead to a ferritic steel. However, the small percentage of nickel enables austenite to form and hence, after quenching, martensite. The result is an alloy with a high corrosion resistance resulting from the high chromium content, with the mechanical properties of a lower chromium content alloy.

Austenitic steels contain 16 to 26 per cent chromium, more than about 6 per cent nickel and a very low percentage of carbon (about 0.1 per cent or less). A common steel of this type has 0.5 per cent carbon, 18 per cent chromium and 8.5 per cent nickel. Steels with this approximate ratio of chromium to nickel are referred to as 18/8 stainless steels. Without the nickel such a steel would be ferritic. However, the effect of the nickel is to give an austenite phase. The stability of this austenite phase increases as the percentage of nickel is increased, until the alloy becomes completely austenitic at all temperatures (Figure 11.16). Such steels cannot be hardened by quenching. They are, however, usually quenched, not to produce martensite but to minimize the formation of chromium carbide as this causes a reduction in the corrosion resistance of the alloy. A very large increase in hardness can be produced by cold working.

Table 11.3 gives details of the mechanical properties of typical stainless steels.

During welding, when temperatures of 500 to 800°C are realized, stainless steels may undergo structural changes which are detrimental to the corrosion

Table 11.3 *Mechanical properties of typical stainless steels*

Stainless steel type	*Composition*	*Condition*	*Yield stress ($MN\ m^{-2}$ or MPa)*	*Tensile strength ($MN\ m^{-2}$ or MPa)*	*Percentage elongation*	*Izod (J)*	*Hardness (HV)*
Ferritic	Carbon 0.06% Chromium 13%	Annealed	280	416	20		170
Martensitic	Carbon 0.12% Chromium 12.5%	Oil-quenched from 1000°C	1190	1850	2.5	7	371
		Oil-quenched from 1000°C, tempered at 750°C	370	570	33	134	172
Cutlery steel	Carbon 0.32% Chromium 13%	Oil-quenched from 980°C, tempered at 180°C		1450	8		600
	Carbon 0.16% Chromium 16.5% Nickel 2.5%	Oil-quenched from 975°C, tempered at 650°C	695	880	22	34	270
Austenitic	Carbon 0.05% Chromium 18% Nickel 8.5% Manganese 0.8%	Annealed	278	618	50		180
		Cold rolled	803	896	30		

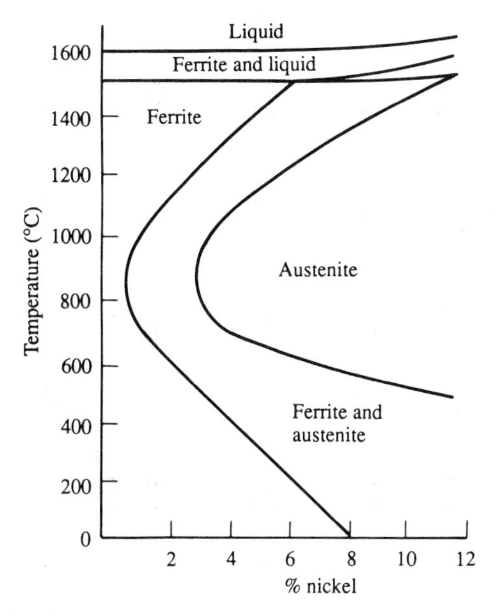

Figure 11.16 The iron–chromium– nickel equilibrium diagram for an 18 per cent chromium alloy

resistance of the material, the other properties being little affected. The effect is known as weld decay and results from the precipitation of chromium-rich carbides at grain boundaries. This removal of chromium from grains to the boundaries decreases the corrosion resistance. One way of overcoming this is to *stabilize* the steel by adding other elements such as niobium and titanium. These have a greater affinity for the carbon than the chromium and so form carbides in preference to the chromium.

A problem that can occur with ferritic steels is the formation of *sigma phase*. This phase is hard and brittle and is formed from ferrite. In the formation a considerable reduction in volume occurs and this can lead to the development of cracks. Sigma phase is shown in the thermal equilibrium diagrams in Figure 11.15. Ferritic steels containing high percentages of chromium can develop sigma phase if held at a temperature of about 475°C for some time. The result of sigma formation is an embrittlement of the alloy. It can be removed by heating to above 650°C. Sigma phase can also be produced in austenitic steels following heating to temperatures in the region of 800°C.

Heat resisting steels

Steels for use at high temperatures must provide good resistance to creep and oxidation and suffer no detrimental changes in structure or properties. There must, therefore, be no such effects as temper embrittlement, sigma phase formation or carbide precipitation. Other properties may also be required, e.g. weldability, machinability and good fatigue resistance.

Carbon steels are limited to about 450°C since at higher temperatures the rate of oxidation can become very high and also there is a marked reduction in stress-bearing capabilities. Low alloy steels have been developed for use at higher temperatures. Such steels include small amounts of elements, such as chromium, vanadium and molybdenum, which combine with the carbon to form carbides. For example, a 1 per cent chromium–½ per cent molybdenum steel has an oxidation limit of about 550°C and is more resistant to creep than a carbon steel. Such a steel is used for steam pipes.

For temperatures in excess of 550°C the oxidation problem can be reduced by increasing the chromium content. For example, a 12 per cent chromium steel has an oxidation limit of about 575°C and is used for steam turbine rotors and blades.

For higher temperatures, austenitic steels can be used. Such steels do not transform to the ferritic structure on cooling to room temperature, but retain an austenitic structure. This is obtained as a result of adding 18 per cent chromium and 8 per cent nickel. Such a steel has an oxidation limit of about 650°C and superior creep resistance, a typical application being for superheater tubes. Enhanced creep resistance can be obtained by incorporating small amounts of niobium, titanium or molybdenum. Steels containing up to 25 per cent chromium and 30 per cent nickel have been developed for use up to about 750°C.

Nickel–chromium alloys (see Chapter 13) can be used for higher temperatures. For example, an 80 per cent chromium–20 per cent nickel alloy has an oxidation limit of about 900°C and good creep properties to quite high temperatures.

For more information reference may be made to Day, M. F., *Materials for High-Temperature Use* (Oxford University Press 1979).

Steels at low temperatures

Most metals have face-centred cubic, body-centred cubic, or hexagonal closepacked crystal structures. The effect of a reduction in temperature on the properties of metals depends on the crystal structure. The strength, ductility and toughness of face-centred cubic metals improves as the temperature decreases. For this reason austenitic stainless steels, along with aluminium, copper and nickel alloys, are widely used at low temperatures. Body-centred cubic metals, however, generally undergo a ductile to brittle transition at some temperature below room temperature. The temperature at which this transition occurs is markedly affected by small changes in the composition of the alloy, grain size, notches and flaws and the rate of loading. Body-centred cubic metals include all plain carbon steels, low and medium alloy steels. Care has thus to be exercised in the use of such steels at low temperatures. Hexagonal close-packed metals have properties intermediate to those of face-centred and body-centred cubic structures.

Coding systems for steels

In Great Britain the standard codes for the specification of steels are specified by the *British Standards Institution*. The following codes are those specified to the system given in BS 970 for wrought steels, published between 1970 and 1972. The system uses six symbols.

The first three digits designate the type of steel.

000 to 199 Carbon and carbon-manganese types, the numbers being 100 times the manganese content.
200 to 240 Free cutting steels, the second and third numbers having an approximate relationship to 100 times the mean sulphur content.
250 Silicon-manganese spring steels.
300 to 499 Stainless and heat resistant valve steels.
500 to 999 Alloy steels.

The fourth symbol is a letter.

A The steel is supplied to a chemical composition determined by chemical analysis.
H The steel is supplied to hardenability specification.
M The steel is supplied to mechanical property specification.
S The material is stainless.

The fifth and sixth digits correspond to 100 times the mean percentage carbon content of the steel. The following examples illustrate the use of the coding.

070M20 A carbon/carbon-manganese type of steel, supplied to mechanical property specification and having 0.20 per cent carbon with 0.70 per cent manganese.
150M19 A carbon/carbon-manganese type of steel supplied to mechanical property specification and having 0.19 per cent carbon with 1.50 per cent manganese.
220M07 A free-cutting steel supplied to mechanical property specification and having 0.7 per cent carbon content. The sulphur content would be about 0.20 per cent.
040A04 A carbon/carbon-manganese type of steel supplied to a chemical composition specification and having 0.04 per cent carbon with 0.40 per cent manganese.

Alloy steels have the first three digits in the code between 500 and 999. These are a few examples from this range of steels.

503M40 A nickel steel having 1.0 per cent nickel and 0.40 per cent carbon, being supplied to a mechanical property specification. The number 503 indicates that the steel is a nickel steel.

526M60 A chromium steel with 0.75 per cent chromium and 0.60 per cent carbon, to a mechanical property specification.

530M40 Also a chromium steel, having 1 per cent chromium with 0.40 per cent carbon to a mechanical property specification.

640M40 A nickel–chromium steel with 1.25 per cent nickel, 0.5 per cent chromium and 0.40 per cent carbon, to a mechanical property specification.

653M31 Also a nickel chromium steel having 3 per cent nickel, 1 per cent chromium and 0.31 per cent carbon, to a mechanical property specification.

A wide range of steels is available. As the specification of a steel is usually according to its tensile strength when in the hardened and tempered condition, a steel can be specified by means of a code letter, the letter indicating the tensile strength range in which the steel falls when in the hardened and tempered condition. The letter is said to refer to the *condition* of the steel. Table 11.4 gives condition codes and examples of steel which, by composition and size of section, fall into the various ranges.

Table 11.4 *Specificity of steels by condition codes*

Condition code	*Tensile strength range (N mm^{-2} or MPa)*
P	550 to 700
Q	629 to 770
R	700 to 850
S	770 to 930
T	850 to 1000
U	930 to 1080
V	1000 to 1150
W	1080 to 1240
X	1150 to 1300
Y	1240 to 1400
Z	1540 minimum

Other countries have other code systems for specifying steels. A common one is that developed by the *American Society of Automotive Engineers*, the code being referred to as a SAE numbering. A four-symbol code is used, the first two numbers indicating the type of steel and the third and fourth numbers indicating 100 times the percentage of carbon content.

1000 series Carbon steels; 10XX is plain carbon with a maximum manganese content of 1.00 per cent, 11XX is resulphurized, 12XX resulphurized and rephosphorized, 13XX has 1.75 per cent manganese, etc.

2000 series	Nickel steels; 23XX are nickel steels having 3.50 per cent nickel, 25XX has 5.00 per cent nickel.
3000 series	Nickel chromium steels; 32XX has 1.75 per cent nickel and 1.07 per cent chromium, 34XX has 3.00 per cent nickel and 0.77 per cent chromium.
4000 series	Chromium molybdenum steels; 40XX has 0.20 or 0.25 per cent molybdenum, 41XX has 0.50 per cent, 0.80 per cent or 0.95 per cent chromium with 0.12 per cent, 0.20 per cent or 0.30 per cent molybdenum.
5000 series	Chromium steels containing up to 1 per cent chromium.
6000 series	Chromium vanadium steels.
7000 series	Steels containing up to 6 per cent tungsten.
8000 series	Steels containing nickel, chromium and molybdenum.
9000 series	Silicon manganese steels.

There are some specially specified steels which have modified SAE numbers.

XXBXX	The 'B' denotes boron intensified steels, the letter being incorporated in the middle of the normal specifications.
XXLXX	The 'L' denotes leaded steels, the letter being incorporated in the middle of the normal specification.

These examples illustrate the use of coding.

1330	A carbon steel having 0.30 per cent carbon and 1.75 per cent manganese.
4012	A steel having 0.12 per cent carbon, 0.20 per cent molybdenum and 0.90 per cent manganese.

11.4 Cast irons

The term *cast iron* arises from the method by which the iron is produced. Pig iron is remelted in a furnace and the properties of the iron modified by the addition of other materials. The resulting iron is then cast, hence the term cast iron. Cast irons have between about 2 and 4 per cent carbon and often significant amounts of silicon, as well as smaller amounts of other elements. Cast irons are used widely because of their ease of melting and hence the production of components by casting.

Grey and white cast irons

The iron–carbon thermal equilibrium diagram (Figure 11.1) shows that at a temperature of about 1147°C a eutectic occurs for alloys with 4.3 per cent carbon. Between 2.0 and 4.3 per cent, increasing the carbon percentage reduces the temperature at which solidification occurs. Cast irons have between 2 and 4 per cent carbon.

Consider the cooling of an iron from the liquid state. When solidification starts to occur the result is the formation of austenite in the liquid. At 1147°C solidification occurs. With very slow cooling austenite plus graphite can be formed. Graphite occurs because the rate of cooling is so slow that at such a temperature cementite is not stable and breaks down to give graphite. As the temperature is further lowered the graphite grows due to precipitation from the austenite. At 723°C the remaining austenite changes to ferrite plus graphite. The result at room temperatue is a structure of ferrite and graphite. This type of cast iron is known as a *grey iron* because of the grey appearance of its freshly fractured surface. This grey iron is referred to as a ferritic form of grey iron.

A faster cooling rate can lead to a structure in which the austenite changes to a mixture of ferrite and pearlite at 723°C, or with an even faster cooling rate, entirely pearlite. The results are structures involving graphite with either ferrite and pearlite or just pearlite. The greater the amount of pearlite in these grey irons the harder the material.

When the iron solidifies from the liquid state the graphite usually forms as flakes. Hence the structure at room temperature has graphite flakes in ferrite, ferrite and pearlite, or pearlite, depending on the rate of cooling (Figure 11.17).

Figure 11.17 A grey cast iron. The structure consists of black graphite flakes in a matrix of pearlite (Courtesy of BCIRA)

With even faster cooling a different type of structure is formed. The solidification at 1147°C gives austenite and cementite. As the temperature is further lowered the cementite grows due to precipitation from the austenite. At 723°C the remaining austenite changes to pearlite. The result at room temperature is a structure of cementite and pearlite (Figure 11.18). This type of cast iron is known as a *white iron* because of the white appearance of its freshly fractured surface. White iron, because of its high cementite content, is hard and brittle. This makes it difficult to machine and hence of limited use. The main use is where a wear-resistant surface is required.

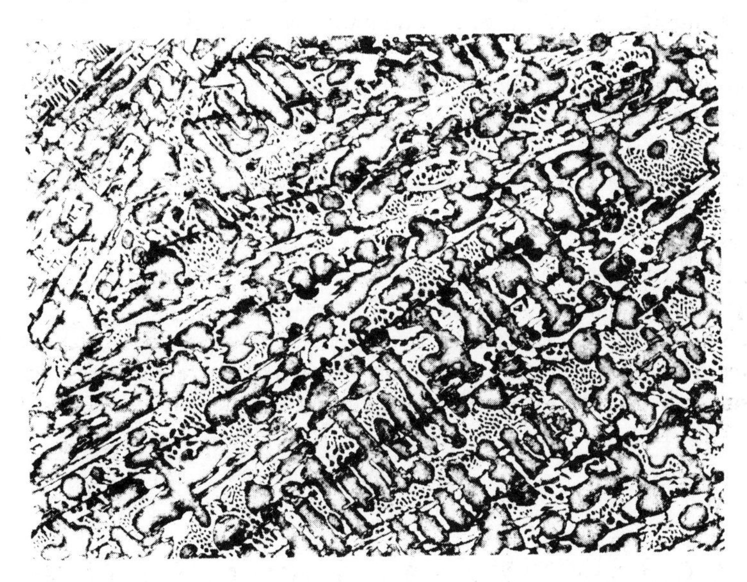

Figure 11.18 A white cast iron. The structure consists of white cementite and dark pearlite (Courtesy of BCIRA)

The tendency of cast iron to solidify with the carbon present in the form of cementite rather than graphite increases as the carbon percentage (or carbon equivalent (see later)) is reduced and as the cooling rate increases. The iron is said to chill if such changes occur. Chilled structures are hard and brittle.

Section sensitivity

A casting will often have sections of varying thickness. The rate of cooling of a part of a casting will depend on the thickness of the section concerned – the thinner the section the more rapidly it will cool. This means that there is likely to be a variation in the properties of different parts of a casting.

Thus, for example, a very thin section might be a white iron, and so it will be very hard and virtually unmachinable. A different section might be grey with white edges. A thicker section might be a fine-grained grey iron while an even thicker section might be, at least internally, a coarse-grained grey iron.

Because the properties of grey iron depend on the rate of cooling and this is influenced by section thickness, the British Standard specifications for grey iron are in terms of the properties of a test piece of specific diameter, 30 mm, when cast. Seven grades are used, the grade number being the value of the tensile strength in MN m^{-2} (MPa). The grades used are 150, 180, 220, 260, 300, 350, 400. Figure 11.19 shows the effect on the tensile strength of the different grades of section thickness. This has to be taken into account in considering the likely strengths at different thicknesses within a casting produced to a particular grade.

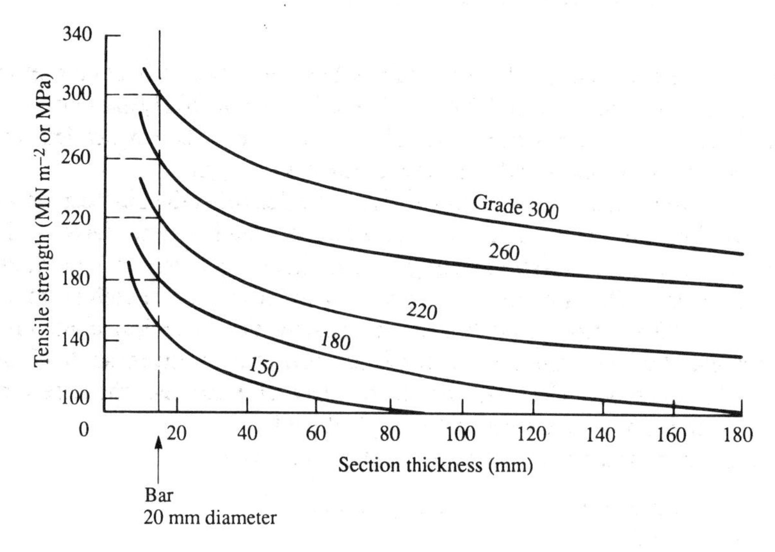

Figure 11.19 Effect of section thickness on tensile strength of grey cast iron (Based on data supplied by BCIRA)

Thus an iron which in the standard test piece bar of 30 mm diameter gives a tensile strength of 220 MN m^{-2} (MPa), i.e. grade 220, would give this same tensile strength in the centre of a casting which is 17 mm thick (the cooling of effectively a flat plate is lower than a bar). However if the thickness were 100 mm the tensile strength would be 147 MN m^{-2} (MPa). This lower value occurs because the centre of this thicker section would be cooling slower than the test piece.

Chemical composition

The tendency of graphite to be produced in a cast iron depends on the carbon content and the rate of cooling. An increase in the carbon content up to 4.3 per cent (the eutectic value) lowers the melting point of the iron, as indicated by the iron–carbon thermal equilibrium diagram (Figure 11.1). This decrease in melting point favours the production of an iron with graphite rather than cementite because the cooling from the liquid state is from a lower temperature and hence the cooling rate is slower.

Some elements when included with the carbon in the iron affect the formation of graphite. Silicon and phosporus affect the composition of the eutectic point and are considered to have a carbon equivalence in the formation of graphite. The combined effect of carbon, silicon and phosphorus is represented by the value of the so-called *carbon equivalent* E_c:

$$E_c = \text{\% carbon} + \frac{\text{\% silicon} + \text{\% phosphorus}}{3}$$

The carbon equivalent value determines how close the composition of an iron is to the eutectic value of 4.3 per cent and hence how likely grey iron is to be produced rather than white iron. The greater the carbon equivalent the lower the tensile strength and hardness of the cast material.

Sulphur often exists in cast iron and has the effect of stabilizing cementite (the opposite effect to silicon). The presence of sulphur thus favours the production of white iron rather than grey iron, hence increasing the hardness and brittleness. The amount of sulphur in an iron has, therefore to be controlled. The addition of small amounts of manganese to an iron containing sulphur enables the sulphur to combine with the manganese to form manganese sulphide. The removal of the free sulphur has the effect of increasing the chance of grey iron being produced.

The following is a typical composition for a grey iron:

Carbon 3.2 to 3.5 per cent (most as graphite)
Silicon 1.3 to 2.3 per cent
Phosphorus 0.15 to 1.0 per cent
Sulphur 0.10 per cent
Manganese 0.5 to 0.7 per cent

The material has a carbon equivalent approaching 4.3 per cent, the eutectic value.

Heat treatment of grey irons

Annealing is used with grey cast irons to provide optimum machinability and remove stresses. Annealing involves heating the cast iron to about 40°C above the A_1 temperature, i.e. about 760°C, soaking at that temperature and then cooling very slowly.

Grey cast iron can be hardened by quenching. Such a treatment is generally followed by tempering. A tempering temperature of 475°C is common.

A stress relieving treatment is often used before a significant amount of machining takes place. Such a treatment involves heating to about 550°C, i.e. below the A_1 temperature.

Malleable cast irons

Malleable cast irons are produced by the heat treatment of white cast irons. Three forms of malleable iron occur: *whiteheart*, *blackheart* and *pearlitic*. Malleable irons have better ductility than grey cast irons and this combined with their high tensile strength makes them useful materials.

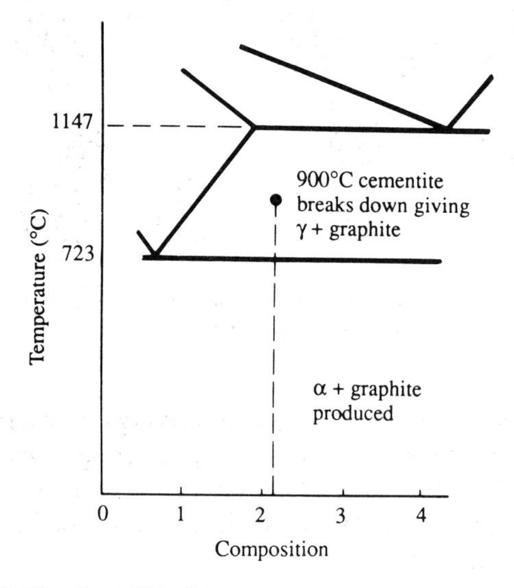

Figure 11.20 Production of blackheart malleable iron

In the blackheart process, white iron castings are heated in a non-oxidizing atmosphere to 900°C and soaked at that temperature for two days or more. This causes the cementite to break down. The result is spherical aggregates of graphite in austenite. The casting is then cooled very slowly, resulting in the austenite changing into ferrite and more graphite (Figure 11.20). Figure 11.21 shows the form of the product.

The whiteheat process also involves heating white iron casting to about 900°C and soaking at that temperature. But in this process the castings are packed in canisters with haematite iron ore. This gives an oxidizing atmosphere. Where the casting is thin, the carbon is oxidized forming a gas and so leaves the casting. In the thicker sections of the casting only the carbon in the surface layers leaves. The result, after very slow cooling, is a ferritic structure in the thin sections of the casting and, in the thick sections, a ferrite outer layer with a ferrite plus pearlite inner core.

Pearlite malleable iron is produced by heating a white iron casting in a

Figure 11.21 Blackheart malleable iron. The structure consists of 'rosettes' of graphite in a matrix of ferrite (Courtesy of BCIRA)

non-oxidizing atmosphere to 900°C and then soaking at that temperature. This causes the cementite to break down and give spherical aggregates of graphite in austenite, as with blackheart iron. However, if more rapid cooling is used a pearlite structure is produced. This pearlitic malleable iron has a higher tensile strength than blackheart iron.

Pearlite malleable irons can be produced by other methods. One method involves adding 1 per cent manganese to the iron. The manganese inhibits the production of graphite. The higher strength pearlitic malleable irons are produced by quenching the iron from 900°C and then tempering.

British Standards specify three grades of blackheart malleable cast irons (BS 3333), five grades of pearlite malleable irons (BS 3333) and two grades of whiteheart malleable irons (BS 3333) and two grades of whiteheart malleable irons (BS 309). The grading format used is to use the letters B, P or A to indicate blackheart, pearlitic or whiteheart, followed by a number to indicate the minimum requirements for section sizes greater than 15 mm of the tensile strength in MN m^{-2} (MPa) and a final number to indicate the percentage elongation. Thus, for example, B 340/12 is a blackheart cast iron with a minimum tensile strength of 340 MN m^{-2} (MPa) and a percentage elongation of 12 per cent.

Nodular cast irons

Nodular iron, or *spheroidal-graphite (SG) iron* or *ductile iron* as it is sometimes called, has the graphite in the iron in the form of nodules or spheres. Magnesium or cerium is added to the iron before casting occurs. The effect of these materials is to prevent the formation of graphite flakes during the slow cooling of the iron – the graphite forms nodules instead. At room temperature the structure of the cast iron is mainly pearlitic with nodules of graphite (Figure 11.22). The resulting material is more ductile than a grey iron.

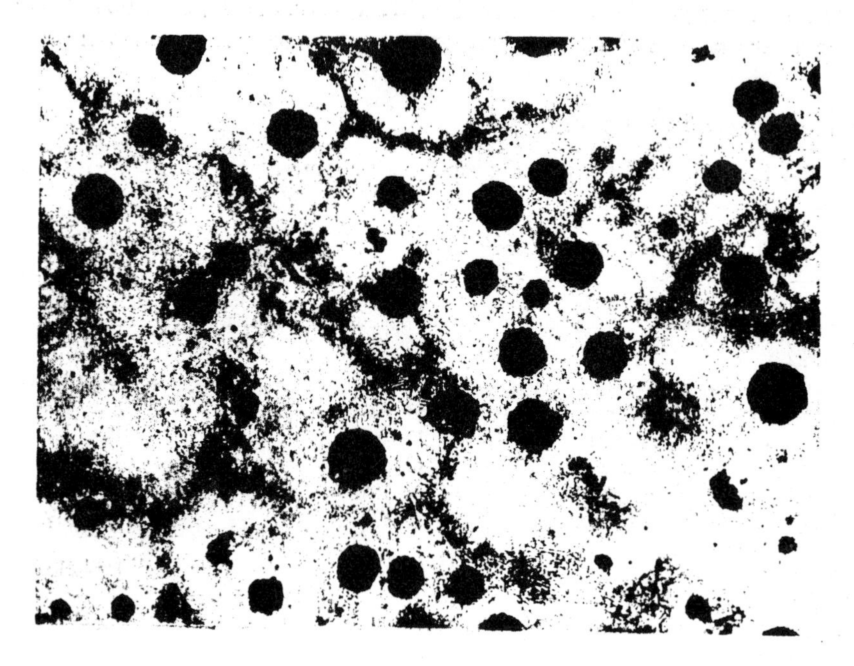

Figure 11.22 Spheroidal graphite iron. The structure consists of dark spheroidal graphite in a matrix of pearlite (Courtesy of BCIRA)

A heat treatment process can be applied to a pearlitic nodular iron to give a microstructure of graphite nodules in ferrite. The treatment is to heat to 900°C, soak at that temperature and then slowly cool. This ferritic form is more ductile, but has less tensile strength, than the pearlitic form.

British Standards (BS 2789) specify six grades of nodular cast irons. The grade form used consists of two numbers, the first representing the minimum tensile strength in MN m^{-2} (MPa) and the second the percentage elongation. Thus grade 420/12 is a nodular cast iron with a minimum tensile strength of 420 MN m^{-2} (MPa) and percentage elongation of 12 per cent.

Austempered ductile irons

Ductile irons are generally used in the as-cast condition or possibly after an annealing or normalizing treatment. The properties can, however, be considerably improved by an austempering heat treatment. This treatment involves heating the casting to 850 to 950°C, the immersing it in a bath of liquid held at a temperature which is usually in the range 235 to 425°C. It is held at this temperature for up to four hours. During this time bainite is formed. Bainite consists of ferrite plates between which, or inside which, short cementite rods have formed. Tensile strength and yield strength are increased without too large a decrease in ductility. Wear resistance is greater than that of other ductile irons and some steels, though not as good as white cast iron.

Properties and uses of cast irons

Table 11.5 shows the types of properties that might be obtained with cast irons.

Table 11.5 *Typical properties of cast irons*

Type of material	*Condition*	*Tensile strength* ($MN\ m^{-2}$)	*Elongation* (%)	*Hardness* (*HB*)
Grey iron	As cast	150–400	0.5–0.7	130–300
White iron	As cast	230–460	0	400
Blackheart	Annealed	290–340	6–12	125–140
Whiteheart	Annealed	270–410	3–10	120–180
Pearlitic malleable	Normalized	440–570	3–7	140–240
Nodular ferritic	As cast	370–500	7–17	115–215
Nodular pearlitic	As cast	600–800	2–3	215–305
Ductile iron	Austempered	800–1600	1–16	250–480

With the exception of the white iron, all the other cast irons listed in Table 11.5 give good to very good machining. Similarly for casting, the white iron gives only fair castings while the other cast irons give good to very good castings. Fair welds can be achieved with the cast irons, apart from the white iron for which welding is poor. White iron has a very high abrasion resistance.

The following are typical uses of the various cast irons:

1 *Grey iron* – water pipes, motor cylinders and pistons, machine castings, crankcases, machine tool beds, manhole covers.
2 *White iron* – wear-resistant parts, such as grinding mill parts, crusher equipment.
3 *Blackheart* – wheel hubs, pedals, levers, general hardware, brake shoes.
4 *Whiteheart* – wheel hubs, bicycle and motorcycle frame fittings.
5 *Pearlitic malleable* – camshafts, gears, couplings, axle housings.

6 *Nodular ferritic* – heavy duty piping.
7 *Nodular pearlitic* – crankshafts.
8 *Austempered ductile iron* – crankshafts, camshafts, railway couplings, machinery guides, plough shares.

In selecting a cast iron for a particular application the following considerations are taken into account:

1 *Grey irons* – very good machinability and stability. Able to damp out vibrations, i.e. considered to have a good damping capacity. Excellent wear resistance due to the graphite giving a self-lubricating effect for metal-to-metal contacts. Relatively poor tensile strength and ductility, but good strength in compression. Not good for shock loads.
2 *White iron* – excellent abrasion resistance. Very hard. Virtually unmachinable so has to be cast to the required shape and dimensions.
3 *Malleable iron* – good machinability and stability. Higher tensile strength and ductility than grey iron. Better for shock loads than grey iron.
4 *Nodular iron* – high tensile strengths with reasonable ductility. Good machinability and wear characteristics, but not as good as grey iron. Better for shock loads than grey iron (about the same as malleable iron).

The ability to damp out vibrations, i.e. the *damping capacity*, referred to above is important as the higher the capacity the less noise there will be as a result of vibrations of a casting. The presence of graphite flakes or nodules enhances the damping capacity.

For more information concerning the selection of cast irons the reader is referred to J. P. Scholes, *The Selection and Use of Cast Irons* (Oxford University Press, 1979).

Problems

1 Sketch and label the steel section of the iron–carbon system, using the terms austenite, ferrite and cementite.
2 How do the structures of *hypoeutectoid* and *hypereutectoid* steels differ at room temperature as a result of being slowly cooled from the austenite state?
3 Describe the form of the microstructure of a slowly-cooled steel having the eutectoid structure.
4 What would be the expected structure of a 1.1 per cent carbon steel if it were cooled slowly from the austenitic state?
5 Explain how the percentage of carbon present in a carbon steel affects the mechanical properties of the steel.
6 Carbon steel is used for the following items. Which type of carbon steel would be most appropriate for each item?

(a) Railway track rails.

(b) Ball bearings.
(c) Hammers.
(d) Reinforcement bars for concrete work.
(e) Knives.

7 Explain the significance of the M_s and M_f temperatures for the hardness possible with a carbon steel.
8 What is the effect of tempering on martensite and hence the hardness of a material?
9 What factors limit the engineering applications of plain carbon steels?
10 In what ways do alloying elements, other than carbon, have an effect on the properties of steel?
11 In what main ways do the following elements affect the properties of steel when they are alloyed with it? (a) manganese, (b) chromium, (c) molybdenum, (d) sulphur, (e) silicon.
12 What elements added to a steel can improve the corrosion properties of that steel?
13 What elements are usually added to a steel to increase its machinability?
14 State two elements that, when added to steel, restrict grain growth.
15 A commonly used structural steel has the following composition: 0.40 per cent carbon, 0.55 per cent manganese, 1.50 per cent nickel, 1.20 per cent chromium, 0.30 per cent molybdenum. What are the effects of the nickel, chromium and molybdenum on the properties of the steel?
16 What are the properties required of (a) tool steels in general, (b) tool steels for use in hot working processes, (c) shock-resistant tool steels?
17 What are stainless steels?
18 Explain how the three forms of stainless steel – ferritic, martensitic and austenitic – are produced and how their properties differ.
19 Explain the significance of sigma phase on the properties of stainless steels.
20 Explain what is meant by weld decay and how it can be prevented.
21 Describe the properties required of heat-resisting steels.
22 State the type of steel and the percentages of the various constituents for the steels specified by the following BS 970 codes.
(a) 040A12, (b) 070M26, (c) 150M19, (d) 210M15, (e) 503A37, (f) 653M31.
23 A steel is specified as having the condition 'R' when at a certain ruling section. What is meant by the condition 'R' and why is the ruling section specified?
24 A chromium-base tool steel is required for hot work. What code letters should be looked for in tool steels quoted to BS 4659 specifications?
25 Explain the effects on the microstructure and properties of cast iron of (a) cooling rate, (b) carbon content, (c) the addition of silicon, manganese, sulphur and phosphorus.
26 Describe the conditions under which (a) grey iron, (b) white iron are produced.

27 In what way does the section thickness of a casting affect the structure of the cast iron?

28 How do the mechanical properties of malleable irons compare with those of grey irons?

29 Describe the way in which the structures of nodular irons and malleable irons differ from those of grey irons?

30 Which types of cast iron have high ductilities?

31 Which types of cast irons have high tensile strengths?

32 Describe the forms of the microstructure of (a) blackheart and (b) whiteheart cast irons.

33 Why is it important to know whether a casting will require any machining before deciding on the material to be used?

34 How does the presence of graphite as flakes or nodules affect the properties of the cast iron?

35 Which type of cast iron would be most suitable for a situation where there was a high amount of wear anticipated?

36 Which types of case iron would you suggest for the following applications? Justify your answers.

(a) Sewage pipe.
(b) Crankshaft in an internal combustion engine.
(c) Brake discs.
(d) Manhole cover.

12

Heat treatment of steels

12.1 Heat treatment

The term *heat treatment* is used to describe a process involving controlled heating and cooling of a material in order to change its microstructure and properties. Vital factors in determining the outcome of heat treatment are the temperature to which the material is heated and the rate at which it is cooled back to room temperature. The following is a brief overview of the various heat treatment processes.

1 *Full annealing* – heating to produce a fully austenitic structure followed by furnace cooling.
2 *Normalizing* – heating to produce a fully austenitic structure followed by air cooling.
3 *Subcritical annealing* – heating to below the eutectoid temperature followed by furnace cooling.

The heat treatment of *annealing* is used to make a steel softer and more ductile, remove stresses in the material and reduce grain size. This treatment involves heating a steel to a temperature high enough to convert the structure of the steel to austenite, holding it there long enough for the change to fully occur, then very slowly cooling to room temperature. This change has to be slow enough for the structure to be converted to ferrite and pearlite. The end result is a steel in as soft a condition as possible. The above annealing process is referred to as *full annealing* and is one of a group of heat treatment processes which produce equilibrium conditions, i.e. the rate of cooling is slow enough for all movements of atoms in the material by diffusion to be completed. The following are brief descriptions of processes used to produce equilibrium structures.

4 *Hardening* – heating to produce a fully austenitic structure followed by rapid cooling, i.e. quenching, to produce a martensitic structure. Slow cooling from the austenitic structure, in which the carbon in the steel is dispersed throughout the structure, involves carbon atoms diffusing out of the austenitic structure to form cementite and hence pearlite, a laminar

structure of cementite and ferrite. If, however, the steel is rapidly cooled, i.e. quenched by, for example, immersion in water, then there is insufficient time for the carbon atoms to diffuse out of the austenitic structure. They become trapped within the structure and the result is that the austenitic structure transforms to a new structure, a *martensitic structure*. This is not an equilibrium structure, hence it does not appear on the thermal equilibrium diagram. The effect of martensite on the properties of the steel is to make it harder and more brittle.

5 *Austempering* – heating to produce a fully austenitic structure followed by quenching in a molten salt bath to produce a bainite structure, followed by cooling. The equilibrium structures produced as a result of the very slow cooling of annealing and the martensitic structures produced by rapid cooling represent the extremes; at an intermediate rate of cooling a structure called *bainite* can be produced. Slow cooling from the austenitic state produces ferrite and cementite; the distribution of these in alternate plates giving pearlite. However, at the intermediate cooling rates the pearlite does not form, instead the structure produced consists of ferrite plates between which, or inside which, short cementite rods form. This is the bainite structure. The result is a structure which is harder than that which would be obtained by annealing, but softer than martensitic structures.

6 *Tempering* – the heating of a previously quenched material to produce equilibrium products from the martensitic structure and so increase ductility. In many applications the hard material produced by quenching, i.e. the martensitic structure, needs to be made more ductile. This can be achieved by tempering. This involves heating the previously quenched material so that some diffusion of carbon atoms can occur and form cementite, leaving the structure behind as ferrite. Because ferrite is more ductile the resulting material is more ductile. The extent to which these changes occur depends on the tempering temperature and the time for which the material is held at that temperature.

7 *Precipitation hardening* – heating to produce a fully austenitic structure followed by a suitable treatment to produce a martensitic structure and then an ageing process to produce precipitates, and so increase hardness. Some steels, e.g. maraging steels, can be hardened by a heat treatment which causes precipitates to be formed, the process being known as precipitation hardening. The steels are heated to the austenitic state and then cooled to an appropriate rate to give a martensitic structure, or this state is achieved after a more complicated heat treatment. Following machining or working, the material is aged. This involves heating to a temperature and holding for a period which is long enough for fine precipitates of intermetallic compounds to be produced.

8 *Flame hardening and induction hardening* – heating the surface layers to the austenitic state and then quenching to give a harder martensitic state for the surface.

9 *Carburizing* – heating to the austenitic state in a carbon-rich atmosphere so that carbon diffuses into the surface layers, then quenching to convert the surface layers to martensite.
10 *Nitriding* – heating the component in a mixture of ammonia and hydrogen so that nitrogen diffuses into the surface layers and hard nitrides are formed.

There are many situations where there is a need for a hard surface to a component without the entire component being made hard. There are essentially two types of process by which this can be achieved – selective heating and quenching of the surface layers to transform the structure to martensite or changing the composition of the surface layers by increasing the carbon content so that a quenching process produces martensite; or introducing nitrogen into the surface to give hard nitrides.

12.2 Microstructural transformations

The iron–carbon diagram shown in Figure 11.1 shows the phases that will be present in iron–carbon alloys, of a range of compositions, at different temperatures if sufficient time is allowed for all the movements of atoms, i.e. diffusion processes, at any temperature to proceed to completion. If a steel is heated so that the temperature only rises slowly then the graph of temperature against time for the steel sample may be like that shown in Figure 12.1. At a particular temperature, in this case 723°C, the graph shows a discontinuity. Though heat is being supplied at a constant rate, the temperature does not rise at a constant rate at this particular temperature. Such a temperature is known as an *arrest point* or *critical point*. For the example given, a hypoeutectoid steel, the temperature of 723°C marks the transformation of the steel from a ferrite-plus-pearlite structure to ferrite-plus-austenite. Critical points occur whenever the structure of the steel changes.

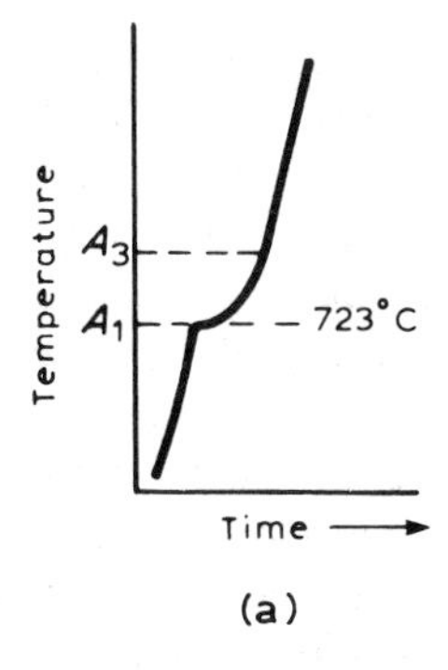

Figure 12.1 Heating curve for a hypoeutectoid steel

The lower critical point, i.e. 723°C, is denoted as A_1 and marks the transformation from a ferrite-plus-pearlite structure to one of ferrite-plus-austenite. The upper critical point is denoted as A_3 and marks the transformation from a ferrite-plus-austenite structure to one of just austenite.

For a hypereutectoid steel the lower critical point A_1 marks the transformation from a steel with a structure of pearlite plus cementite to one of austenite plus cementite. The upper critical point marks the transformation from the austenite-plus-cementite structure to austenite and is denoted by A_{cm}.

Cooling curves give critical points slightly different from those produced by heating, cooling giving lower values than heating. The cooling critical points are generally denoted by the inclusion of the letter 'r', e.g. Ar_1, and the heating points by the letter 'c', e.g. Ac_1. Figure 12.2 shows the graph of the critical point temperatures plotted against the percentage of carbon in the steel. The graph is, in fact, essentially just the iron–carbon diagram of Figure 11.1.

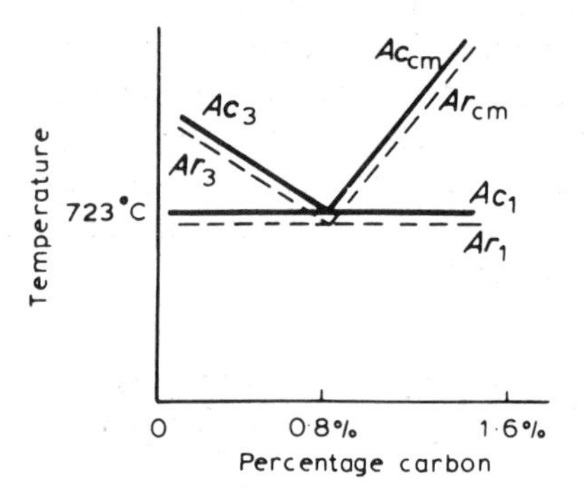

Figure 12.2 Critical points

The time scales for the iron–carbon diagram in Figure 11.1 or the critical points in Figure 12.2 are relatively long; long enough for all diffusion processes to be completed, i.e. reach equilibrium. Faster temperature changes can, however, lead to the diffusion processes not being completed, i.e. not reaching equilibrium. The faster changes can thus lead to the microstructure being in a non-equilibrium state, and so not that forecast by the iron–carbon diagram.

12.3 TTT diagrams

The effect of time and temperature on the microstructure of a steel can be expressed in terms of a *time-temperature transformation (TTT) diagram*. Such a diagram is sometimes known as an *isothermal transformation (IT) diagram* or *Bain S curves*. The diagram shows the rate at which austenite is transformed, at a given temperature, and the products produced by the transformation and any further transformations. Such a diagram is specific to just one particular steel. Thus steels with different percentages of carbon give different TTT diagrams, as do steels in which the other alloying elements are different.

One way of obtaining a TTT diagram for a particular steel is to take a small sample of the steel and heat it to a temperature at which it is completely transformed into austenite. The sample is then quickly transferred to a bath of liquid at some predetermined temperature. The time for which the sample is at this temperature is measured. This is the time during which the austenite transformation is occurring and the temperature of this change is that of the bath of liquid. This austenite transformation is taking place at a constant temperature, hence the term isothermal transformation (isothermal means constant temperature).

After the required time the sample is quickly quenched in cold water. This halts the transformation that was occurring and has the effect of converting the remaining austenite into martensite. The amount of martensite produced can be determined by microscopic examination of the sample. Hence the amount of the sample that has not been transformed from austenite, at the temperature of the bath, can be determined. This type of investigation is carried out for a number of different times at the particular bath temperature (Figure 12.3). The way in which the microstructure of the steel changes with time can be represented by the line drawn in Figure 12.4. By repeating the above sequence of operations for different temperatures a composite diagram can be built up, the TTT diagram (Figure 12.5).

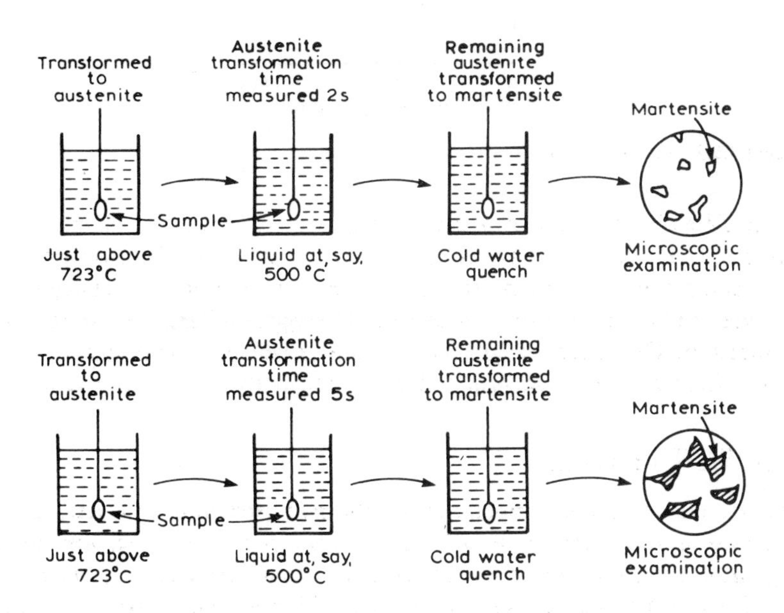

Figure 12.3 Obtaining two points on the TTT diagram at one particular temperature, 500°C

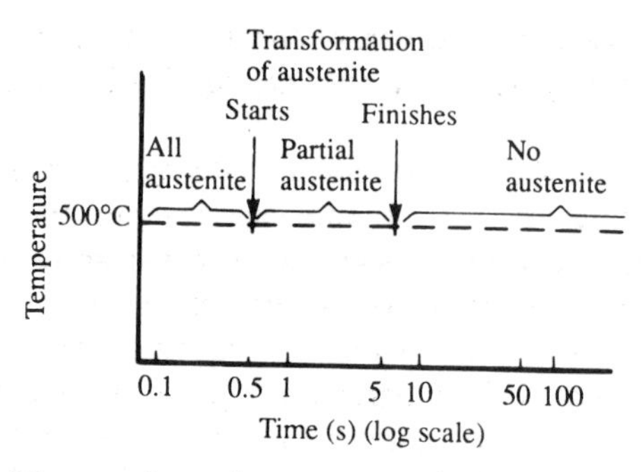

Figure 12.4 The transformation at one particular temperature

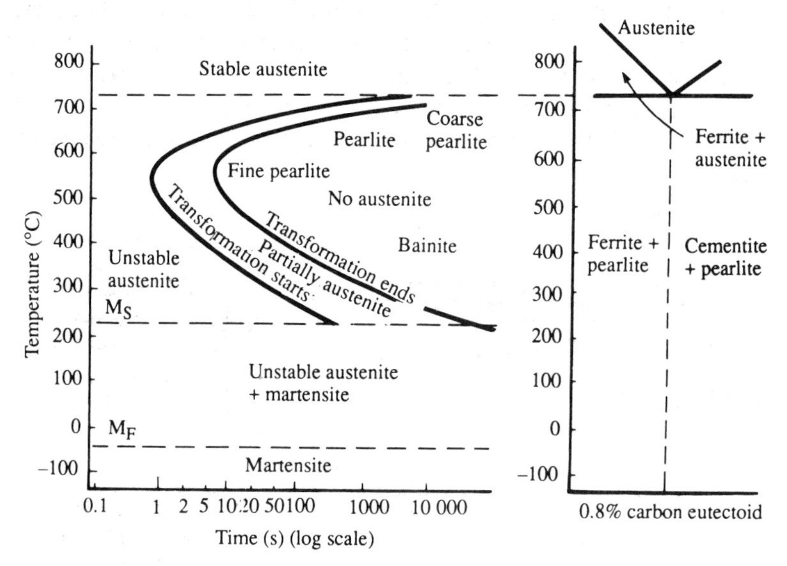

Figure 12.5 The TTT diagram for a 0.8 per cent plain carbon steel and the thermal equilibrium diagram

Above 723°C, for the steel chosen for Figure 12.5, austenite is the only stable phase. For temperatures below this the austenite is unstable and given sufficient time suffers a transformation. From 723°C down to about 566°C, the lower the temperature the shorter the time taken for the austenite to begin to transform. On completion of the transformation the microstructure is pearlite, the higher temperature giving a coarser pearlite than the lower temperature. From 566°C down to about 215°C the time taken for the austenite to begin to transform increases. The product of the transformation is known as *bainite*. Bainite, like pearlite, is a mixture of ferrite and cementite, however while pearlite consists of alternate layers of the two constituents, bainite consists of a feathery-like structure. From 215°C down to −20°C the unstable austenite transforms to martensite. At this lower temperature the structure is just martensite. For comparison with the conditions which would occur with very slow cooling, the relevant part of the thermal equilibrium diagram is also included in Figure 12.5.

In the region 215°C to −20°C the amount of martensite formed is practically independent of time, being determined mainly by the temperature of the steel. The temperature at which martensite begins to form is denoted by M_s and the temperature at which the transformation is complete by M_f.

The TTT diagram in Figure 12.5 is for the eutectoid composition, i.e. 0.8 per cent carbon. Different TTT diagrams are produced for other carbon contents. Figure 12.6 shows the diagram for a 0.45 per cent carbon steel, also the relevant part of the thermal equilibrium diagram. The TTT diagram for the 0.45 per cent carbon steel differs from that of the 0.8 per cent carbon steel in that there is an additional region due to ferrite transformations. Also, with the lower percentage carbon the TTT diagram has effectively been shifted to the left, i.e. the transformations begin more quickly and are completed more quickly at any given temperature. For example, compare the transformations at 500°C for the 0.45 per cent carbon steel (Figure 12.7) with those given in Figure 12.4 for the 0.8 per cent carbon steel.

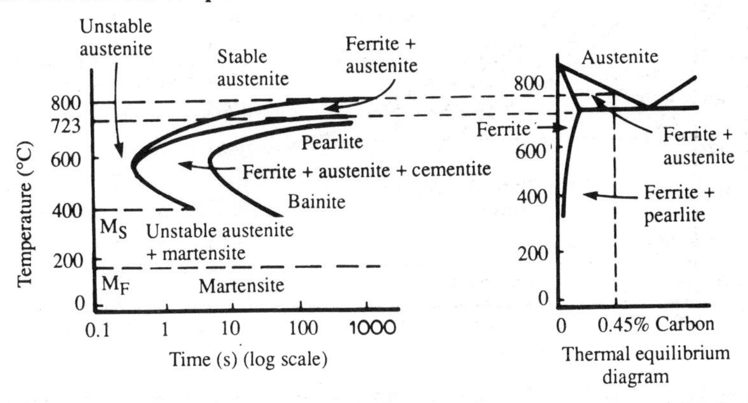

Figure 12.6 The TTT diagram for a 0.45 per cent plain carbon steel and the thermal equilibrium diagram

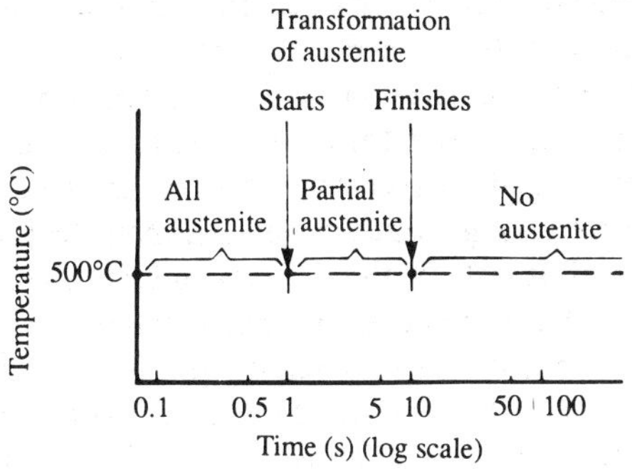

Figure 12.7 The transformations at one particular temperature for a 0.45 per cent plain carbon steel

The lower the percentage of carbon in a plain carbon steel the more to the left the TTT diagram is shifted. This means a shorter time is needed at any particular temperature for the transformation to pearlite, or bainite.

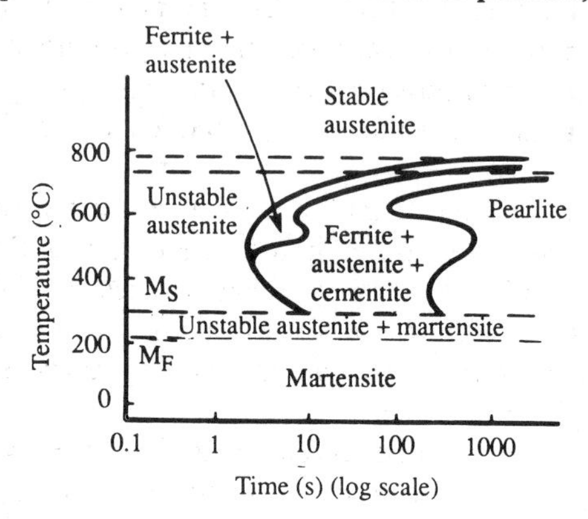

Figure 12.8 The TTT diagram for a 0.4 per cent carbon, 0.8 per cent chromium steel

The TTT diagram is also changed by other elements added to a carbon steel. Figure 12.8 shows the TTT diagram for an alloy steel having 0.4 per cent carbon and 0.8 per cent chromium. The effect of the chromium is to shift the TTT diagram to the right, i.e. increase the time needed at any particular temperature for the transformation to pearlite or bainite (compare Figure 12.9 with 12.7). Nickel, molybdenum and manganese all have similar effects to that shown by the chromium additive to the steel.

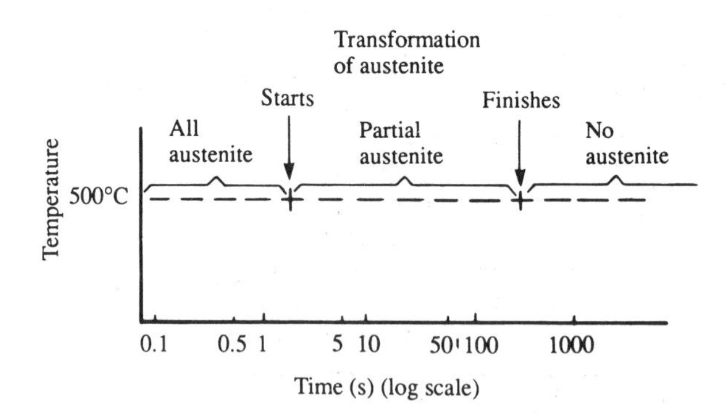

Figure 12.9 The transformations at one particular temperature for a 0.4 per cent carbon, 0.8 per cent chromium steel

Quenching steels

A hard, strong, steel can be produced by heating the steel to a temperature greater than the A_3 critical temperature so that the microstructure is entirely changed to austenite, then cooling quickly so that the austenite is transformed into martensite. The treatment thus involves continuous cooling. The cooling rate depends on the medium used for the quenching, water giving a faster cooling rate than oil.

The TTT diagrams are derived from data obtained for transformations at constant temperature. However, the diagrams are only slightly modified if continuous cooling is considered, the effect being to displace the TTT diagram to the right and downward. Also the continuous curves frequently show gaps where no transformation occurs on cooling. Such a diagram is referred to as a *continuous cooling transformation curve*.

Figure 12.10 shows the continuous cooling transformation curve for a 0.8 per cent carbon steel, i.e. the eutectoid composition. Superimposed on the diagram are graphs of temperature against time that could occur for the steel when quenched in different media, i.e. different cooling rate graphs. With fast cooling (curve 1) from the stable austenite temperature the only transformation that occurs is from the austenite to martensite. With a slightly slower cooling rate (curve 2) a partial transformation to pearlite occurs and the final microstructure is thus martensite and pearlite. With an even slower cooling rate (curve 3) the austenite changes entirely to pearlite and the final structure is fine pearlite, no martensite being produced. With the yet slower cooling rate

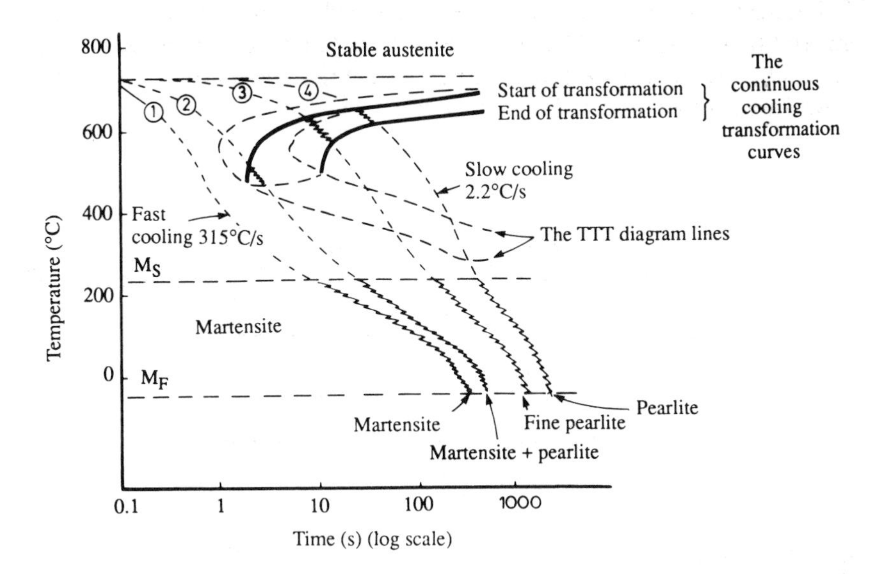

Figure 12.10 Cooling curves superimposed on the continuous cooling transformation curves for a 0.8 per cent plain carbon steel

(curve 4) the austenite changes entirely to pearlite and again no martensite is produced.

For the microstructure to be entirely martensite, and hence show the maximum hardness, the cooling rate must be such that the only transformation occurring is from austenite to martensite. For the structure to be entirely pearlite the cooling rate must be such that the cooling rate line passes through the start and end of transformation lines. The structure will be partly pearlite and partly martensite if the cooling rate line passes into the transformation region but does not pass completely through it from the start to the end of transformation lines.

If a steel is being quenched to give an entirely martensitic microstructure then the cooling rate line must not pass through the transformation region. The minimum cooling rate that will do this is shown in Figure 12.11. This minimum cooling rate is called the *critical cooling rate*. Cooling rates faster than this will give a completely martensite structure but rates slower than this will not.

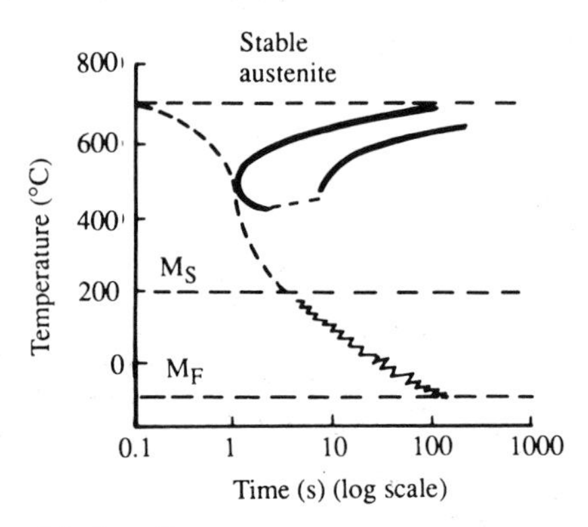

Figure 12.11 The critical cooling rate

The critical cooling rate thus depends on the form of the continuous cooling transformation curve (and hence the TTT diagram). Figure 12.6 shows the TTT diagram for a 0.45 per cent plain carbon steel. The result of having the lower percentage of carbon, than that giving Figure 12.5, has been to shift the diagram to the left. This means that the critical cooling rate for the 0.45 per cent carbon steel is higher than that for the 0.8 per cent carbon steel. The more the TTT diagram is shifted to the left the faster the cooling rate needed to obtain a completely martensitic structure. In fact, by about 0.3 per cent carbon the TTT diagram has been shifted so far to the left that the cooling rate required for such a change is too fast to be possible and so such steels cannot be hardened by quenching.

A problem with the quenching of plain carbon steels to give a martensitic structure is that the high cooling rates needed inevitably mean water-quenching. But such rapid quenching can lead to cracking and distortion. The effect of adding elements, other than carbon, to steel is to shift the TTT curve, and hence the continuous cooling transformation curve, to the right. This means that the critical cooling rate is less and thus oil or salt quenching, or even air cooling, may be used and still result in a martensitic structure.

The rate of cooling that occurs for a piece of steel when quenched in, say, water depends on the size of the piece of steel. The bigger the piece of steel the slower its rate of cooling. The quenching medium has thus to be chosen with a particular size piece of steel in mind. Figure 12.12 shows the continuous cooling transformations that occur for specific quenching media as a function of the diameters of the bars being quenched. Thus for a 10 mm diameter bar, air cooling will result in a microstructure involving ferrite, pearlite and a small amount of bainite. With oil quenching the result is bainite plus martensite. With water quenching the microstructure is only martensite. Thus for this particular diameter bar water quenching must be used if a completely martensitic structure is required. The microstructures given by the diagram are those at the centre of the quenched bar and those at the surface may differ due to the surface cooling at a different rate.

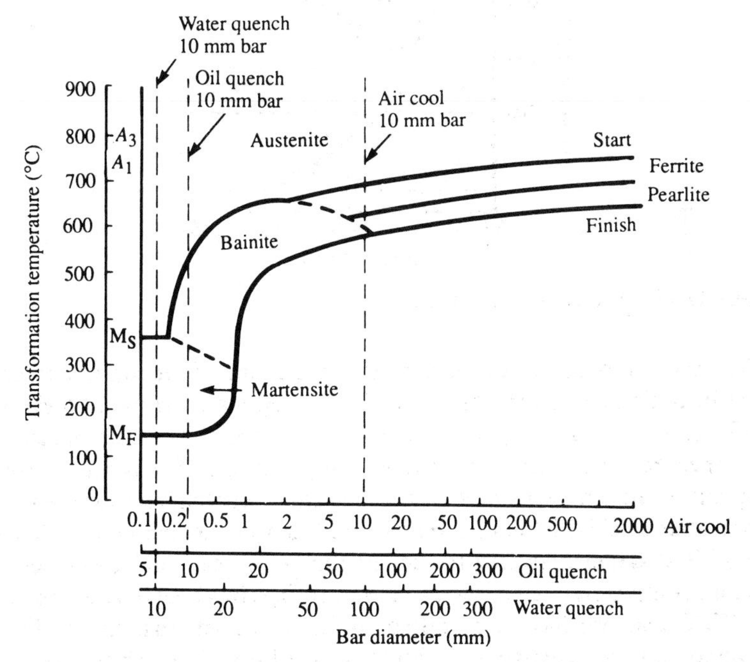

Figure 12.12 Continuous cooling transformations as a function of bar diameter and quenching media for a 0.38 per cent carbon, 0.70 per cent manganese plain carbon steel (Based on a diagram in *Atlas of Continuous Cooling Transformation Diagrams for Engineering Steels*, British Steel Corporation, 1977)

Figure 12.12 was for a plain carbon steel, Figure 12.13 is however for a steel with a similar percentage of carbon but with chromium and molybdenum alloying elements. As will be apparent from a comparison of the two diagrams, a much less severe quenching can be used and still a martensitic structure obtained. For the 10 mm diameter bar air cooling will suffice.

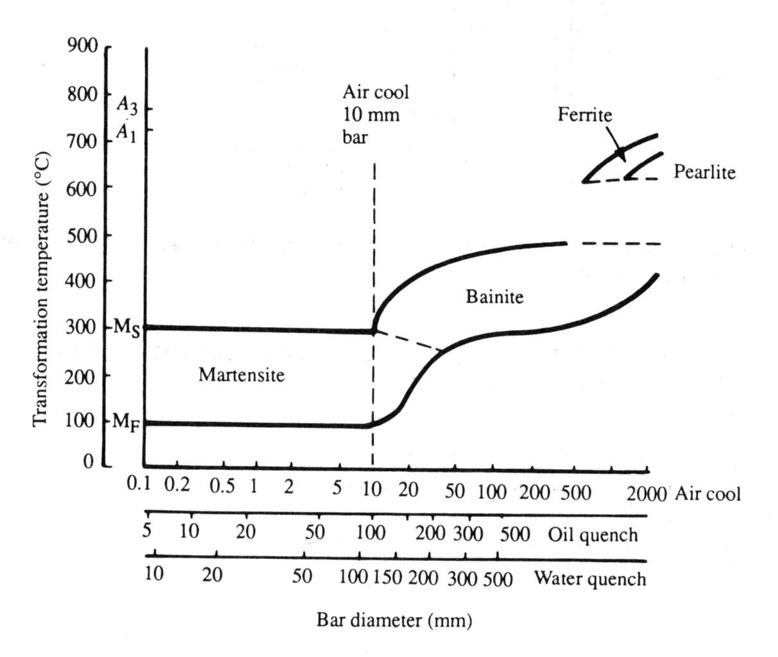

Figure 12.13 Continuous cooling transformations as a function of bar diameter and quenching media for a 0.40 per cent, 1.50 per cent nickel, 1.20 per cent chromium, 0.30 per cent molybdenum alloy steel (Based on a diagram in *Atlas of Continuous Cooling Transformation Diagrams for Engineering Steels*, British Steel Corporation, 1977)

Hardness

The maximum hardness that can be produced with any carbon steel is that occurring when the steel has a completely martensitic structure. Figure 12.14 shows how the hardness of such a structure depends on the percentage of carbon in the steel. It also shows the hardness that would be produced if the steel had been given a slow quench, air cooled, and an entirely pearlitic structure had been produced.

When a carbon steel is quenched to room temperature, though the cooling rate may be fast enough to convert all the austenite to martensite, the entire transformation may be incomplete. The temperature at which austenite begins

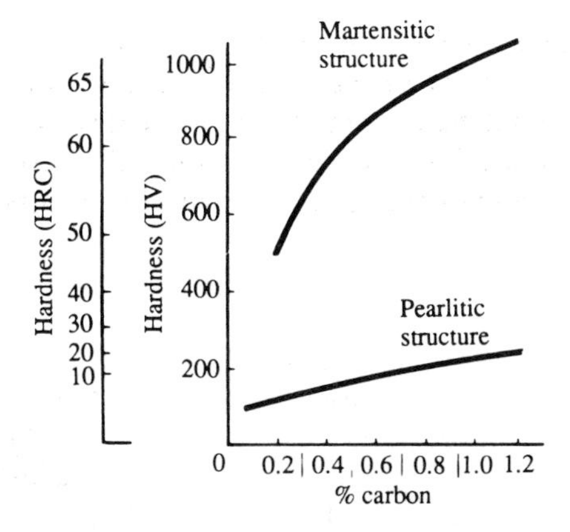

Figure 12.14 The hardness of carbon steel with martensitic or pearlitic structures as a function of carbon content

to transform to martensite M_s depends on the carbon content of the steel (see Figure 11.10 and Table 12.1).

Table 12.1 *Relation of M_s to carbon content*

Carbon content	*M_s/°c*
0.2	490
0.4	420
0.8	250
1.2	150

The temperature at which the transformation from austenite to martensite is complete M_F is about 215°C below the M_s temperatures. Thus for the 0.2 per cent carbon steel the martensite will be complete by 275°C, the 0.8 per cent carbon steel by 35°C. Both these are above room temperature. However the M_F temperature for the 1.2 per cent carbon steel is −65°C and thus such a steel will consist at room temperature of martensite plus austenite. This has the effect of reducing the overall hardness of that steel to below that which it would have attained if completely martensitic.

12.4 Hardenability

When a block of steel is quenched the surface can show a different rate of cooling to that of the inner core of the block. Thus the formation of martensite may differ at the surface from the inner core. This means a difference in hardness. Figure 12.15 shows how the hardness varies with depths for a

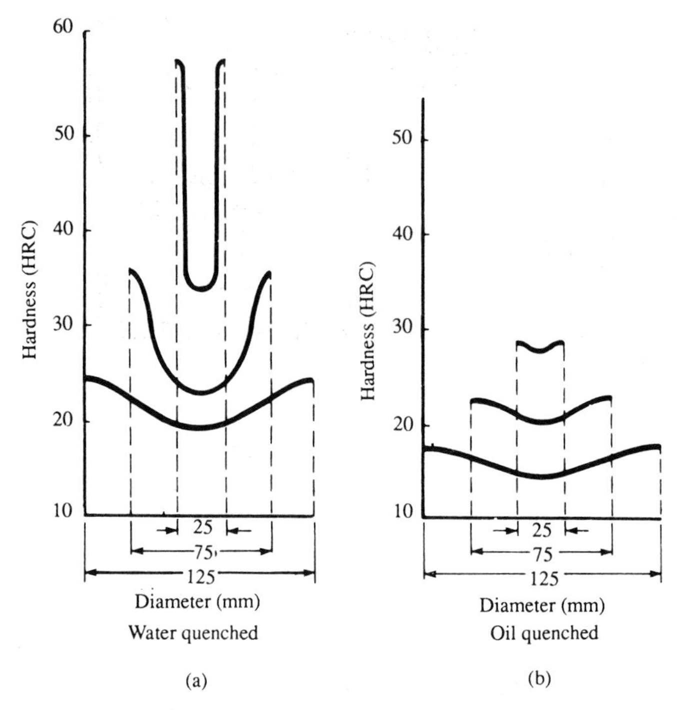

Figure 12.15 Variation of hardness with depth in quenched bars of different diameters, for a 0.48 per cent plain carbon steel

number of different diameter bars when quenched in water and in oil.

With the water quenching, the hardness in the inner core is quite significantly different from the surface hardness. The larger diameter bars also show lower surface and core hardness as their increased mass has resulted in a lower overall rate of cooling. With the oil quenching the cooling rates are lower than those of the water-cooled bars and thus the hardness values are lower. The hardness value for a fully martensitic 0.48 per cent plain carbon steel is about 60 HRC. Thus, as the values in Figure 12.15 indicate, the quenched bars are not entirely martensitic.

The term *hardenability* is used as a measure of the depth of hardening introduced into a steel section by quenching (not to be confused with hardness). Hardenability is measured by the response of a steel to a standard test. The *Jominy test* involves heating a standard test piece of the steel to its austenitic state, fixing it in a vertical position and then quenching the lower end by means of a jet of water (Figure 12.16). This method of quenching results in different rates of cooling along the length of the test piece. When the test piece is cool, after the quenching, a flat portion is ground along one side of the test piece, about 0.4 mm deep, and hardness measurements made along the length of the test piece. Figure 12.17 shows the types of result that are produced. The

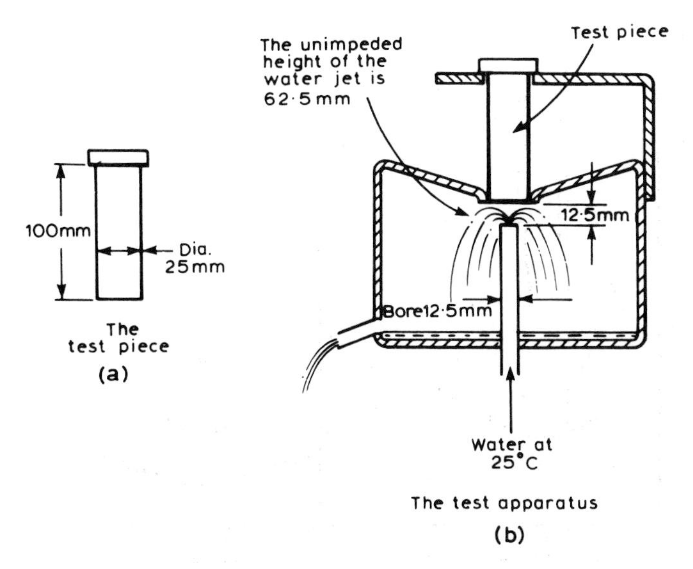

Figure 12.16 The Jominy test (BS 4437)

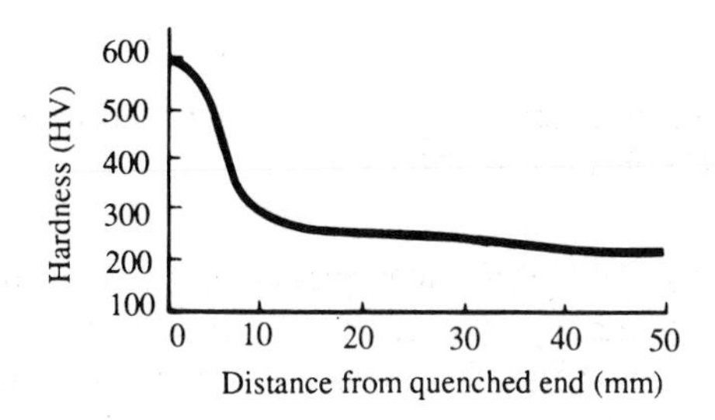

Figure 12.17 Results of the Jominy test for a 0.40 per cent plain carbon steel

hardness is greatest at the end on which the cold jet of water played and least at the other end of the bar.

The significant point about the Jominy test results is not that they give the hardness at different distances along the test piece but that they give the hardness at different cooling rates. Each distance along the test piece corresponds to a different rate of cooling (Figure 12.18). The important point however is that this applies to both points on the surface and points inside any sample of the steel, provided we know the cooling rates at those points. This applies regardless of the quenching medium used. Figure 12.19, for example, shows how the cooling rate at the centre of circular cross-section bars at 700°C is related to their diameter. As the cooling rates are related to distances along the Jominy test piece these distances are also given in the figure.

Thus for the Jominy test result given in Figure 12.17, the hardness at 20 mm

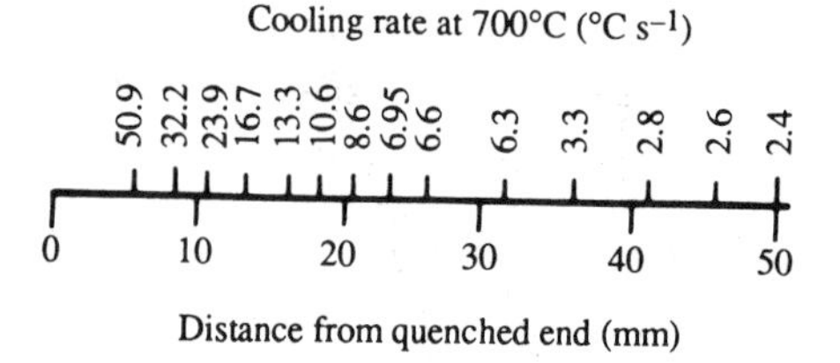

Figure 12.18 Cooling rates at different distances from the quenched end of the Jominy test piece

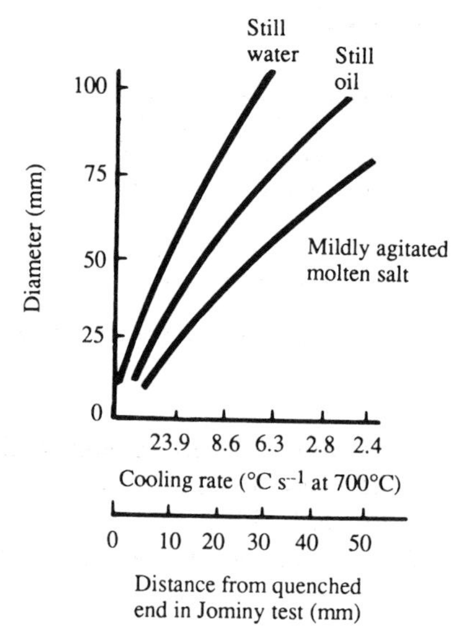

Figure 12.19 Cooling rates at the centres of different diameter bars at 700°C in different media

from the quenched end of the test piece is about 230 HV. This means that, using Figure 12.19, a circular cross-section bar of diameter about 75 mm would have this hardness at its centre when quenched in still water from 700°C. If still oil had been the quenching medium the hardness at the centre of a 60 mm diameter bar would have been 230 HV.

Figure 12.20 shows, in a similar manner to Figure 12.19, how the cooling rate is related to the diameter of circular cross-section bars for points on the surface. The Jominy test result of 230 HV at 20 mm from the quenched end of the test piece means, using Figure 12.20, a circular cross-section bar of diameter 50 mm would have this hardness at its surface when quenched from 700°C in mildly agitated molten salt.

Figure 12.21 shows Jominy test results for two different steels. In order to

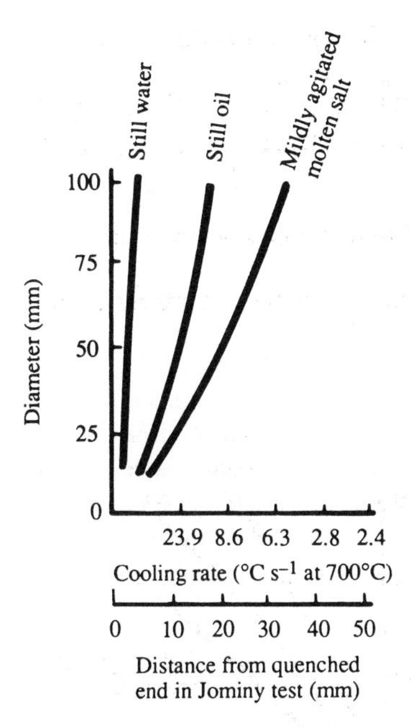

Figure 12.20 Cooling rates at the surfaces of different diameter bars at 700°C in different media

enable the significance of the results to be seen in terms of the hardness at the centre of different diameter bars, the cooling rates have been transposed into diameter values by means of Figure 12.19. The alloy steel can be said to be have better hardenability than the plain carbon steel. The plain carbon steel cannot

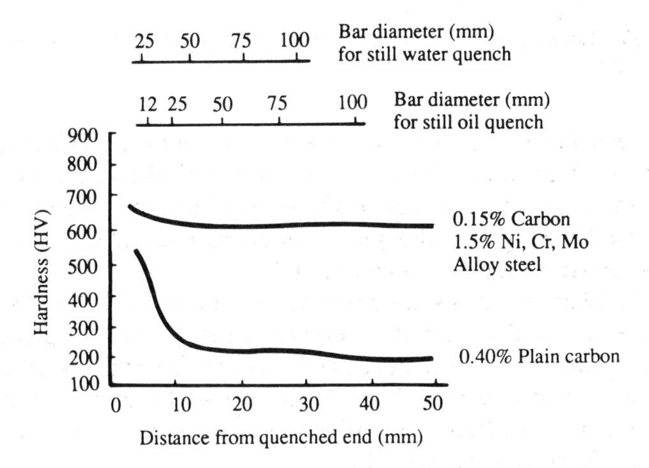

Figure 12.21 Results of Jominy tests

be used in a diameter greater than about 25 mm with a still water quench if the bar is to be fully hardened. However the alloy steel is fully hardened for bars with diameters in excess of 100 mm.

Hardenability and continuous cooling curves

When a bar of material is cooled the centre of the bar undergoes a slower rate of cooling than the surface. The greater the diameter of the bar the greater the difference between these two rates of cooling. Figure 12.22 shows how the spread of cooling rates between the centre of a bar and its surface can be represented on a continuous cooling transformation curve in order to determine the structure across the section. For the example given in Figure 12.22 the central region of the bar has been partially transformed to pearlite or bainite and thus is not as hard as the surface layers which have been completely transformed to martensite. The greater the cross-section size of the bar the greater the spread of cooling rates and thus the more difficult it is to completely harden the bar.

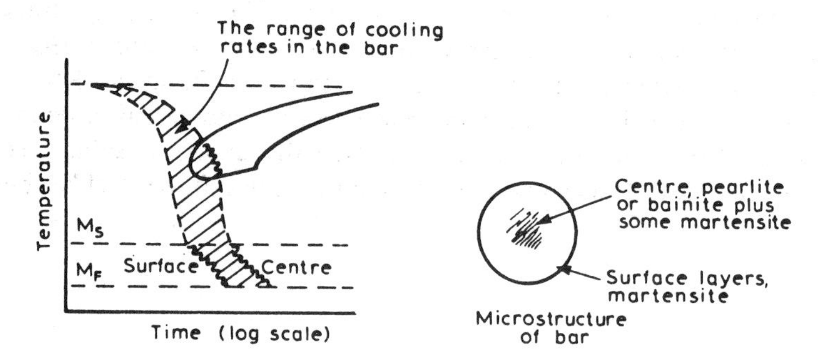

Figure 12.22 Quenching a bar

Ruling section

If you look up the mechanical properties of a steel in the data supplied by the manufacturer or other standard tables you will find that different values of the mechanical properties are quoted for different limiting ruling sections. The

Table 12.2 *Effect on properties of limiting ruling section*

Steel	*Condition*	*Limiting ruling section/mm*	*Tensile strength /N mm^{-2}*	*Minimum elongation (%)*
070 M 55	Hardened and tempered	19	850 to 1000	12
		63	770 to 930	14
		100	700 to 850	14

limiting ruling section is the maximum diameter of round bar at the centre of which the specified properties may be obtained. Table 12.2 shows an example.

The reason for the difference of mechanical properties for different size bars of the same steel is that during the heat treatment different rates of cooling occur at the centres of such bars due to their differences in sizes (see Figure 12.15). This results in differences in microstructure and hence differences in mechanical properties.

12.5 Tempering

Martensite is a hard and brittle substance. Thus a steel which has been quenched so that its entire microstructure is martensitic will be not only hard but probably very brittle, but it will not be tough. The toughness of such a steel can be improved by tempering. The gain in ductility is however balanced by a reduction in strength and hardness.

Tempering is the name given to the process in which a steel, hardened as a result of quenching, is reheated to a temperature below the A_1 temperature in order to modify the structure of the steel. Martensite is a highly stressed supersaturated solid solution of carbon in a distorted ferrite lattice. Heating such a structure enables the carbon atoms trapped in the ferrite lattice to diffuse out of that lattice and form fine cementite. The amount of carbon that diffuses out of the martensite lattice, and hence the amount of softer ferrite structure produced, depends on the temperature to which the steel has been

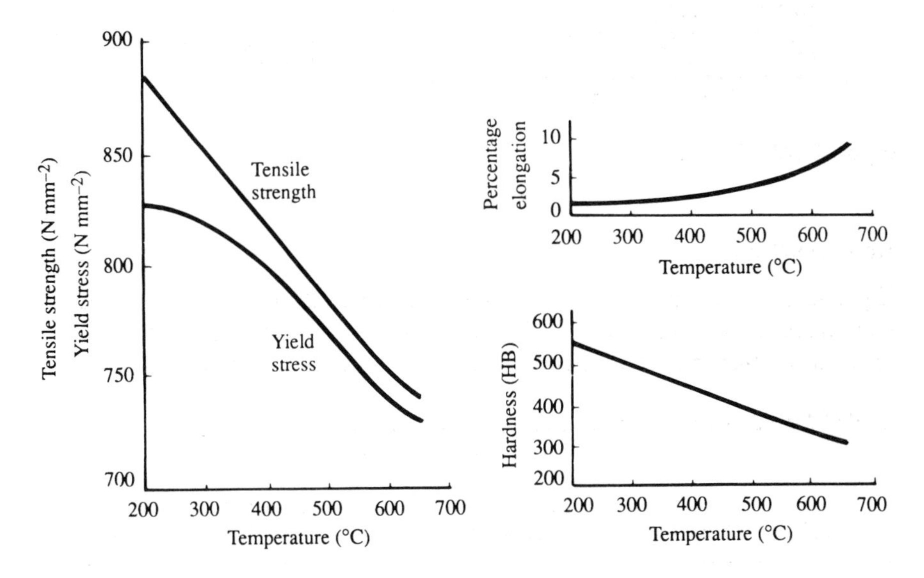

Figure 12.23 The effect of tempering on the properties of a steel (0.40 per cent carbon, 0.70 per cent manganese, 1.8 per cent nickel, 0.80 per cent chromium, 0.25 per cent molybdenum)

heated and the time for which it is held at that temperature. Thus the mechanical properties of the steel can be controlled by the tempering process.

Figure 12.23 shows how, for an oil-quenched steel, the tempering temperature affects the hardness, tensile strength, yield stress and percentage elongation. The higher the tempering temperature the lower the tensile stress, yield stress and hardness but the higher the percentage elongation.

A graph of hardness against tempering temperature is known as the *tempering curve* and often forms part of the data supplied by a steel manufacturer. Figure 12.24 is a typical example, for a tool steel. With this steel, tools for light shock applications where maximum wear resistance is required, e.g. thin sheet punching dies, are tempered at 190°C to 250°C. This gives a hardness of 60 to 63 HRC. Heavy duty applications such as heavy shear blades would be tempered at between 540°C and 560°C and have a hardness of 52 to 56 HRC. For this particular steel the manufacturer recommends that *double tempering* be used. This involves heating to the required tempering temperature, holding at that temperature for at least 60 minutes, then allowing the steel to cool in air before repeating the entire operation.

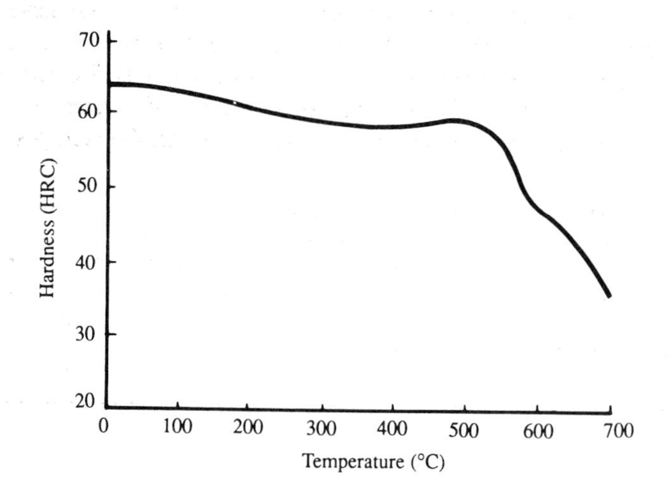

Figure 12.24 Tempering curve for a tool steel KEA 180 (1.55 per cent carbon, 12.0 per cent chromium, 0.85 per cent molybdenum, 0.28 per cent vanadium) (Courtesy of Sanderson Kayser Ltd)

Austempering

The *austempering* process involves heating a steel to the austenite state, i.e. above A_3, and then quenching it in a bath held at a temperature above M_s. The steel then remains at that temperature until the austenite is completely transformed into bainite, after which it is allowed to cool. The initial quenching treatment must be at a cooling rate greater than or equal to the critical cooling rate so that no pearlite is formed. The final cooling rate for the bainite structure can be at any rate as no further changes take place. Figure 12.25 shows the above sequence of events on a TTT diagram.

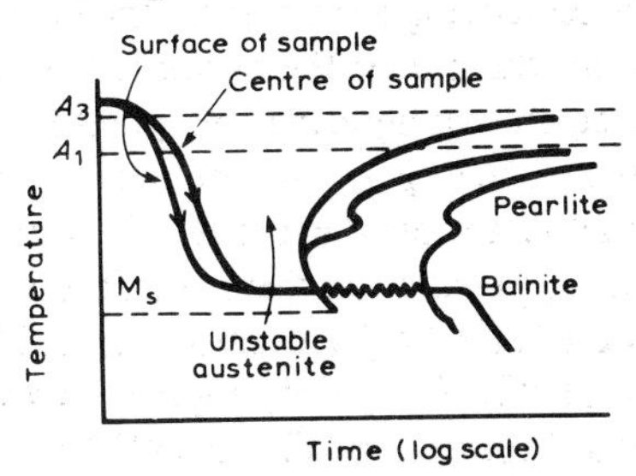

Figure 12.25 Austempering

Bainite structures are softer than martensite structures but have better ductility and impact toughness. Because the steel does not suffer a severe quench treatment it is less likely to crack and distort. Table 12.3 shows how the properties of an austempered steel compare with those of the same steel (0.95 per cent carbon steel) when quenched to martensite and then tempered.

Table 12.3 *The effect on properties of austempering*

Properties	*Austempered*	*Quenched and tempered*
Hardness/HV	545	560
Percentage elongation	11	1
Impact test/J	58	16

Martempering

Martempering is a hardening treatment that consists of heating a steel to its austenitic state, i.e. above A_3, and then quenching it to a temperature just above M_s. The quenching must give a cooling rate faster than the critical cooling rate. The steel is then held at the temperature sufficiently long for the entire piece of steel to come to the same temperature, without any transformation to bainite occurring. It is then cooled in air to change the austenite to martensite. No transformation product other than martensite should occur in this process. Figure 12.26 shows the above sequence of events on a TTT diagram. After such a treatment the steel is then tempered in the usual way.

The effect of such a treatment is to minimize cracking and distortion, the

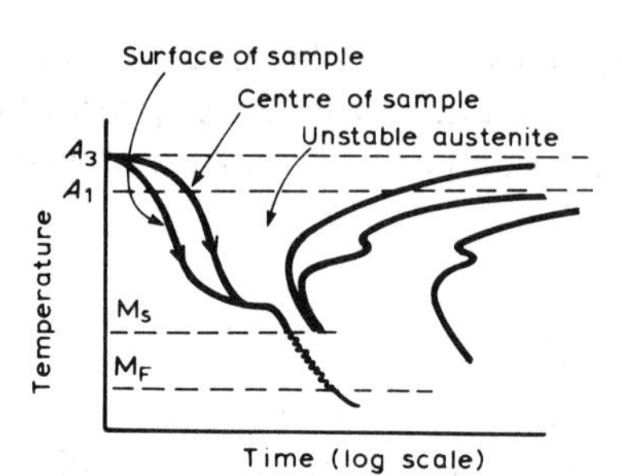

Figure 12.26 Martempering

thermal shock of the quenching is reduced. The hardness and ductility are generally similar to those obtained by the direct quenching to the martensitic state followed by tempering. The impact toughness may however be better.

12.6 Precipitation hardening

A number of different treatments can be used with alloy steels to harden them as a result of producing fine precipitates of carbides, nitrides or intermetallic compounds. The hardening is a consequence of the precipitates hindering the movement of dislocations within the metal.

Maraging steel is an example of steel which is precipitation hardened. The steel has a low carbon content, less than 0.03 per cent, a high nickel content, of approximately 20 per cent, and other elements such as cobalt, titanium and molybdenum. The steel is heated to above the A_3 temperature and the structure becomes fully austenitic. It is then air cooled to give a martensitic structure, which because of the low carbon content of the alloy is relatively soft and easily worked. Following machining and working, the steel is then heated to about 500°C, i.e. below A_1, and held at this temperature for two or three hours. During this time precipitates of intermetallic compounds form.

There are three forms of stainless steel which can be precipitation hardened: martensitic, semi austenitic and austenitic. The martensitic stainless steel has a low carbon content, a high chromium content of approximately 13 to 17 per cent, a nickel content of approximately 4 to 8 per cent and other elements such as copper, aluminium and molybdenum. Following heating to the austenitic state the alloy is air cooled to give a soft martensitic state. Machining and working can take place before the alloy is aged by heating to about 500°C. This results in fine precipitates being produced. The semi-austenitic steel has a chromium content of approximately 15 to 17 per cent, nickel approximately 7 per cent and other elements. Following heating the austenitic state and air cooling the alloys remain austenitic. Either refrigeration or heating to about 700°C is necessary to form a martensitic structure prior to precipitation hardening like the martensitic steel. The austenitic steel has typically 17 per cent chromium, 10 per cent nickel and 0.25 per cent phosphorus. Ageing of such an alloy at about 680°C results in a carbide precipitate being produced and this, together with the lattice strain produced by the phosphorus, results in a degree of hardening.

12.7 Surface hardening

There are many situations where wear or severe stress conditions may indicate the need for a hard surface to a component without the entire component being made hard and possibly brittle. The various methods used for surface hardening can, in the main, be grouped into two categories:

1 Selective heating of the surface layers.
2 Changing the composition of the surface layers.

One selective heating method is called *flame hardening*. This method involves heating the surface of a steel with an oxy-acetylene flame and then immediately quenching the surface in cold water (Figure 12.27). The heating transforms the structure of the surface layers to austenite and the quenching changes this austenite to martensite; the result is a hard layer of martensite in the surface. The depth of hardening depends on the heat supplied per unit surface per unit time. Thus the faster the burner is moved over the surface the less the depth of hardening. This is because less of the steel has its structure transformed into austenite and hence martensite. The temperatures used in the method are typically of the order of 850°C or more, i.e. above the A_3 temperature.

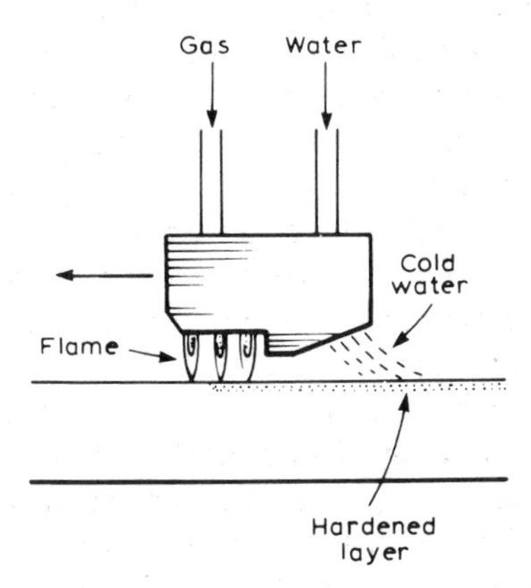

Figure 12.27 A burner with cooling water for flame hardening

Another method involving selective heating is called *induction hardening*. This method involves placing the steel component within a coil through which a high frequency current is passed (Figure 12.28). This alternating current

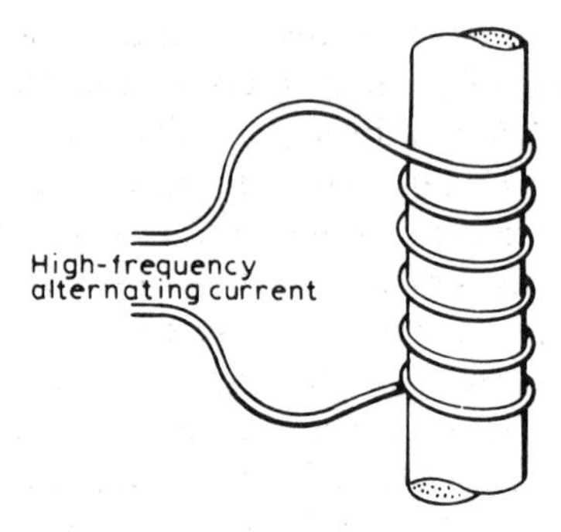

Figure 12.28 Induction hardening of a tube or rod

induces another alternating current to flow in the surface layers of the steel component. This induction is called electromagnetic induction. The induced currents heat the surface layers. The temperatures so produced cause the surface layers to change to austenite. The cooling following the passage of the current transforms the austenite to martensite. The depth of heating produced by this method, and hence the depth of hardening, is related to the frequency of the alternating current used. The higher the frequency the less the hardened depth, but the more rapid the temperature rise (see Table 12.4).

Table 12.4 *The effect of frequency on depth of hardening*

Frequency used/kHz	*Depth of hardening/mm*
3	4.0 to 5.0
10	3.9 to 4.0
450	0.5 to 1.1

The form of the induction coil used depends on the shape of the component being hardened, also the size of the area to be hardened.

The composition of the steel to be surface hardened by selective heating has to be chosen with certain criteria in mind. Before the surface hardening treatment, the steel must have its inner core with the right mechanical properties. These properties will be unaffected by the surface treatment. This means that the composition of the steel and its heat treatment have to be chosen. The heat treatment has also to be chosen with the selective heating process in mind. The selective heating has to be able to change the microstructure of the steel to austenite in a very short amount of time. Hardened and tempered steels respond well but coarse annealed steels do not. In addition the composition of the steel must be such that the quenching part of the selective heating process will produce martensite and so harden the steel. This tends to mean carbon contents of 0.4 per cent or more.

One surface hardening process that involves changing the composition of the surface layers is *carburizing*. This method involves increasing the carbon content of the surface layers, followed by a quenching process to convert the surface layers into martensite. This process is normally carried out on a steel containing less than about 0.2 per cent carbon, the carburizing treatment being used to give about 0.7 to 0.8 per cent carbon in the surface layers. This wide difference in carbon content is needed because the quenching treatment following the carburizing will affect both the inner core and the surface layers and it is only the surface layers that are to be hardened, the inner core should remain soft and tough. Figure 12.29 shows the results of Jominy tests for a 0.18 per cent carbon (0.46 per cent manganese, 1.68 per cent silicon, 0.03 per cent chromium, 0.21 per cent molybdenum) steel for its core where the percentage of carbon remains unchanged and for the surface where the percentage of carbon has been increased to 0.7 per cent by carburizing. Thus with a fast rate of cooling the carburized surface may have a hardness of the order of 60 HRC while the inner core is about 15 HRC.

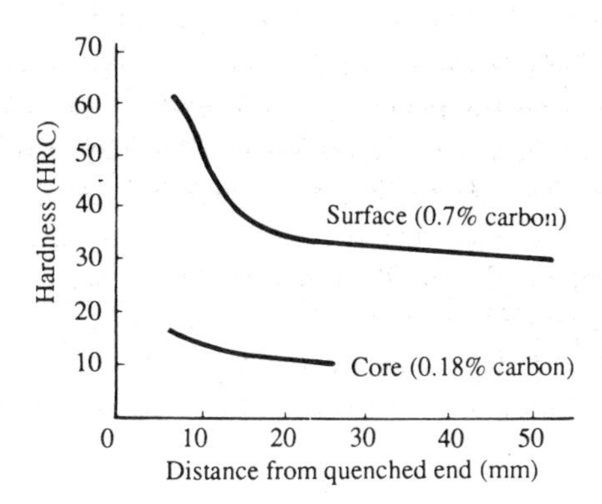

Figure 12.29 Jominy test results for a carburized steel

There are a number of carburizing methods. With *pack carburizing* the steel component is heated to above the A_3 temperature while in a sealed metal box which contains charcoal and barium carbonate. The oxygen present in the box reacts with the carbon to produce carbon monoxide. This carbon-rich atmosphere in contact with the hot steel results in carbon diffusing into the surface austenitic layers. Pack carburizing tends to be used mainly for large components or where a thick surface layer has to be hardened.

In *gas carburizing* the component is heated to above the A_3 temperature in a furnace in an atmosphere of carbon-rich gas. The result is that carbon diffuses into the surface austenitic layers. Gas carburizing is the most widely used method of carburizing.

Salt bath carburizing, or *cyaniding* as it is often known, involves heating the component in a bath of suitable carbon-rich salts. Sodium cyanide is mainly used. The carbon from the molten salt diffuses into the component. In addition there is also diffusion of some nitrogen into the component. Both carbon and nitrogen can result in a microstructure which can be hardened. The method tends to produce relatively thin, hardened layers with high carbon content. This occurs because the carburizing takes place very quickly. One of the problems with this method is the health and safety hazard posed by the poisonous cyanide. Another problem is the removal of salt from the hardened component after the treatment. This can be particularly difficult with threaded parts or blind holes.

Carburizing may result in the production of a large grain structure due to the time for which the material is held at temperature in the austenitic state. Though the final product may be hard there may be poor impact properties due to the large grain size. A heat treatment might thus be used to refine the grains. A two stage process is required as the carbon content of the surface is significantly different from that of the inner core of the material.

The first stage involves a heat treatment to refine the grains in the core. The component is heated to just above the A_3 temperature for the carbon content of the core (Figure 12.30). For a core with 0.2 per cent carbon this is about 870°C. The component is then quenched in oil. The result is a fine grain core, but the surface layers are rather coarse martensite. The martensite is refined by heating to above the A_1 temperature for the carbon content of the surface layers; if this is 0.9 per cent carbon then the temperature would be about 760°C. The component is then water-quenched. This second stage treatment has little effect on the core but refines the martensite in the outer layers. The treatment may then be followed by a low temperature tempering, e.g. about 150°C, to relieve internal stresses produced by the treatment.

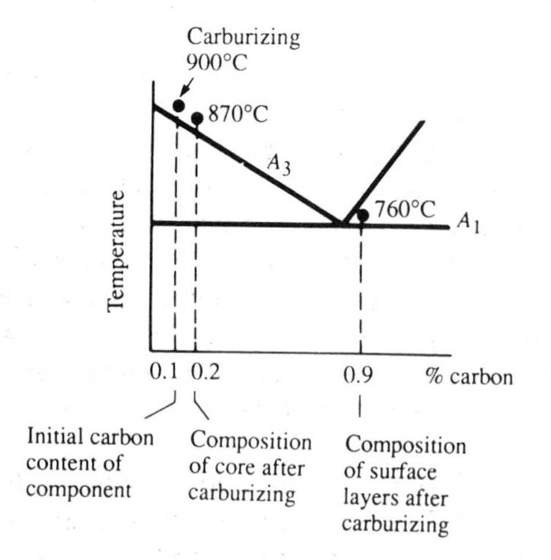

Figure 12.30 Heat treatment for a carburized steel

Nitriding involves changing the surface composition of a steel by diffusing nitrogen into it. Hard compounds, nitrides, are produced. The process is used with those alloy steels that contain elements that form stable nitrides, e.g. aluminium, chromium, molybdenum, tungsten and vanadium. Prior to the nitriding treatment the steel is hardened and tempered to the properties required of the core. The tempering temperature does, however, need to be in the region 560°C to 750°C. The reason for this is that the nitriding process requires a temperature up to about 530°C, and this must not be greater than the tempering temperature, as the nitriding process would temper the steel and so change the properties of the core.

Unlike carburizing, nitriding is carried out at temperatures below the stable austenitic state. The process consists of heating a component in an atmosphere of ammonia gas and hydrogen, the temperatures being of the order of 500 to 530°C. The time taken for the nitrogen to react with the elements in the surface of the steel is often as much as 100 hours. The depth to which the nitrides are formed in the steel depends on the temperature and the time allowed for the reaction. Even with such long times the depth of hardening is unlikely to exceed about 0.7 mm. After the treatment the component is allowed to cool slowly in the ammonia/hydrogen atmosphere. With most nitriding conditions a thin white layer of iron nitrides is formed on the surface of the component. This layer adversely affects the mechanical properties of the steel, being brittle and generally containing cracks. It is therefore removed by mechanical means or chemical solutions.

Because with nitriding no quenching treatments are involved, cracking and distortion are less likely than with other surface hardening treatments. Very high surface hardnesses can be obtained with special alloys. The hardness is retained at temperatures up to about 500°C, whereas that produced by carburizing tends to decrease and the surface become softer at temperatures of the order of 200°C. The capital cost of the plant is however higher than that associated with pack carburizing.

Carbonitriding is the name given to the surface hardness process in which both carbon and nitrogen are allowed to diffuse into a steel when it is in the austenitic-ferritic condition. The component is heated in an atmosphere containing carbon and ammonia and the temperatures used are about 800°C to 850°C. The nitrogen inhibits the diffusion of carbon into the steel and, with the temperatures and times used being smaller than with carburizing, this leads to relatively shallow hardening. Though this process can be used with any steel that is suitable for carburizing it is used generally only for mild steels or low alloy steels.

Ferritic nitrocarburizing involves the diffusion of both carbon and nitrogen into the surface of a steel. The treatment involves a temperature below the A_1 temperature when the steel is in a ferritic condition. A very thin layer of a compound of iron, nitrogen and carbon is produced at the surfaces of the steel. This gives excellent wear and anti-scuffing properties. The process is

Table 12.5 *Surface hardening treatments*

Process	*Temperature /°C*	*Case depth /mm*	*Case hardness /HRC*	*Main user*
Pack carburizing	810–1100	0.25–3	45–65	Low carbon and carburizing alloy steels. Large case depths, large components.
Gas carburizing	810–980	0.07–3	45–65	Low carbon and carburizing alloy steels. Large numbers of components.
Cyaniding	760–870	0.02–0.7	50–60	Low carbon and light alloy steels. Thin case.
Nitriding	500–530	0.07–0.7	50–70	Alloy steels. Lowest distortion.
Carbonitriding	700–900	0.02–0.7	50–60	Low carbon and low alloy steels.
Flame hardening	850–1000	Up to 0.8	55–65	0.4 to 0.7% carbon steels, selective heating.
Induction hardening	850–1000	0.5–5	55–65	0.4 to 0.7% carbon steels, selective heating.

mainly used on mild steel in the rolled or normalized condition.

Table 12.5 compares the main surface hardening treatments. In comparing the process the term *case depth* is used, which is defined graphically (Figure 12.31) in terms of a graph of carbon or nitrogen content against depth under the surface of the steel. A straight line is drawn on the graph such that it passes through the surface carbon content value and so that the area between the line and the core carbon content C_c is the same as that between the actual carbon content graph curve and the core carbon content C_c. Where this line meets the core carbon content line gives the case depth.

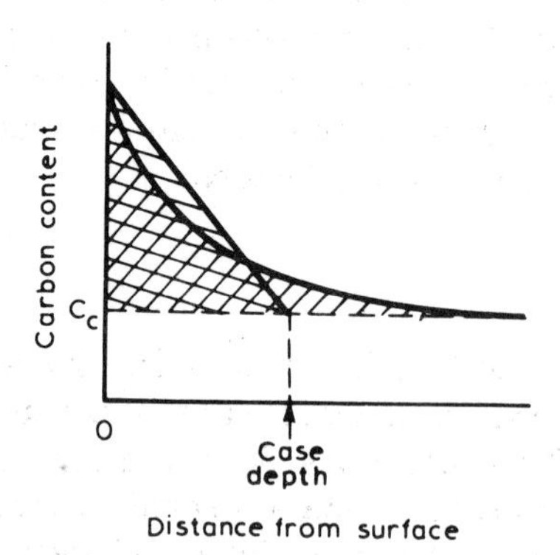

Figure 12.31 Defining case depth

Wear resistance treatments

In addition to the surface hardening treatments, there are a number of other treatments which can be applied to surfaces to reduce wear.

1 *Siliconizing* – this process is used with low carbon, medium carbon and unalloyed steels. Silicon dissolves in iron at a temperature of about 1000°C to form a solid solution. Surfaces exposed to the silicon acquire a low coefficient of friction and can retain some lubricant. Adhesive wear is thus reduced. The process, however, involves high temperatures and the surface is not suitable where high pressures are involved, e.g. ball bearings.
2 *Sulfinuz* – this process can be used with all ferrous metals and titanium alloys. The material is heated in a salt bath to about 540 to 600°C. The treatment introduces carbon, nitrogen and sulphur into the surfaces. The result is good scuffing resistance with some reduction in the coefficient of friction.
3 *Sulf BT* – this process can be used with all ferrous metals, except 13 per cent chromium stainless steels. The material is heated in a salt bath to about 180 to 200°C. The treatment introduces sulphur into the surfaces. The low temperature process minimizes distortion and the surfaces have good scuffing resistance.
4 *Noskuf* – this process is used with case-hardening and direct-hardening steels. It is applied after carburizing and involves heating the material in a salt bath at about 700 to 760°C. The process adds nitrogen and carbon to the surface layers and gives good scuffing resistance.
5 *Phosphating* – this process is used with all ferrous metals. A film is produced on the surface of the material by either chemical or electrochemical treatments at about 40 to 100°C. The porous nature of the surface helps to retain lubricants and resists scuffing. The treatment is less effective than nitriding or Sulf BT in improving wear resistance.
6 *Boriding* – this process is used with low carbon, medium carbon and low alloy steels. Boron is allowed to diffuse into the surfaces at temperatures of the order of 900 to 1000°C. An iron boride compound is produced which gives the surface a good resistance to abrasive wear. The surface is very hard and is difficult to grind or polish; diamond tipped tools and silicon carbide or alumina wheels have to be used.

12.8 Annealing

The term *full annealing* is used for the treatment that involves heating a steel to its austenitic state before subjecting it to very slow cooling. For carbon steels with less than about 0.8 per cent carbon the temperature to which the steel is heated for its microstructure to become austenitic is about 40°C above the A_3 temperature. The result of such a treatment is a microstructure of ferrite and pearlite. This gives a very soft steel. Figure 12.32 shows the relevant austenizing temperature region on the iron–carbon diagram and Figure 12.33, the temperature–time relationship for the process with reference to the continuous cooling transformation curve.

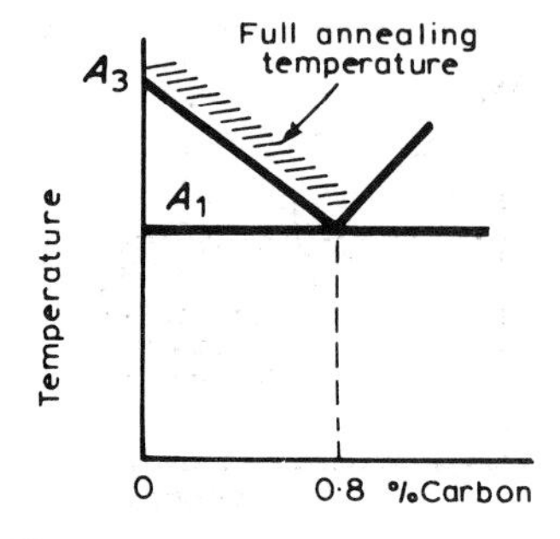

Figure 12.32 The iron–carbon diagram and the temperatures used for full annealing

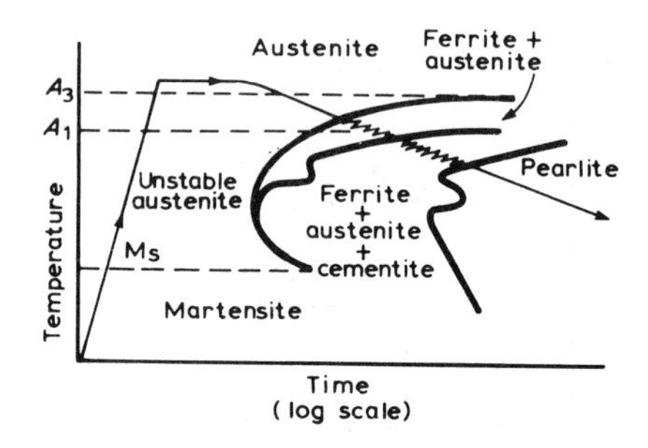

Figure 12.33 The temperature–time relationship for full annealing

For a carbon steel with more than 0.8 per cent carbon, the steel is heated to about 40°C above the A_1 temperature before being slowly cooled. The result of such a treatment is to produce a microstructure of austenite plus cementite at the temperature just above A_1. After cooling, the structure is pearlite plus excess cementite in dispersed spheroidal form. The overall result is a soft steel. The reason for not heating this steel to above the A_3 temperature is that slow cooling of such a steel results in a network of cementite surrounding pearlite. This has the effect of making the steel relatively brittle.

Sub-critical annealing, generally referred to as *process annealing*, is used often during cold-working processes with low-carbon steels, less than about 0.3 per cent carbon, where the material has to be made more ductile for the process to continue. The process involves heating the steel to a temperature just below the A_1 temperature, holding it at that temperature for a while and then allowing the material to cool in air. This is a faster rate of cooling than that employed with full annealing, where the material is cooled in the furnace. This process leads to no change in microstructure, no austenite being produced. The result is however a recrystallization. Prior to this treatment the crystals may have been deformed by the cold working process, afterwards a new crystal structure occurs with no deformation. The effect of this treatment is to give a

Table 12.6 *The effect of annealing on properties for a 0.15 per cent plain carbon steel*

Condition	*Hardness/HV*	*Percentage elongation*
Cold worked	187	22
Process annealed	162	25

reduction in hardness and an increase in percentage elongation (see Table 12.6).

Figure 12.34 shows the region of the iron–carbon diagram relevant to process annealing.

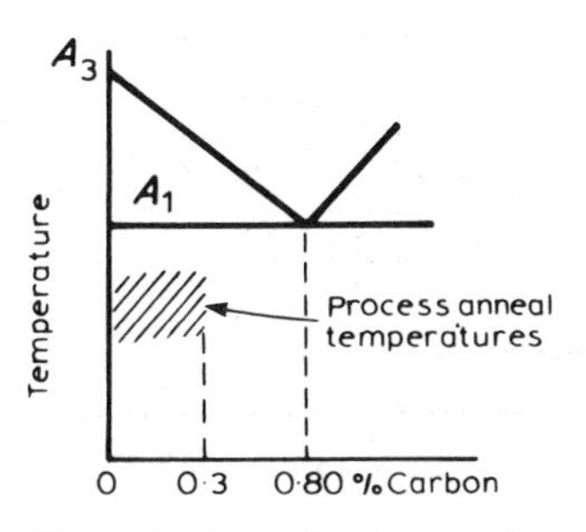

Figure 12.34 The iron–carbon diagram and process annealing

Figure 12.35 The iron–carbon diagram and spheroidizing annealing

When sub-critical annealing is applied to steels having percentages of carbon higher than 0.3 per cent, the effect of the heating is to cause the cementite to assume spherical shapes. Because of this the process is referred to as *spheroidizing annealing*. Figure 12.35 shows the relevant part of the iron–carbon diagram for this treatment. The process results in a reduction in strength and hardness but an increase in ductility and machinability. The spherical pieces of cementite act as chip breakers during machining and so improve machinability.

The term *normalizing* is used to describe an annealing process which involves heating a steel to about 40°C above the A_3 temperature and then allowing the steel to cool in air. The process is thus similar to full annealing in that the microstructure is transformed to austenite but differs in that a faster rate of cooling is used (Figure 12.36). The result of such a treatment is a ferrite and

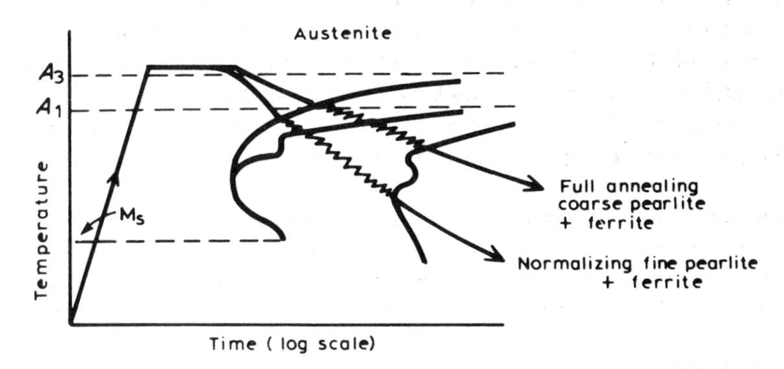

Figure 12.36 The temperature–time relationship for full annealing and normalizing

pearlite microstructure, the hardness and strength being slightly greater than that which would have occurred with full annealing.

12.9 Selecting heat treatment

Heat treatment can be defined as the controlled heating and cooling of metals for the purpose of altering their mechanical properties. While heat treatment is often used to strengthen or harden a steel there are numerous occasions when heat treatment is used to prepare the material for fabrication, e.g. machining or restoring ductility so that deformation can continue. Full annealing, process annealing, spheroidizing annealing and normalizing are examples of such treatments. Strengthening and hardening treatments for steels are quenching followed by tempering, austempering and martempering. There is also the possibility of surface hardening, without changing the core properties of the steel concerned.

Suppose a steel has to be softened for some fabrication process. The choice of heat treatment will depend on what hardness is required of the component for the fabrication process. If only a relatively small decrease in hardness and increase in ductility are required and the steel has a low percentage of carbon, then process annealing may be chosen. This would have the advantage of being a faster process than full annealing and so cause less of a production hold-up.

Suppose a steel shaft has to be hardened so that the surface is at a particular hardness and the central core of the shaft has also to be a particular hardness. Consideration has to be given to the hardenability of the steel concerned and whether the required hardness distribution across the shaft section can be obtained by quenching in some particular media followed by tempering.

The selection of a heat treatment depends on the mechanical properties required and the type of steel concerned. Thus in the case of the heat treatment to soften a piece of steel for fabrication, process annealing may be chosen for a low carbon steel but if the steel has more than about 0.3 per cent carbon a different treatment will have to be chosen. Similarly the hardening of the steel shaft depends on the hardenability of the steel concerned. Changing the type of steel used could significantly change the hardenability and hence the method used to produce the required hardness distribution across the shaft section. The size of the shaft also affects the treatment used.

Problems

1 Explain what is meant by the term critical points and, in this context, the symbols A_1, A_3 and A_{cm}.
2 What is shown by a TTT diagram?
3 Explain how a TTT diagram is obtained.
4 Sketch typical TTT diagrams for plain carbon steels having (a) the eutectoid composition, (b) less carbon than the eutectoid composition.

5 What is the effect on the TTT diagram of a carbon steel of the addition of chromium to the alloy?
6 What are continuous cooling transformation curves?
7 What is meant by the critical cooling rate? How is the critical cooling rate determined by the continuous cooling transformation curve?
8 Explain, with the aid of a continuous cooling transformation curve, how the structures produced by cooling a plain carbon steel are determined by the rate of cooling. Consider cooling rates both greater and less than the critical cooling rate.
9 Why must water quenching be used to harden plain carbon steels but some alloy steels can be hardened by air cooling?
10 Figure 12.37 is the TTT diagram for a cold work tool steel (KE 162) which is hardened by either air cooling or an oil quench. Figure 12.38 is the TTT diagram for a medium carbon chrome vanadium steel (KE 896) which is hardened by quenching in oil except in the case of sections thicker than about 60 mm which require quenching in water to achieve maximum hardness. How can an examination of the TTT diagrams disclose the fact that the KE 162 steel requires a slower rate of cooling than the KE 896 steel? Why do the thicker sections of KE 896 require a faster rate of cooling than thinner sections?

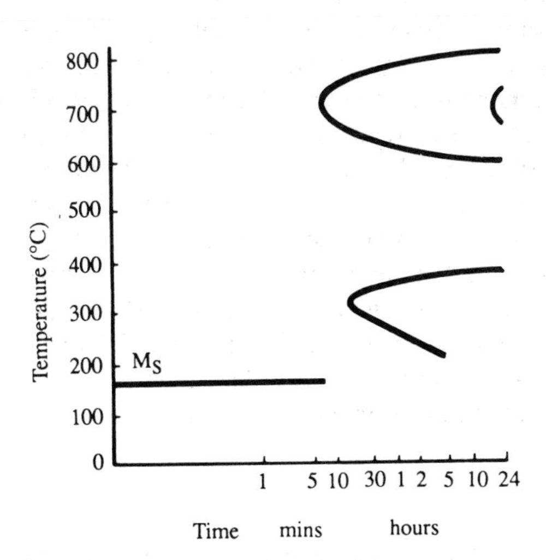

Figure 12.37 The TTT diagram for KEA 162 steel (1.0 per cent carbon, 0.65 per cent manganese, 5.0 per cent chromium, 1.0 per cent molybdenum, 0.20 per cent vanadium) (Courtesy of Sanderson Kayser Ltd)

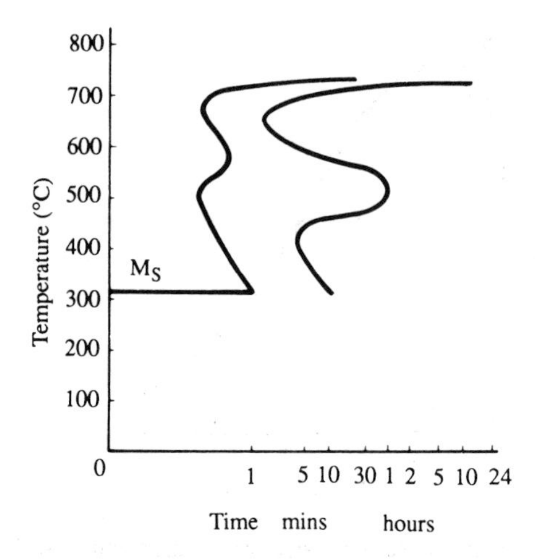

Figure 12.38 The TTT diagram for KE 896 steel (0.50 per cent carbon, 0.65 per cent manganese, 1.1 per cent chromium, 0.20 per cent vanadium) (Courtesy of Sanderson Kayser Ltd)

11 Explain the term hardenability.
12 Describe the Jominy test.
13 Explain how Jominy test results can be used to determine the hardness distribution that would occur across a section of a steel component when it is quenched.
14 Use Figure 12.21 for this question. (a) What would be the hardness at the centre of a 12 mm diameter bar of 0.40 per cent plain carbon steel when it is quenched in still water? (b) What would the answer to (a) have been if still oil had been used for the quenching? (c) For the alloy steel given in the figure, what would be the maximum diameter of bar that could be quenched in still water if the hardness at the centre of the bar had to be 600 HV or more?
15 Explain, using a continuous cooling transformation curve, why the structure at the centre of a bar may differ from that at the surface.
16 Explain the term 'limiting ruling section'.
17 Explain what the tempering process is and why the mechanical properties of a tempered steel depend on the tempering temperature.
18 With the tempering curve given in Figure 12.24 for the KEA 180 steel, what tempering temperature should be used if the steel is to have a hardness of 45 HRC?
19 Figure 12.39 shows the tempering curve for a 1.4 per cent plain carbon steel. What would be the hardness of such a steel after tempering at 300°C?
20 The tempering curve in Figure 12.39 was for a plain carbon steel that had been heated to 1200°C, water quenched and then cooled to a subzero

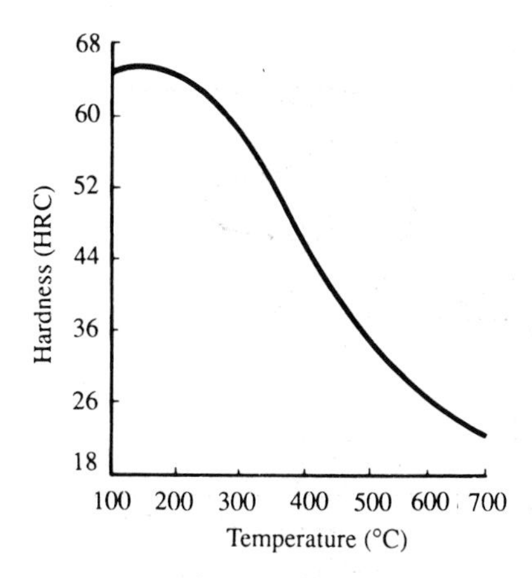

Figure 12.39 Tempering curve for a 1.4 per cent plain carbon steel

temperature. (a) What was the purpose of the heating to 1200°C? (b) Why was the steel cooled to below zero?

21 Describe the following treatment processes. (a) Full annealing, (b) Process annealing, (c) Spheroidizing annealing, (d) Normalizing.

22 In what ways do the processes listed in Question 21 differ?

23 Explain, using a TTT diagram, how full annealing and normalizing can lead to different microstructures.

24 Explain using TTT diagrams the (a) austempering and (b) martempering processes.

25 Explain how flame hardening leads to the surface hardening of a steel.

26 Compare the surface hardening processes of nitriding and carburizing.

27 Which surface hardening treatment would be most appropriate for an alloy steel which cannot be heated above 550°C?

28 What factors have to be considered in determining the type of quenching to be used to harden a steel?

29 A steel with 0.6 per cent carbon is to be surface hardened. Which type of process, selection heating or carburizing, would be suitable?

13

Non-ferrous alloys

13.1 The range of alloys

The term *ferrous alloys* is used for those alloys having iron as the base element, e.g. cast iron and steel. The term *non-ferrous alloys* is used for those alloys which do not have iron as the base element, e.g. alloys of aluminium. The following are some of the non-ferrous alloys in common use in engineering:

Aluminium alloys Aluminium alloys have a low density, good electrical and thermal conductivity, high corrosion resistance. Typical uses are metal boxes, cooking utensils, aircraft bodywork and parts.

Copper alloys Copper alloys have good electrical and thermal conductivity, high corrosion resistance. Typical uses are pump and valve parts, coins, instrument parts, springs, screws. The names brass and bronze are given to some forms of copper alloys.

Magnesium alloys Magnesium alloys have a low density, good electrical and thermal conductivity. Typical uses are castings and forgings in the aircraft industry.

Nickel alloys Nickel alloys have good electrical and thermal conductivity, high corrosion resistance, can be used at high temperatures. Typical uses are pipes and containers in the chemical industry where high resistance to corrosive atmospheres is required, food processing equipment, gas turbine parts. The names Monel, Inconel and Nimonic are given to some forms of nickel alloys.

Titanium alloys Titanium alloys have a low density, high strength, high corrosion resistance, can be used at high temperatures. Typical uses are in aircraft for compressor discs, blades and casings, in chemical plant where high resistance to corrosive atmospheres is required.

Zinc alloys Zinc alloys have good electrical and thermal conductivity, high corrosion resistance, low melting points. Typical uses are as car door handles, toys, car carburettor bodies – components that in general are produced by die casting.

Non-ferrous alloys have, in general, these advantages over ferrous alloys:

1 Good resistance to corrosion without special processes having to be carried out.
2 Most non-ferrous alloys have a much lower density and hence lighter weight components can be produced.
3 Casting is often easier because of the lower melting points.
4 Cold working processes are often easier because of the greater ductility.
5 Higher thermal and electrical conductivities.
6 More decorative colours.

Ferrous alloys have these advantages over non-ferrous alloys:

1 Generally greater strengths.
2 Generally greater stiffness, i.e. larger values of Young's modulus.
3 Better for welding.

13.2 Aluminium

Pure aluminium has a density of 2.7×10^3 kg m^{-3}, compared with that of 7.9×10^3 kg m^{-3} for iron. Thus for the same size component the aluminium version will be about one-third of the mass of an iron version. Pure aluminium is a weak, very ductile, material. It has an electrical conductivity about two-thirds that of copper but weight for weight is a better conductor. It has a high thermal conductivity. Aluminium has a great affinity for oxygen and any fresh metal in air rapidly oxidizes to give a thin layer of the oxide on the metal surface. This surface layer is not penetrated by oxygen and so protects the metal from further attack. The good corrosion resistance of aluminium is due to this thin oxide layer on its surface.

Aluminium of high purity (99.5 per cent aluminium, or greater) is too weak a material to be used in any other capacity than a lining for vessels. It is used in this way to give a high corrosion resistant surface.

Aluminium of commercial purity, 99.0 to 99.5 per cent aluminium, is widely used as a foil for sealing milk bottles, thermal insulation, and kitchen foil for cooking. The presence of a relatively small percentage of impurities in aluminium considerably increases the tensile strength and hardness of the material.

The mechanical properties of aluminium depend not only on the purity of the aluminium but also upon the amount of work to which it has been subject. The effect of working the material is to fragment the grains. This results in an increase in tensile strength and hardness and a decrease in ductility. By controlling the amount of working different degrees of strength and hardness can be produced. These are said to be different *tempers*. The properties of aluminium may thus, for example, be referred to as that for the annealed condition, the half-hard temper and the fully hardened temper.

In material specifications the temper of aluminium and its non-heat treatable alloys is indicated as follows:

British Standards	*Condition*
M	As manufactured, e.g. as rolled
O	In the annealed or soft condition
H	Work hardened, the degree of work hardening being indicated by numbers from 1 to 8
H2	Quarter work hardened
H4	Half work hardened
H6	Three-quarters work hardened
H8	Fully work hardened

American Standards	*Condition*
F	As fabricated, e.g. rolled
O	In the annealed or soft condition
H	Strain hardened – the letter is followed by two numbers, the first indicating the combination of processes used to achieve the temper and the second the degree of hardness
H1	Cold worked only
H2	Cold worked and partially annealed
H3	Cold worked and then stabilized by a low temperature anneal
Hx2	Quarter hard
Hx4	Half hard
Hx8	Fully hard
Hx9	Extra hard temper

Table 13.1 shows typical properties of aluminium.

Sintered aluminium product

Sintered aluminium product (SAP) is formed by sintering aluminium to powder which has a thin coating of aluminium oxide. Aluminium oxide, (often referred to as alumina) is a hard, refractory material. When the powder is compacted in the sintering process, the surface oxide becomes dispersed throughout the aluminium to give a dispersion strengthened metal (see Chapter 16 for more details). At room temperature SAP has a tensile strength of about 400 MN m^{-2} (MPa), compared with 100 MN m^{-2} (MPa) or less for aluminium.

Table 13.1 *Typical properties of aluminium*

Composition (%)	*Condition*	*Tensile strength ($MN\ m^{-2}$ or MPa)*	*Elongation (%)*	*Hardness (HB)*
99.99	Annealed (O)	45	60	15
	Half hard (H4)	82	24	22
	Full hard (H8)	105	12	30
99.8	Annealed (O)	66	50	19
	Half hard (H4)	99	17	31
	Full hard (H8)	134	11	38
99.5	Annealed (O)	78	47	21
	Half hard (H4)	110	13	33
	Full hard (H8)	140	10	40
99	Annealed (O)	87	43	22
	Half hard (H4)	120	12	35
	Full hard (H8)	150	10	42

Aluminium alloys

Aluminium alloys can be divided into two groups, wrought alloys and cast alloys. Each of these can be divided into two further groups:

1 Those alloys which are not heat treatable.
2 Those alloys which are heat treated.

The term *wrought material* is used for a material that is suitable for shaping by working processes, e.g. forging, extrusion, rolling. The term *cast material* is used for a material that is suitable for shaping by a casting process.

Non-heat treatment wrought alloys
The non-heat treatable wrought alloys of aluminium do not significantly respond to heat treatment but have their properties controlled by the extent of the working to which they are subject. A range of tempers is thus produced. Common alloys in this category are aluminium with manganese or magnesium. A common aluminium–manganese alloy has 1.25 per cent manganese, the effect of this manganese being to increase the tensile strength of the aluminium. The alloy still has a high ductility and good corrosion properties. This leads to uses such as kitchen utensils, tubing and corrugated sheet for building. Aluminium–magnesium alloys have up to 7 per cent magnesium. The greater the percentage of magnesium, the greater the tensile strength (Figure 13.1). The alloy still has good ductility. It has excellent corrosion resistance and thus finds considerable use in marine environments, e.g. constructional materials for boats and ships.

Heat treatable wrought alloys

The heat treatable wrought alloys can have their properties changed by heat

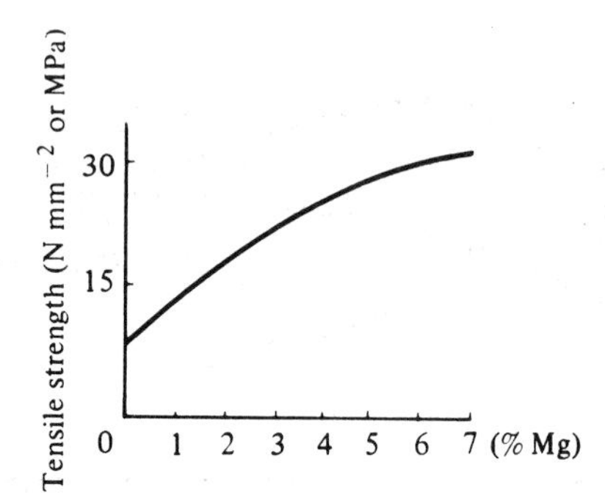

Figure 13.1 The effect on the tensile strength of magnesium content in annealed aluminium–magnesium alloys

treatment. Copper, magnesium, zinc and silicon are common additions to aluminium to give such alloys. Figure 13.2 shows the thermal equilibrium diagram for aluminium–copper alloys. When such an alloy, say 3 per cent copper–97 per cent aluminium, is slowly cooled, the structure at about 540°C is a solid solution of the α phase. When the temperature falls below the solvus temperature a copper–aluminium compound is precipitated. The result at room temperature is α solid solution with this copper–aluminium compound precipitate ($CuAl_2$). The precipitate is rather coarse, but this structure of the alloy can be changed by heating to about 500°C, soaking at that temperature, and then quenching, to give a supersaturated solid solution, just α phase with no precipitate. This treatment, known as *solution treatment*, results in an unstable situation. With time a fine precipitate will be produced. Heating to, say, 165°C for about ten hours hastens the production of this fine precipitate

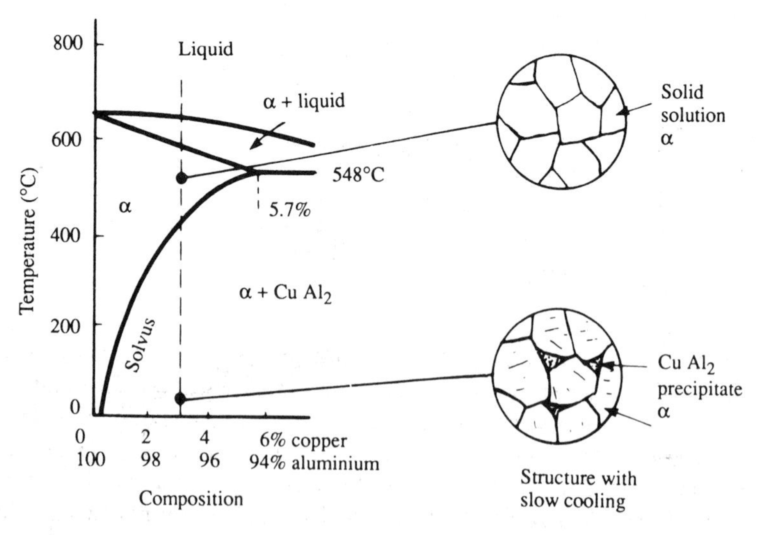

Figure 13.2 Thermal equilibrium diagram for aluminium–copper alloys

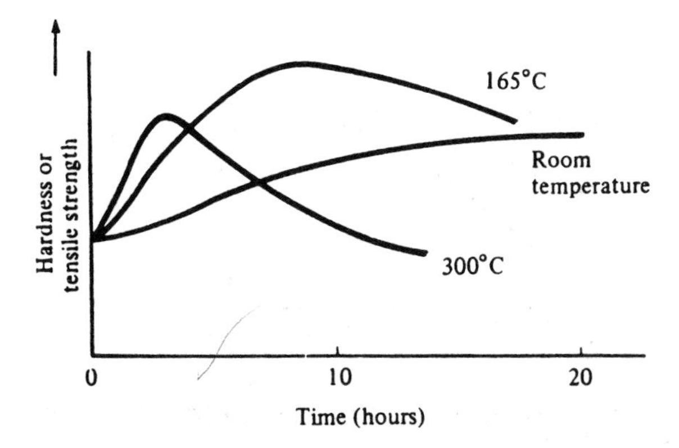

Figure 13.3 The effects of time and temperature on hardness and strength for an aluminium–copper alloy

(Figure 13.3). The microstructure with this fine precipitate is both stronger and harder. The treatment is referred to as *precipitation hardening*.

A common group of heat treatable wrought alloys is based on aluminium with copper. Thus one form has 4.0 per cent copper, 0.8 per cent magnesium, 0.5 per cent silicon and 0.7 per cent manganese. This alloy is known as Duralumin. The heat treatment process used is solution treatment at 480°C, quenching and then precipitation hardening either at room temperature for about four days or at 165°C for ten hours. This alloy is widely used in aircraft bodywork. The presence of the copper does, however, reduce the corrosion resistance and thus the alloy is often clad with a thin layer of high-purity aluminium to improve the corrosion resistance (Figure 13.4).

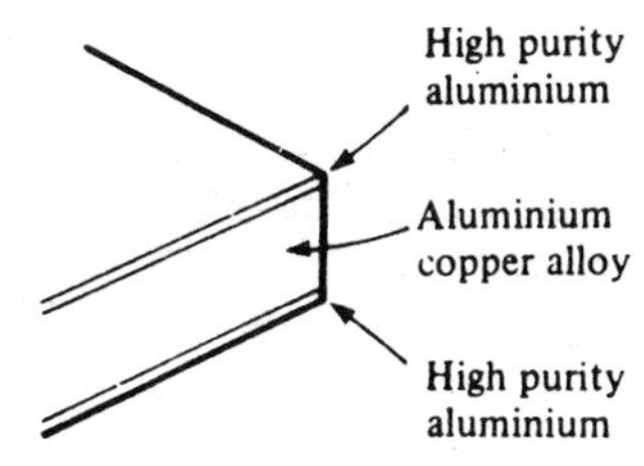

Figure 13.4 A clad duraluminium sheet

The precipitation hardening of an aluminium–copper alloy is due to the precipitate of the aluminium–copper compound. The age hardening of aluminium–copper–magnesium–silicon alloys is due to the precipitates of both an aluminium–copper compound $CuAl_2$ and an aluminium–copper–magnesium compound $CuAl_2Mg$. Other heat treatable wrought alloys are based on aluminium with magnesium and silicon. The age hardening with this alloy is due to the precipitate of a magnesium–silicon compound, Mg_2Si. A typical alloy has the composition 0.7 per cent magnesium, 1.0 per cent silicon and 0.6 per cent manganese. This alloy is not as strong as the duralumin but has greater ductility. It is used for ladders, scaffold tubes, container bodies, structural members for road and rail vehicles. The heat treatment is solution treatment at 510°C with precipitation hardening by quenching followed by precipitation hardening of ten hours at about 165°C. Another group of alloys is based on aluminium–zinc–magnesium–copper, e.g. 5.5 per cent zinc, 2.8 per cent magnesium, 0.45 per cent copper, 0.5 per cent manganese. These alloys have the highest strength of the aluminium alloys and are used for structural applications in aircraft and spacecraft.

Non-heat treatable cast alloys

An alloy for use in the casting process must flow readily to all parts of the mould and on solidifying it should not shrink too much and any shrinking should not result in fractures. In choosing an alloy for casting, the type of casting process being used needs to be taken into account. In sand casting the mould is made of sand bonded with clay or a resin. The cooling rate with such a method is relatively slow. A material for use by this method must give a material of suitable strength after a slow cooling process. With die casting the mould is made of metal and the hot metal is injected into the die under pressure. This results in a fast cooling. A material for use by this method must develop suitable strength after fast cooling.

A family of aluminium alloys that can be used in the 'as cast' condition, i.e. no heat treatment is used, has aluminium with between 9 and 13 per cent silicon. These alloys can be used for both sand and die casting. Figure 13.5

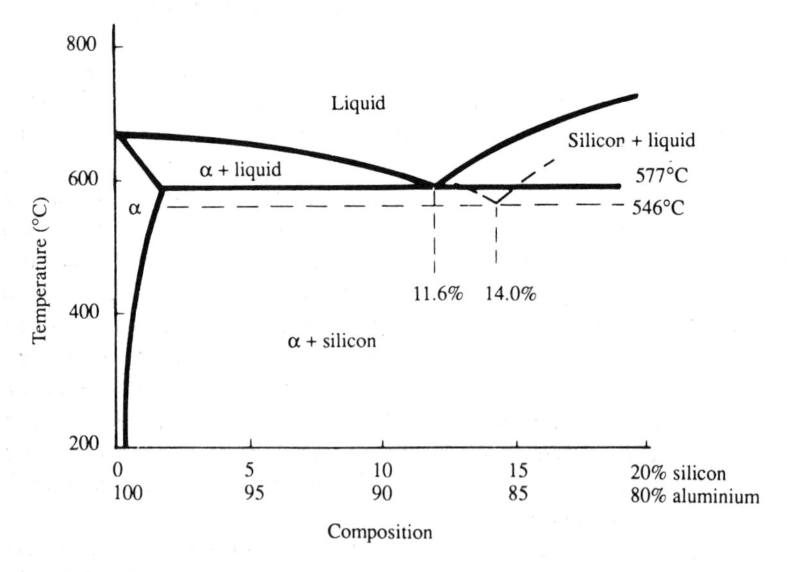

Figure 13.5 The aluminium–silicon thermal equilibrium diagram

shows the thermal equilibrium diagram for aluminium–silicon alloys. The addition of silicon to aluminium increases its fluidity, between about 9 to 13 per cent giving a suitable fluidity for casting purposes. The eutectic for aluminium–silicon alloys has a composition of 11.6 per cent silicon. An alloy of this composition changes from the liquid to the solid state without any change in temperature, and alloys close to this composition solidify over a small temperature range (Figure 13.6). This makes them particularly suitable for die casting where a quick change from liquid to solid is required in order that a rapid ejection from the die can permit high output rates.

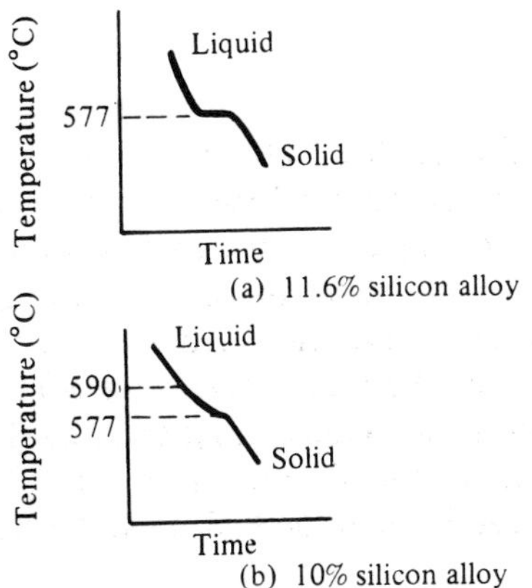

Figure 13.6 Cooling curves (a) 11.6 per cent silicon alloy, (b) 10 per cent silicon alloy

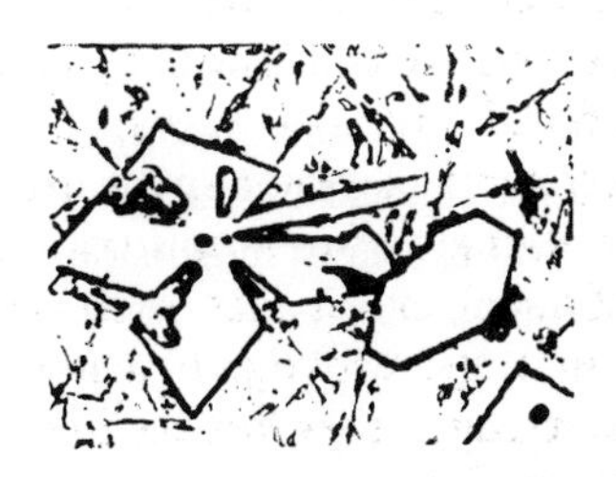

Figure 13.7 A 13 per cent cast silicon alloy showing silicon with eutectic (From Rollason, E. C., *Metallurgy for Engineers*, Edward Arnold)

For the eutectic composition the microstructure shows a rather coarse eutectic structure of α phase and silicon. For an alloy having more silicon than the 11.6 per cent eutectic value, the microstructure consists of silicon crystals in eutectic structure (Figure 13.7). The coarse eutectic structure, together with the presence of the embrittling silicon crystals, results in rather poor mechanical properties for the casting. The structure can, however, be made finer and the silicon crystal formation prevented by a process known as *modification*. This involves adding about 0.005 to 0.15 per cent metallic sodium to the liquid alloy before casting. This produces a considerable refinement of the eutectic structure and also causes the eutectic composition to change to about 14.0 per cent silicon. This displacement of the eutectic point is indicated on Figure 13.5 by the dashed line. Thus for a silicon content below 14 per cent, the structure, as modified, has α phase crystals in a finer eutectic structure (Figure 13.8). The result is an increase in both tensile strength and ductility and so a much better casting material.

The aluminium–silicon alloy is widely used for both sand and die casting, being used for many castings in cars, e.g. sumps, gear boxes and radiators. It is

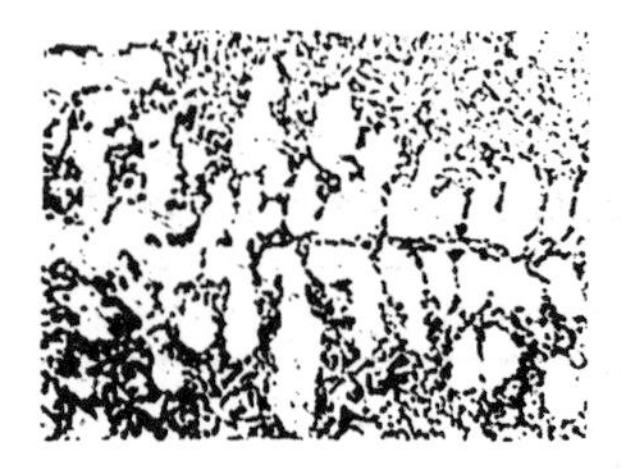

Figure 13.8 The same material as in Figure 13.7, but modified by sodium, showing aluminium with a fine eutectic (From Rollason, E. C., *Metallurgy for Engineers*, Edward Arnold)

also used for pump parts, motor housing and a wide variety of thin walled and complex castings.

Other cast alloys that are not heat treated are aluminium–silicon–copper alloys, e.g. 5.0 per cent silicon and 3.0 per cent copper, and aluminium–magnesium–manganese alloys, e.g. 4.5 per cent magnesium and 0.5 per cent manganese. The silicon–copper alloys can be both sand and die cast, the magnesium–manganese alloys are however only suitable for sand casting. They have excellent corrosion resistance and are often used in marine environments.

Heat treatable cast alloys

The addition of copper, magnesium and other elements to aluminium alloys, either singly or in some suitable combination, can enable the alloy to be heat treated. Thus an alloy having 5.5 per cent silicon and 0.6 per cent magnesium can be subjected to solution treatment followed by precipitation hardening to give a high-strength casting material. Another heat treatable casting alloy has 4.0 per cent copper, 2.0 per cent nickel and 1.5 per cent magnesium.

Properties and specification of aluminium alloys

Table 13.2 shows typical properties of aluminium alloy. In specifications non-heat treatable alloys have their temper indicated in the same way as aluminium. Such alloys have properties which are affected by work hardening. Heat treatable alloys are specified as follows:

British Standards	*Condition*
M	As manufactured
O	In the annealed condition
TB	Solution treated, quenched and naturally aged
TB7	Solution treated and stabilized (cast metals)
TD	Solution treated, quenched, cold worked and naturally aged
TE	Cooled from a hot working process and artificially aged

Table 13.2 *Typical properties of aluminium alloys*

Composition	*Condition*	*Tensile strength (MN m^{-2} or MPa)*	*Elongation (%)*	*Hardness (HB)*
Wrought, non-heat treated alloys				
1.25% Mn	Annealed (O)	110	30	30
	Hard (H8)	180	3	50
2.25% Mg	Annealed (O)	180	22	45
	¾ hard (H6)	250	4	70
5.0% Mg	Annealed (O)	300	16	65
	¼ hard (H2)	340	8	80
Wrought, heated treated alloys				
4.0% Cu, 0.8% Mg, 0.5% Si, 0.7% Mn	Annealed (O)	180	20	45
	Solution treated, precipitation hardened (TF)	430	20	100
4.3% Cu, 0.6% Mg, 0.8% Si, 0.75% Mn	Annealed (O)	190	12	45
	Solution treated, precipitation hardened (TF)	450	10	125
0.7% Mg, 1.0% Si, 0.6% Mn	Annealed (0)	120	15	47
	Solution treated, precipitation hardened (TF)	300	12	100
5.5% Zn, 2.8% Mg, 0.45% Cu, 0.5% Mn	Solution treated, precipitation hardened (TF)	500	6	170
Cast, non-heat treatment alloys				
12% Si	Sand cast (M)	160	5	55
	Die cast (M)	185	7	60
5% Si, 3% Cu	Sand cast (M)	150	2	70
	Die cast (M)	170	3	80
4.5% Mg, 0.5% Mn	Sand cast (M)	140	3	60
Cast, heat treated alloys				
5.5% Si, 0.6% Mg	Sand cast, solution treated, precipitation hardened (TF)	235	2	85
4.0% Cu, 2% Ni, 1.5% Mg	Sand cast, solution treated, precipitation hardened (TF)	275	1	110

TF	Solution treated, quenched and artificially aged
TF7	Solution treated, quenched and stabilized
TH	Solution treated, quenched, cold worked and artificially aged

TS	Thermally treated to improve dimensional stability (cast metals)
American Standards	*Condition*
T	Heat treated, numbers are then appended to indicate the form of treatment
T2	Annealed casting
T3	Solution treated, cold worked and naturally aged
T4	Solution treated and naturally aged
T5	Artificially aged
T6	Solution treated and artificially aged
T8	Solution treated, work hardened and artificially aged
T9	Solution treated, artificially aged and work hardened

Surface treatments of aluminium

Aluminium develops an oxide surface layer in air. This layer can be thickened by an electrolytic process, the treatment being known as *anodizing*. The freshly formed anodic layer is porous and has to be sealed, by immersion in boiling water or a special solution, as maximum protection against atmospheric corrosion is required. The porous film can be coloured by pigments for decorative purposes. The anodic layer is hard since aluminium oxide (alumina) is a ceramic material.

13.3 Copper

Copper has a density of 8.93×10^3 kg m^{-3}. It has very high electrical and thermal conductivity and can be manipulated readily by either hot or cold working. Pure copper is very ductile and relatively weak. The tensile strength and hardness can be increased by working; this does, however, decrease the ductility. Copper has good corrosion resistance. This is because there is a surface reaction between copper and the oxygen in the air which results in the formation of a thin protective oxide layer.

Very pure copper can be produced by an electrolytic refining process. An impure slab of copper is used as the anode while a pure thin sheet of copper is used as the cathode. The two electrodes are suspended in a warm solution of dilute sulphuric acid (Figure 13.9). The passage of an electric current through the arrangement causes copper to leave the anode and become deposited on the cathode. The result is a thicker, pure copper cathode, while the anode effectively disappears; the impurities have fallen to the bottom of the container. The copper produced by this process is often called *cathode copper* and has a purity greater than 99.99 per cent. It is used mainly as the raw material for the production of alloys, though there is some use as a casting material.

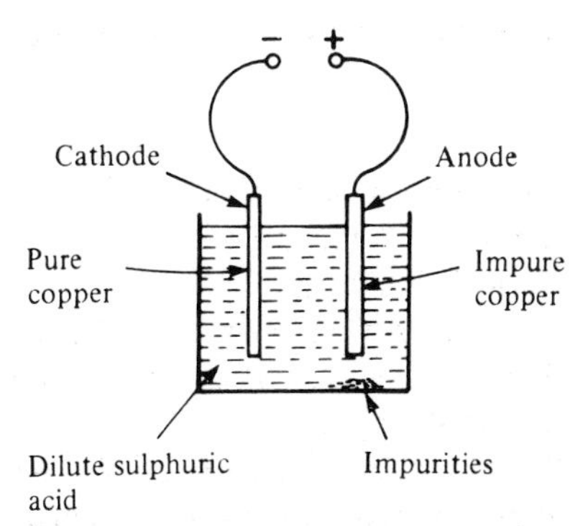

Figure 13.9 Basic arrangement for the electrolytic refining of copper

Electrolytic tough pitch high-conductivity copper is produced from cathode copper which has been melted and cast into billets, and other suitable shapes, for working. It contains a small amount of oxygen, present in the form of cuprous oxide, which has little effect on the electrical conductivity of the copper. This type of copper should not be heated in an atmosphere where it can combine with hydrogen because the hydrogen can diffuse into the metal and combine with the cuprous oxide to generate steam. This steam can exert sufficient pressure to cause cracking of the copper.

Fire refined tough pitch high-conductivity copper is produced from impure copper. In the fire refining process, the impure copper is melted in an oxidizing atmosphere. The impurities react with the oxygen to give a slag which is removed. The remaining oxygen is partially removed by poles of green hardwood being thrust into the liquid metal, the resulting combustion removes oxygen from the metal. The resulting copper has an electrical conductivity almost as good as the electrolytic tough pitch high-conductivity copper.

Oxygen-free high-conductivity copper can be produced if, when cathode copper is melted and cast into billets, there is no oxygen present in the atmosphere. Such copper can be used in atmospheres where hydrogen is present.

Another method of producing oxygen-free copper is to add phosphorus during the refining. The effect of small amounts of phosphorus in the copper is a very marked decrease in the electrical conductivity, of the order of 20 per cent. Such copper is known as *phosphorus deoxidized copper* and it can give good welds, unlike the other forms of copper.

The addition of about 0.5 per cent arsenic to copper increases its tensile strength, especially at temperatures of about 400°C. It also improves its corrosion resistance but greatly reduces the electrical and thermal conductivities. This type of copper is known as *arsenical copper*.

Electrolytic tough pitch high-conductivity copper finds use in high-grade electrical applications, e.g. wiring and busbars. Fire-refined tough pitch high-conductivity copper is used for standard electrical applications. Tough pitch copper is also used for heat exchangers and chemical plant. Oxygen-free high-conductivity copper is used for high-conductivity applications where hydrogen may be present, electronic components and as the anodes in the electrolytic refining of copper. Phosphorus deoxidized copper is used in chemical plant where good weldability is necessary and for plumbing and general pipework. Arsenical copper is used for general engineering work, being useful to temperatures of the order of 400°C.

Table 13.3 shows typical properties of the various forms of copper.

Table 13.3 *Typical properties of the various types of copper*

Composition	*Condition*	*Tensile strength ($N\ mm^{-2}$ or MPa)*	*Elongation (%)*	*Hardness (HB)*
Electrolytic tough-pitch high conductivity copper				
99.90 min 0.05 oxygen	Annealed	220	50	45
	Hard	400	4	115
Fire refined tough-pitch high conductivity copper				
99.85 min 0.05 oxygen	Annealed	220	50	45
	Hard	400	4	115
Oxygen-free high-conductivity copper				
99.95 min	Annealed	220	60	45
	Hard	400	6	115
Phosphorus deoxidized copper				
99.85 min 0.013–0.05 P	Annealed	220	60	45
	Hard	400	4	115
Arsenical copper				
99.20 min 0.05 oxygen, 0.3–0.5 As	Annealed	220	50	45
	Hard	440	4	115

Copper alloys

The most common elements with which copper is alloyed are zinc, tin, aluminium and nickel. The copper–zinc alloys are referred to as brasses, the copper–tin alloys as tin bronzes, the copper–aluminium alloys as aluminium bronzes and the copper–nickel alloys as cupronickels, though where zinc is also present they are called nickel silvers. A less common copper alloy involves copper and beryllium.

The copper–nickel thermal equilibrium diagram is rather simple as the two metals are completely soluble in each other in both the liquid and solid states. The copper–zinc, copper–tin and copper–aluminium thermal equilibrium diagrams are however rather complex. In all cases, the α phase solid solutions have the same types of microstructure and are ductile and suitable for cold working. When the amount of zinc, tin or aluminium exceeds that required to saturate the α solid solution, a β phase is produced. The microstructures of these β phases are similar and alloys containing this phase are stronger and less ductile. They cannot be readily cold worked and are hot worked or cast.

Further additions lead to yet further phases which are hard and brittle.

Brasses

The *brasses* are copper–zinc alloys containing up to about 43 per cent zinc. Figure 13.10 shows the relevant part of the thermal equilibrium diagram. Brasses with between 0 and 35 per cent zinc solidify as α solid solutions, usually cored (Figure 13.11). These brasses have high ductility and can readily be cold worked. *Gilding brass*, 15 per cent zinc, is used for jewellery because it has a colour resembling that of gold and can so easily be worked. *Cartridge brass*, 30 per cent zinc and frequently referred to as *70/30 brass*, is used where high ductility is required with relatively high strength. It is called cartridge brass because of its use in the production of cartridge and shell cases. The brasses in the 0 to 30 per cent zinc range all have their tensile strength and hardness increased by working, but the ductilities decrease.

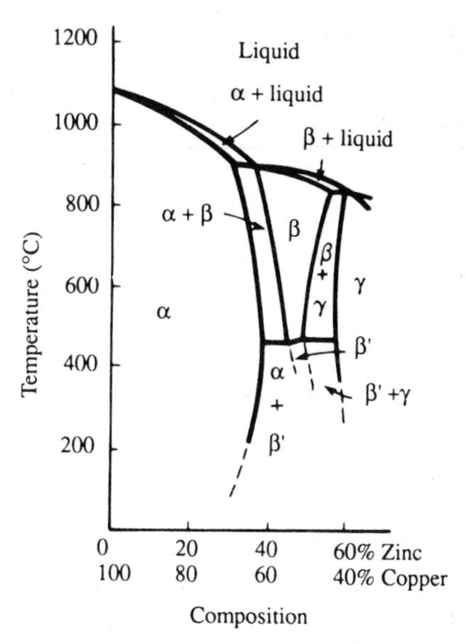

Figure 13.10 Thermal equilibrium diagram for copper–zinc alloys

Brasses with between 35 and 46 per cent zinc solidify as a mixture of two phases (Figure 13.12). Between about 900°C and 453°C the two phases are α and β. At 453°C this β phase transforms to a low temperature modification referred to as β′ phase. Thus at room temperature the two phases present are α

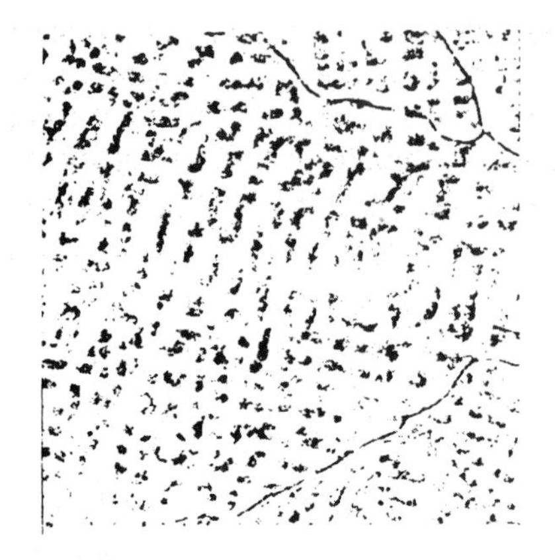

Figure 13.11 A 70 per cent copper–30 per cent zinc alloy – the structure is heavily cored

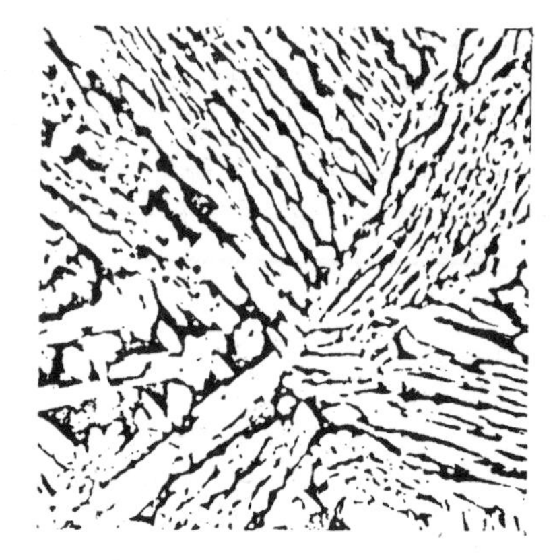

Figure 13.12 A 60 per cent copper–40 per cent zinc alloy. This consists of light α phase in a matrix of dark β′.

and β′ The presence of the β′ phase produces a drop in ductility but an increase in tensile strength to the maximum value for a brass (Figure 13.13). These brasses are known as *alpha-beta* or *duplex brasses*. They are not cold worked but have good properties for hot forming processes, e.g. extrusion. This is because the β phase is more ductile than the β′ phase and hence the combination of α plus β gives a very ductile material. The hot working should take place at temperatures in excess of 453°C. The name *Muntz metal* is given to a brass with 60 per cent copper–40 per cent zinc.

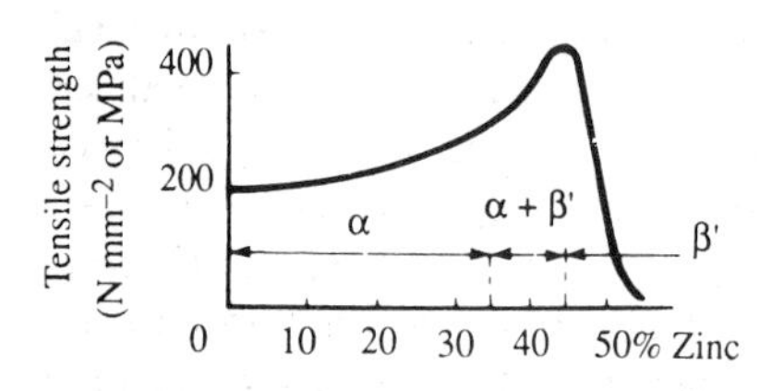

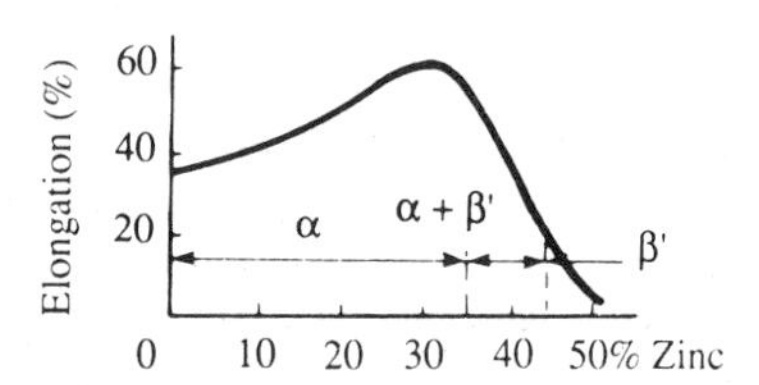

Figure 13.13 Strength and ductility for copper–zinc alloys

The addition of lead to Muntz metal improves considerably the machining properties, without significantly changing the strength and ductility. *Leaded Muntz metal* has 60 per cent copper, 0.3 to 0.8 per cent lead and the remainder zinc.

Copper–zinc alloys containing just the β′ phase have little industrial application. The presence of γ phase in a brass results in a considerable drop in strength and ductility, a weak brittle product being obtained.

Tin bronzes

Copper–tin alloys are known as *tin bronzes*. Figure 13.14 shows the thermal equilibrium diagram for such alloys. The dashed lines on the diagram indicate the phase that can occur with extremely slow cooling. The structure that normally occurs with up to about 10 per cent tin is predominantly α solid solution. Higher percentage tin alloys will invariably include a significant amount of the δ phase. This is a brittle intermetallic compound, the α phase being ductile.

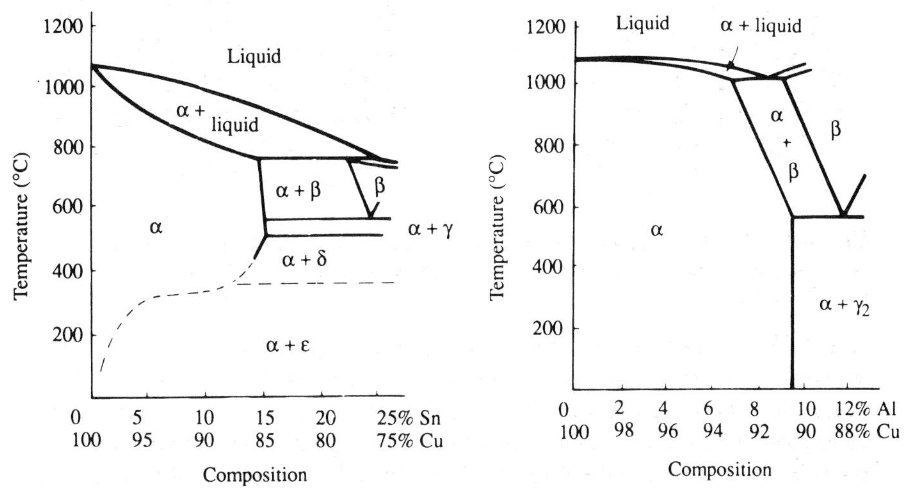

Figure 13.14 Thermal equilibrium diagram for tin bronzes

Figure 13.15 Thermal equilibrium diagram for aluminium bronzes

Bronzes that contain up to about 8 per cent tin are α bronzes and can be cold worked. In making bronze, oxygen can react with the metals and lead to a weak alloy. Phosphorus is normally added to the liquid metals to act as a deoxidizer. Some of the phosphorus remains in the final alloy. This type of alloy is known as *phosphor bronze*. These alloys are used for springs, bellows, electrical contact, clips, instrument components. A typical phosphor bronze might have about 95 per cent copper, 5 per cent tin and 0.02 to 0.40 per cent phosphorus.

The above discussion refers to wrought phosphor bronzes. Cast phosphor bronzes contain between 5 and 13 per cent tin with as much as 0.5 per cent phosphorus. A typical cast phosphor bronze used for the production of bearings and high grade gears has about 90 per cent copper, 10 per cent tin and a maximum of 0.5 per cent phosphorus. This material is particularly useful for bearing surfaces; it has a low coefficient of friction and can withstand heavy loads. The hardness of the material occurs by virtue of the presence of both δ phase and a copper–phosphorus compound.

Casting bronzes that contain zinc are called *gunmetals*. This reduces the cost of the alloy and also makes unnecessary the use of phosphorus for deoxidation as this function is performed by the zinc. *Admiralty gunmetal* contains 88 per cent copper, 10 per cent tin and 2 per cent zinc. This alloy finds general use for marine components, hence the word 'Admiralty'.

Aluminium bronzes

Copper–aluminium alloys are known as *aluminium bronzes*. Figure 13.15 shows the thermal equilibrium diagram for such alloys. Up to about 9 per cent aluminium gives *alpha bronzes*, such alloys containing just the α phase. Alloys with up to about 7 per cent aluminium can be cold worked readily. *Duplex alloys* with about 10 per cent aluminium are used for casting. Aluminium bronzes have high strength, and good resistance to corrosion and wear. These corrosion and wear properties arise because of the thin film of aluminium oxide formed on the surfaces. Typical applications of such materials are high-strength and highly corrosion-resistant items in marine and chemical environments, e.g. pump casings, gears, valve parts.

Beryllium bronzes

Copper alloyed with small percentages of beryllium can be precipitation heat treated to give alloys with very high tensile strengths, such alloys being known as *beryllium bronzes*, or *beryllium copper*. The alloys are used for high-conductivity, high-strength electrical components, springs, clips and fastenings.

Silicon bronzes

Silicon bronzes contain 4 to 5 per cent silicon, in some cases with small amounts of iron or manganese. The alloys have a high corrosion resistance, good strength and weldability.

Cupronickels

Alloys of copper and nickel are known as *cupronickels*, though if zinc is also present they are referred to as *nickel silvers*. Copper and nickel are soluble in each other in both the liquid and solid states, they thus form a solid solution whatever the proportions of the two elements. They are thus α phase over the entire range and suitable for both hot and cold working over the entire range. The alloys have high strength and ductility, and good corrosion resistance. The 'silver' coinage in use in Britain is a 75 per cent copper–25 per cent nickel alloy. The addition of 1 to 2 per cent iron to the alloys increases their corrosion resistance. A 66 per cent copper–30 per cent nickel–2 per cent manganese–2 per cent iron alloy is particularly resistant to corrosion, and erosion, and is used for components immersed in moving sea water.

Nickel silvers have a silvery appearance and find use for items such as knives, forks and spoons. The alloys can be cold worked and usually contain about 20 per cent nickel, 60 per cent copper and 20 per cent zinc.

For more details regarding copper and its alloys the reader is referred to E. G. West, *The Selection and Use of Copper-rich Alloys* (Oxford University Press).

Properties and specifications of copper alloys

Table 13.4 shows typical properties of copper alloys. With British Standards

Table 13.4 *Typical properties of copper alloys*

Composition (%)	*Condition*	*Tensile strength ($N\ mm^{-2}$ or MPa)*	*Elongation (%)*	*Hardness (HB)*
Brasses				
90 copper, 10 zinc	Annealed	280	48	65
80 copper, 20 zinc	Annealed	320	50	67
70 copper, 30 zinc	Annealed	330	70	65
	Hard	690	5	185
60 copper, 40 zinc	Annealed	380	40	75
Tin bronzes				
95 copper 5 tin, 0.02–0.40 phosphorus	Annealed	340	55	80
	Hard	700	6	200
91 copper, 8–9 tin, 0.02–0.40 phosphorus	Annealed	420	65	90
	Hard	850	4	250
Gunmetal				
88 copper, 10 tin, 2 zinc	Sand cast	300	20	80
Aluminium bronzes				
95 copper, 5 aluminium	Annealed	370	65	90
	Hard	650	15	190
88 copper, 9.5 aluminium, 2.5 iron	Sand cast	545	30	110
Cupronickels				
87.5 copper, 10 nickel, 1.5 iron, 1 manganese	Annealed	320	40	155
75 copper, 25 nickel, 0.5 manganese	Annealed	360	40	90
	Hard	600	5	170
Nickel silver				
64 copper, 21 zinc, 15 nickel	Annealed	400	50	100
	Hard	600	10	180
Beryllium bronzes				
98 copper, 1.7 beryllium, 0.2 to 0.6 cobalt and nickel	Solution treated, precipitation hardened	1200	3	370
Silicon bronzes				
95 copper, 2.7 to 3.5 silicon, 0.7 to 1.5 manganese	Annealed	380	60	75
	Hard	630	14	180

copper and its alloys are classified by one or two letters followed by a number. The letters used indicate the form of alloy concerned.

British Standard designation	*Alloy*
C	Copper
CZ	Brass
CN	Copper–nickel
PB	Phosphor bronze
CA	Copper–aluminium bronze
NS	Nickel–silver
CS	Copper–silicon
CB	Copper–beryllium

The classification used in the USA distinguishes between wrought and cast alloys, numbers from C100 to C799 designating wrought alloys and C800 to C999 cast alloys. The first number following the C indicates the form of alloy.

	Copper Development Association designation	*Alloy*
Wrought alloys:	C1xx	Coppers and high-copper alloys
	C2xx	Brasses
	C3xx	Leaded brasses
	C4xx	Tin brasses
	C5xx	Phosphor bronzes
	C6xx	Copper–aluminium, copper–silicon, and some copper–zinc alloys
	C7xx	Copper–nickel and copper–nickel-zinc
Cast alloys:	C8xx	Cast coppers, high coppers, brasses, manganese bronze, copper–zinc–silicon.
	C9xx	Cast copper–tin, copper–tin–lead, copper–tin–nickel, copper–aluminium–iron, copper–nickel–iron, copper–nickel–zinc

Corrosion and copper

Pure copper exposed to the atmosphere, e.g. roofing, acquires a green colouration due mainly to a reaction of the copper with sulphur in the atmosphere. The removal of copper by this means is very slow and indeed the colouration is generally one of the architectural aims. Below about 250°C oxide growth on copper is insignificant.

Copper alloys are susceptible to a number of forms of corrosion. Brasses, particularly with more than about 15 per cent zinc, are susceptible to dezincification and stress corrosion (see Chapter 9 for more details). High zinc brasses, however, are more resistant to erosion than low zinc ones, e.g. Muntz metal is used for tubing if there are high water velocities. Bronzes have better resistance to stress corrosion and erosion than brasses. Cupronickels are very resistant to such forms of corrosion.

Applications of copper alloys
There are a considerable number of copper alloys. The following indicates the type of selection that could be made.

Electrical conductors	Electrolytic tough-pitch, high-conductivity copper
Tubing and heat exchangers	Phosphorus deoxidized copper is generally used. Muntz metal, cupronickel or naval brass (62 per cent copper–37 per cent zinc–1 per cent tin) is used if the water velocities are high
Pressure vessels	Phosphorus deoxidized copper, copper-clad steel or aluminium bronze
Bearings	Phosphor bronze. Other bronzes and brasses with some lead content are used in some circumstances
Gears	Phosphor bronze. For light duty gunmetals, aluminium bronze or die cast brasses may be used
Valves	Aluminium bronze
Springs	Phosphor bronze, nickel silver, basis brass are used for low cost springs. Beryllium bronze is the best material.

13.4 Magnesium

Magnesium has a density of 1.7×10^3 kg m^{-3} and thus a very low density compared with other metals. It has an electrical conductivity of about 60 per cent of that of copper, as well as a high thermal conductivity. It has a low tensile strength, needing to be alloyed with other metals to improve its strength. Under ordinary atmospheric conditions magnesium has good corrosion resistance, which is provided by an oxide layer that develops on the surface of the magnesium in air. However, this oxide layer is not completely impervious, particularly in air that contains salts, and thus the corrosion resistance can be low under adverse conditions. Magnesium is only used generally in its alloy form, the pure metal finding little application.

Magnesium alloys

Because of the low density of magnesium, the magnesium-base alloys have low densities. Thus magnesium alloys are used in applications where lightness is the primary consideration, e.g. in aircraft and spacecraft. Aluminium alloys have higher densities than magnesium alloys but can have greater strength. The strength-to-weight ratio for magnesium alloys is, however, greater than that of aluminium alloys. Magnesium alloys also have the advantage of good machinability and weld readily.

Magnesium–aluminium–zinc alloys and magnesium–zinc–zirconium are the main two groups of alloys in general use. Small amounts of other elements are also present in these alloys. The composition of an alloy depends on whether it is to be used for casting or working, i.e. as a wrought alloy. The cast alloys can often be heat treated to improve their properties.

A general-purpose wrought alloy has about 93 per cent magnesium–6 per cent aluminium–1 per cent zinc–0.3 per cent manganese. This alloy can be forged, extruded and welded, and has excellent machinability. A high-strength wrought alloy has 96.4 per cent magnesium–3 per cent zinc–0.6 per cent zirconium. A general-purpose casting alloy has about 91 per cent magnesium–8 per cent aluminium–0.5 per cent zinc–0.3 per cent manganese. A high-strength casting alloy has 94.8 per cent magnesium–4.5 per cent zinc–0.7 per cent zirconium. Both these casting alloys can be heat treated.

Table 13.5 shows typical properties of magnesium alloys.

Table 13.5 *Properties of typical wrought and cast magnesium alloys*

Composition (%)	*Condition*	*Tensile strength ($MN\ m^{-2}$ or MPa)*	*Elongation (%)*	*Hardness (HB)*
Wrought alloys				
93 magnesium, 6 aluminium, 1 zinc, 0.3 manganese	Forged	290	8	65
96.4 magnesium, 3 zinc, 0.6 zirconium	Extruded	310	8	70
Cast alloys				
91 magnesium, 8 aluminium, 0.5 zinc, 0.6 zirconium	As cast	140	2	55
	Heat treated	200	6	75
94.8 magnesium, 4.5 zinc, 0.7 zirconium	Heat treated	230	5	70

13.5 Nickel

Nickel has a density of 8.88×10^3 kg m^{-3} and a melting point of 1455°C. It possesses excellent corrosion resistance, hence it is used often as a cladding on a steel base. This combination allows the corrosion resistance of the nickel to be realized without the high cost involved in using entirely nickel. Nickel has good tensile strength and maintains it at quite elevated temperatures. Nickel can be both cold and hot worked, has good machining properties and can be joined by welding, brazing and soldering.

Nickel is used in the food processing industry in chemical plant, and in the petroleum industry, because of its corrosion resistance and strength. It is also used in the production of chromium-plated mild steel, the nickel forming an intermediate layer between the steel and the chromium. The nickel is electroplated on to the steel.

Nickel alloys

Nickel is used as the base metal for a number of alloys with excellent corrosion resistance and strength at high temperatures. One group of alloys is based on nickel combined with copper. The alloys containing about two-thirds nickel and one-third copper are called *Monels*. Monel 400 has 66.5 per cent nickel and 31.5 per cent copper. It has high strength, toughness and weldability. It is highly resistant to sea water, alkalis, many acids and superheated steam, hence its use for marine fixtures and fasteners, food processing equipment and chemical engineering plant components. Monel K-500 has a basic Monel composition but includes 3.0 per cent aluminium and 0.6 per cent titanium. These enable age hardening precipitates to be formed and so give higher strength, while still maintaining the excellent corrosion properties. All the Monels retain their strength and toughness to temperatures of 500°C. Figure 13.16 shows how the properties vary with temperature for Monel K-500.

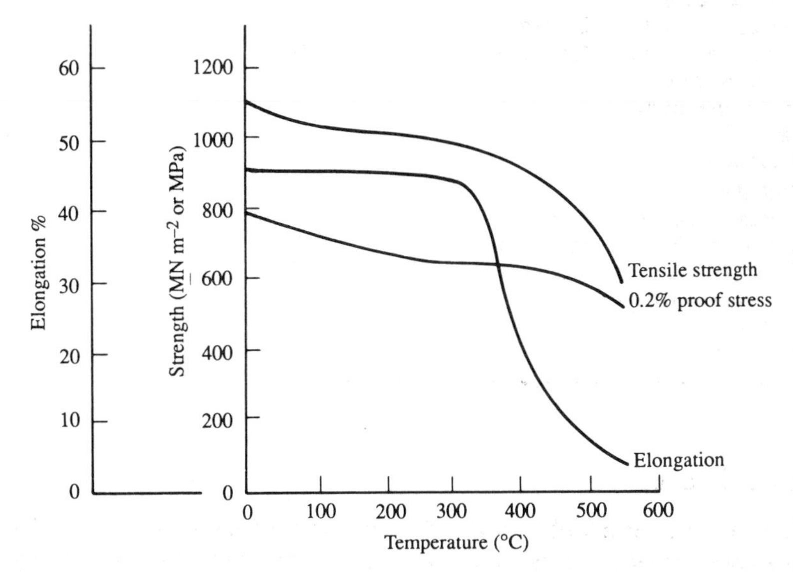

Figure 13.16 High temperature properties of Monel K-500 after age hardening

Another common group of nickel alloys are based on nickel combined with chromium. *Inconel 600* contains 76.5 per cent nickel, 15.5 per cent chromium and 8 per cent iron. The alloy has a high strength and excellent resistance to corrosion at both normal and high temperatures. The alloy is not heat treatable but can be work hardened. Figure 13.17 shows how the properties vary with temperature. *Inconel 601* contains 61.6 per cent nickel, 23 per cent chromium, 14 per cent iron and 1.4 per cent aluminium. The high chromium content of this alloy gives it good corrosion resistance in many environments and this, together with the aluminium, also gives it high temperature oxidation resistance. At high temperatures the aluminium, chromium and nickel oxides form an extremely protective and adherent film on the metal surface.

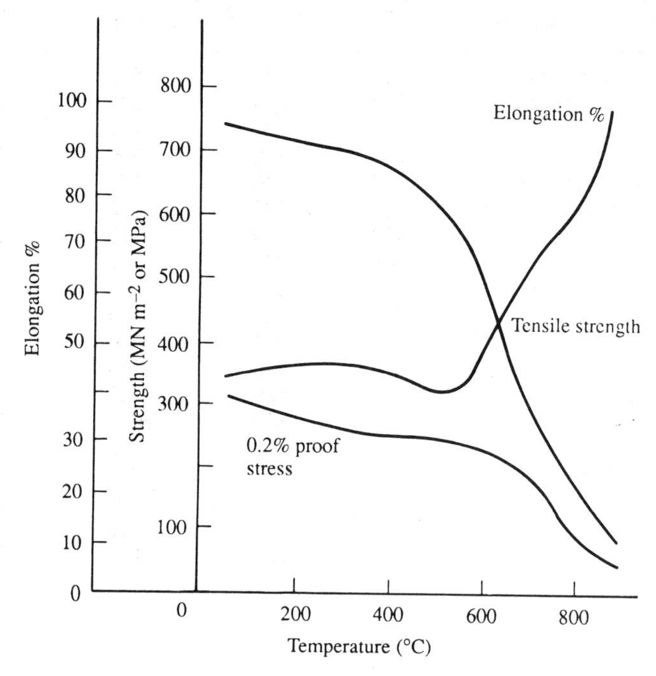

Figure 13.17 High temperature properties of Inconel 600

The *Nimonic* series of alloys are based on nickel alloys with about 20 per cent chromium. They have high strengths and good creep resistance at high temperatures and are used in gas turbines for discs and blades. Nimonic 80 is essentially a 76.5 per cent nickel–20 per cent chromium solid solution with 2.25 per cent titanium and 1 per cent aluminium forming precipitates. The result is an alloy that can be precipitation hardened. Nimonic 80 was developed in 1941 and has since been followed by many other Nimonic and related alloys with yet better properties at high temperatures.

Properties of nickel alloys

Table 13.6 shows typical properties of a range of nickel alloys. Nickel alloys have good high and low temperature strength, coupled with good corrosion resistance in many environments. Unfortunately, the alloys are expensive and thus their use tends to be restricted to applications where other, cheaper, materials cannot be used because of inadequate properties.

Table 13.6 *Typical properties of nickel alloys*

Alloy and % composition	*Condition*	*Temperature (°C)*	*0.2% proof stress (MN m^{-2})*	*Tensile strength (MN m^{-2})*	*Elongation (%)*
Monel 400	Hot rolled	20	230	560	45
Cu 30, Fe 1.5, Mn 1.0		200	200	540	50
		400	220	460	52
		600	120	260	30
Monel K-500	Heat treated	20	340	680	45
Cu 29, Al 2.8, Ti 0.5		200	290	650	40
		400	260	600	30
		600	290	460	5
Inconel 600	Hot rolled	20	250	590	50
Cr 16, Fe 6		400	185	560	50
		600	150	530	10
Nimonic 80A	Heat treated	20	740	1240	24
Cr 20, T1 2.0, Al 1.5		400	680	1150	26
		600	620	1080	20
		800	490	620	24
Nimonic 115	Heat treated	20	860	1230	27
Cr 14, Co 13, Mo 3		600	790	1100	20
T1 2, Al 0.5		800	760	1020	19
		1000	200	420	26

13.6 Titanium

Titanium has a relatively low density, 4.5×10^3 kg m^{-3}, just over half that of steel. It has a relatively low strength when pure but alloying gives a considerable increase in strength. Because of the low density of titanium its alloys have a high strength-to-weight ratio. Also, it has excellent corrosion resistance. However, titanium is an expensive metal, its high cost reflecting the difficulties experienced in the extraction and formation of the material; the ores are quite plentiful.

Titanium can exist in two crystal forms, α which is a hexagonal close-packed structure and β which is a body-centred cubic. In pure titanium the α structure is the stable phase up to 883°C and is transformed into the β form above this temperature.

Commercially, pure titanium ranges in purity from 99.0 to 99.5 per cent, the main impurities being iron, carbon, oxygen, nitrogen and hydrogen. Such material is lower in strength than titanium alloys but more corrosion resistant and less expensive. The properties of the commercially pure titanium are largely determined by the oxygen content. Table 13.7 shows the composition and properties of such materials. Because of excellent corrosion resistance, commercially pure titanium is used for chemical plant components, surgical implants, marine and aircraft engine parts, etc.

Table 13.7 *Composition and properties of commercially pure annealed titanium*

Composition (%)	*Temperature* (°C)	*Tensile strength* ($MN\ m^{-2}$)	*Yield stress* ($MN\ m^{-2}$)	*Elongation* (%)
99.5 Ti, 0.18 O, 0.08 C,	20	330	240	30
0.20 Fe, 0.02 N, 0.015 H	300	150	95	32
99.2 Ti, 0.20 O, 0.08 C,	20	440	350	28
0.25 Fe, 0.03 N, 0.015 H	300	200	120	35
99.1 Ti, 0.30 O, 0.08 C,	20	520	450	25
0.25 Fe, 0.05 N, 0.015 H	300	240	140	34
99.0 Ti, 0.40 O, 0.08 C,	20	670	590	20
0.50 Fe, 0.05 N, 0.015 H	300	310	170	37

Titanium alloys

Titanium alloys can be grouped into a number of categories according to the phases present in their structure. The addition of elements such as aluminium, tin, oxygen or nitrogen results in the enlargement of the α phase region on the thermal equilibrium diagram, such elements being referred to as α-stabilizing elements. Other elements such as vanadium, molybdenum, silicon and copper enlarge the β phase region and are known as β-stabilizing elements. There are other elements that are sometimes added to titanium and which are not either α or β stabilizers. Zirconium is used to contribute solid solution strengthening.

The alloys are grouped into four categories, each category having distinctive properties.

1 α-titanium alloys

These are composed entirely of α phase. An example of such an alloy is 92.5 per cent titanium, 5 per cent aluminium and 2.5 per cent tin. Both aluminium and tin are α stabilizers. Such alloys have the hexagonal close-packed structure of titanium. The all α-phase alloys are strong and maintain their strength at high temperatures but are difficult to work. The alloys have good weldability and are used where high temperature strength is required, e.g. steam turbine blades.

2 Near α titanium alloys

These are composed of almost all α phase with a small amount of β phase dispersed throughout the α. This is achieved by adding small amounts, about 1 to 2 per cent, of β-stabilizing elements such as molybdenum and vanadium to what is otherwise an α-stabilized alloy. An example of such an alloy is 90 per cent titanium, 8 per cent aluminium, 1 per cent molybdenum and 1 per cent vanadium. This alloy is normally used in the annealed condition. Two forms of annealing are used – mill annealing and duplex annealing. Mill annealing involves heating the alloy to 790°C for eight hours and then furnace cooling. Duplex annealing involves mill annealing followed by reheating to 790°C for quarter of an hour and then air cooling. The result of such annealing is β particles dispersed throughout an α matrix. Solution heat treatment and ageing can be used to increase the strength but unfortunately it makes the alloy susceptible to stress corrosion cracking in salt water environments. The alloy, in the annealed condition, is used for airframe and jet engine parts which require high strengths, good creep resistance and toughness up to temperatures of about 850°C. The alloy has good weldability.

3 α β titanium alloys

These contain sufficient quantities of β-stabilizing elements for there to be appreciable amounts of β phase at room temperature. Such an alloy is 90 per cent titanium, 6 per cent aluminium and 4 per cent vanadium. The aluminium stabilizes the α phase while the vanadium stabilizes the β phase. The α-β alloys can be solution treated, quenched and aged for increased strength. The microstructure of the alloys depends on their composition and heat treatment. For example, a fast cooling rate, such as quenching in cold water, from a temperature where the material was all β produces a martensitic structure with

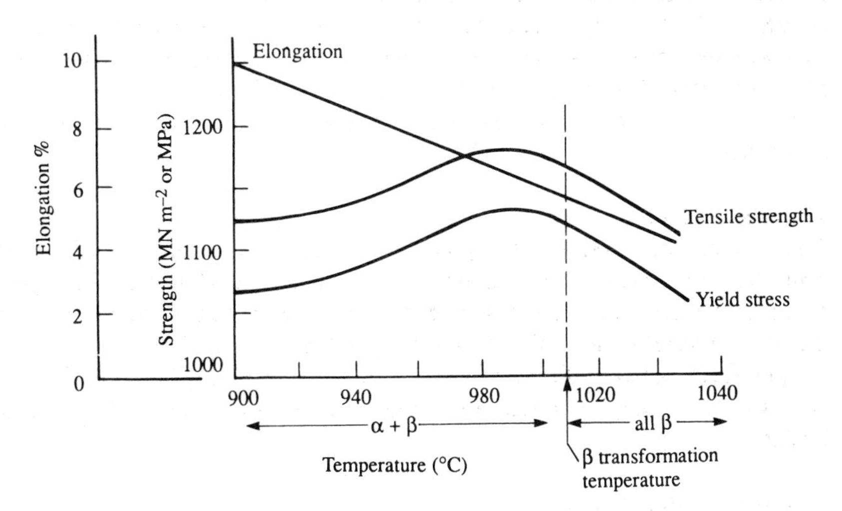

Figure 13.18 Effect of the solution heat treatment temperature on the properties of 90 per cent Ti, 6 per cent Al, 4 per cent V alloy after ageing for 8 h at 593°C

some increase in hardness. Ageing can then produce some further increase in strength as a result of β precipitates occurring. Figure 13.18 shows the effect on the resulting properties of quenching from different temperatures. The maximum strength is obtained by solution treatment to a temperature just below the β transformation temperature.

The 90 per cent titanium, 6 per cent aluminium, 4 per cent vanadium alloy can be readily welded, forged and machined. At 482°C the alloy shows a metallurgical change in structure and there is a change in its mechanical properties, a decrease in strength and an increase in ductility. The alloy is used for both high and low temperatue applications, e.g. rocket motor cases, turbine blades and cryogenic vessels.

4 β titanium alloys

When sufficiently high amounts of β-stabilizing elements are added to titanium the resulting structure can be made entirely β at room temperature after quenching, or in some cases air cooling. Unlike α-titanium alloys, β-titanium alloys are readily cold worked in the solution treated and quenched condition and can subsequently be aged to give very high strengths. In the high-strength condition the alloys have low ductilities. They can also suffer

Table 13.8 *Composition and properties of titanium alloys*

Composition (%)	*Condition*	*Temperature (°C)*	*Tensile strength ($MN\ m^{-2}$)*	*Yield stress ($MN\ m^{-2}$)*	*Elongation (%)*
α alloy					
92.5 Ti, 5 Al, 2.5 Sn.	Annealed	20	1050	820	16
		300	570	450	18
Near α alloy					
90 Ti, 8 Al, 1 V.	Duplex Annealed	20	1015	970	15
		300	800	630	20
		400	750	570	20
		500	630	520	25
α–β alloy					
90 Ti, 6 Al, 4 V	Annealed	20	1000	940	14
		300	730	660	14
		400	680	580	18
		500	540	430	35
	Solution treated and aged	20	1190	1100	10
		300	870	710	10
		400	810	630	12
		500	660	490	22
β alloy					
77 Ti, 13 V, 11 Cr, 3 Al.	Solution treated and aged	20	1240	1190	8
		300	890	840	19

from poor fatigue performance. The alloys are thus not as widely used as the α-β alloys.

A typical β titanium alloy has 77 per cent titanium, 13 per cent vanadium, 11 per cent chromium and 3 per cent aluminium. The alloy is usually used in the solution treated, quenched and aged condition in order to obtain the very high tensile strength. The alloy is used for aerospace components, honeycomb panels and high strength fasteners.

Properties of titanium alloys
Table 13.8 shows the composition and typical properties of a range of commonly used titanium alloys. In considering such alloys account needs to be taken of the possibility of stress corrosion cracking in salt water environments for near α alloys and the poor fatigue performance of β alloys.

13.7 Zinc

Zinc has a density of 7.1×10^3 kg m^{-3}. Pure zinc has a melting point of only 419°C and is a relatively weak metal. It has good corrosion resistance, due to the formation of an impervious oxide layer on the surface. Zinc is frequently used as a coating on steel in order to protect that material against corrosion, the product being known as galvanized steel.

Zinc alloys

The main use of zinc alloys is for die casting. They are excellent for this purpose by virtue of their low melting points and the lack of corrosion of dies used with them. The two alloys in common use for this purpose are known as alloy A and alloy B. *Alloy A*, the more widely used of the two, has the composition of 3.9 to4.3 per cent (max) aluminium, 0.03 per cent (max) copper, 0.03 to 0.06 per cent (max) magnesium, the remainder being zinc. *Alloy B* has the composition 3.9 to 4.3 per cent (max) aluminium, 0.75 to 1.25 per cent (max) copper, 0.03 to 0.06 per cent (max) magnesium, with the remainder being zinc. Alloy A is the more ductile, alloy B has the greater strength.

The zinc used in the alloys has to be extremely pure so that little, if any, other impurities are introduced into the alloys, typically the required purity is 99.99 per cent. The reason for this requirement is that the presence of very small amounts of cadmium, lead or tin renders the alloy susceptible to intercrystalline corrosion. The products of this corrosion cause a casting to swell and may lead to failure in service.

After casting, the alloys undergo a shrinkage which takes about a month to complete; after that there is a slight expansion. A casting can be *stabilized* by annealing at 100°C for about six hours.

Zinc alloys can be machined and, to a limited extent, worked. Soldering and welding are not generally feasible.

Zinc alloy die castings are widely used in domestic appliances, for toys, car parts such as door handles and fuel pump bodies, optical instrument cases.

The properties that might be obtained with zinc die-casting alloys are shown in Table 13.9.

Table 13.9 *Properties of zinc casting alloys*

Composition (%)	*Condition*	*Tensile strength* ($MN\ m^{-2}$ *or MPa*)	*Elongation* (%)	*Hardness* (*HB*)
Alloy A	As cast	285	10	83
Alloy B	As cast	330	7	92

13.8 Wear resistance treatments

The following are wear resistance treatments applied to the surfaces of non-ferrous metals (see Section 12.7 for the comparable treatments for ferrous alloys).

Delsun

This process is used with brasses and bronzes. The treatment involves electrolytic deposition of an alloy containing mainly tin, antimony and cadmium and is followed by a diffusion treatment in a molten salt bath at about 400°C during which the deposited material fully integrates with the brass or bronze. The process increases surface hardness and so improves abrasive resistance. It also gives good corrosion resistance in saline environments.

Trical

This process is used with aluminium bronze. The treatment involves electrolytic deposition of a metallic alloy on the surface and is followed by a diffusion treatment in a neutral atmosphere or a salt bath during which the deposited material fully integrates with the aluminium bronze. The result is a hardened zone overlaid by a thin soft anti-adhesion layer which gives good resistance to wear.

Zinal

This process is used with aluminium alloys. The treatment involves electrolytic deposition of a metallic alloy followed by a diffusion treatment during which the deposited material fully integrates with the aluminium alloy. The surface hardness is improved and increases the ability of the surface to retain lubricants. Wear resistance and protection against seizure are thus improved.

Tidurdan

This process is applied to titanium. It involves electrolytic deposition followed by a diffusion treatment during which the deposited material fully integrates with the titanium. The result is a reduction in the coefficient of friction and a reduction in scuffing.

Conversion coatings

Changes can be produced in the surface layers of some metals, e.g. aluminium, magnesium and copper, by chemical or electro-chemical reactions. Such treatments can improve wear and corrosion resistance (see section 9.3).

Problems

1 What is the effect on the strength and ductility of aluminium of (a) the purity of the aluminium, (b) the temper?
2 Describe the effect on the strength of aluminium of the percentage of magnesium alloyed with it.
3 Describe the solution treatment and precipitation hardening processes for aluminium–copper alloys.
4 Describe the features of aluminium–silicon alloys which make them suitable for use with die casting.
5 What is the effect of heat treatment on the properties of an aluminium alloy, such as a 4.3 per cent copper–0.6 per cent magnesium–0.8 per cent silicon–0.75 per cent manganese?
6 Ladders are often made from an aluminium alloy. What are the properties required of that material in this particular use?
7 Explain with the aid of a thermal equilibrium diagram how the addition of a small amount of sodium to the melt of an aluminium–silicon alloy changes the properties of alloys with, say, 12 per cent silicon.
8 What are the differences between the following forms of copper: electrolytic tough-pitch, high-conductivity copper; fire-refined, tough-pitch, high-conductivity copper; oxygen-free, high-conductivity copper; phosphorus-deoxidized copper; arsenical copper.
9 Which form of copper should be used in an atmosphere containing hydrogen?
10 What is the effect of cold work on the properties of copper?
11 What is the effect of the percentage of zinc in a copper–zinc alloy on its strength and ductility?
12 What brass composition would be most suitable for applications requiring

(a) maximum tensile strength, (b) maximum ductility, (c) the best combined tensile strength and ductility?

13 In general, what are the differences in (a) composition, (b) properties of α phase and duplex alloys of copper?

14 The 'silver' coinage used in Britain is made from a cupronickel alloy. What are the properties required of this material for such a use?

15 What are the constituent elements in (a) brasses, (b) phosphor bronzes and (c) cupronickels?

16 The name Muntz metal is given to a 60 per cent copper–40 per cent zinc alloy. Some forms of this alloy also include a small percentage of lead. What is the reason for the lead?

17 What percentage of aluminium would be likely to be present in an aluminium bronze that is to be cold worked?

18 What is meant by the term strength-to-weight ratio? Magnesium alloys have a high strength-to-weight ratio; of what significance is this in the uses to which the alloys of magnesium are put?

19 How does the corrosion resistance of magnesium alloys compare with other non-ferrous alloys?

20 What are the general characteristics of the nickel–copper alloys known as Monel?

21 Though titanium alloys are expensive compared with other non-ferrous alloys, they are used in modern aircraft such as *Concorde*. What advantages do such alloys possess which outweighs their cost?

22 What problems can arise when impurities are present in zinc die casting alloys?

23 What is the purpose of 'stabilizing annealing' for zinc die-casting alloys?

24 Describe the useful features of zinc die-casting alloys which makes them so widely used.

25 This question is based on the aluminium–silicon equilibrium diagram given in Figure 13.5.
   (a) At what temperature will the solidification of a 85 per cent aluminium–15 per cent silicon start? At what temperature will it be completely solid?
   (b) The above alloy is modified by the addition of a small amount of sodium. At what temperature will solidification start? At what temperature will it be completely solid?

26 This question is based on the copper–zinc thermal equilibrium diagram given in Figure 13.10.
   (a) What is the temperature at which a 85 per cent copper–15 per cent zinc alloy begins to solidify?
   (b) The pouring temperature used for casting is about 100°C above the liquidus temperature. What is the pouring temperature for the above alloy?

27 This question is based on the copper–zinc thermal equilibrium diagram given in Figure 13.10.

(a) Which of the following brasses would you expect to be just α solid solution? (i) 10 per cent zinc–90 per cent copper, (ii) 20 per cent zinc–80 per cent copper, (iii) 40 per cent zinc–60 per cent copper.

(b) How would you expect the properties of the above brasses to differ? How are the differences related to the phases present in the alloys?

28 This question is based on the copper–nickel thermal equilibrium diagram given in Figure 3.5.

(a) How does the copper–nickel thermal equilibrium diagram differ from that of copper–zinc?

(b) Over what range of compositions will copper–nickel alloys be α solid solution?

29 The following group of questions is concerned with justifying the choice of a particular alloy for a specific application.

(a) Why are zinc alloys used for die casting?
(b) Why are magnesium alloys used in aircraft?
(c) Why are milk bottle caps made of an aluminium alloy?
(d) Why are titanium alloys extensively used in high-speed aircraft?
(e) Why are domestic water pipes made of copper?
(f) Why is cartridge brass used for the production of cartridge cases?
(g) Why are the ribs of hand gliders made of an aluminium alloy?
(h) Why are nickel alloys used for gas-turbine blades?
(i) Why is brass used for cylinder lock keys?
(j) Why are kitchen pans made of aluminium or copper alloys?
(k) Why is copper used for gaskets?
(l) Why are electrical cables made of copper?
(m) Why are cupronickels used for tubes in desalination plants?

30 Compare the properties of ferrous and non-ferrous alloys, commenting on the relative ease of processing.

PART FOUR

Non-metallic materials

14

Forming processes with non-metallic materials

14.1 The main polymer-forming processes

Many polymer-forming processes are essentially two stage, the first stage being the production of the polymer in a powder, granule or sheet form and the second stage being the shaping of this material into the required shape. The first stage can involve the mixing with the polymer of suitable additives and even other polymers in order that the finished material should have the required properties. The additives may be in the form of solids, liquids or gases. Thus solid additives such as cork dust, paper pulp, chalk or carbon black are added to thermosetting polymers to reduce the brittleness of the material. Liquid additives may be used to improve the flow characteristics of the material during processing. Gas additives enable foamed or expanded materials to be produced.

Second-stage processes generally involve heating the powder, granule or sheet material until it softens, shaping the softened material to the required shape and then cooling it. The main types of process are:

1 Extrusion
2 Moulding
3 Casting
4 Calendering
5 Foaming
6 Machining

The choice of process will depend on a number of factors, such as:

1 The quantity of items required.
2 The size of the items.
3 The rate at which the items are to be produced.
4 The requirements for holes, inserts, enclosed volumes, threads.
5 The type of material being used.

Items required in a continuous length are generally extruded while items required in large quantities, particularly small items, are generally moulded, the injection moulding process being used. Casting and forming are relatively slow processes, unlike injection moulding which is much faster. Calendering is used for the production of sheet plastic. Forming involves the shaping of sheet plastic.

Thermoplastic materials can be softened and resoftened indefinitely by the application of heat, provided the temperature is not so high as to cause decomposition. Because they flow readily with the application of heat they are particularly suitable for processing by extrusion and injection moulding; they can also be readily formed. Polyethylene, polyvinyl chloride, polystyrene, polyamide (nylon), polycarbonate, cellulose acetate and polytetrafluoroethylene are examples of thermoplastics.

Thermosetting materials undergo a chemical change when they are subject to heat which cannot be changed by further heating. Moulding and casting are processes often used with such materials. Typical thermosetting resins are phenol formaldehyde, urea formaldehyde, melamine formaldehyde, unsaturated polyesters and epoxides.

Extrusion involves the forcing of the molten polymer through a die. The process is comparable with the squeezing of toothpaste out of its tube. Figure 14.1 shows the basic form of the extrusion process. The polymer is fed into a screw mechanism which takes the polymer through a heated zone and forces it out through the die. In the case of an extruded product such as curtain rail the product is obtained by just cooling the extruded material. If thin film or sheet is the required product, a die may be used which gives an extruded cylinder of material. This cylinder while still hot is inflated by compressed air to give a sleeve of thin film. Another way of obtaining film or sheet is to use a slit die and cool the extruded product by allowing it to fall vertically into some cooling system.

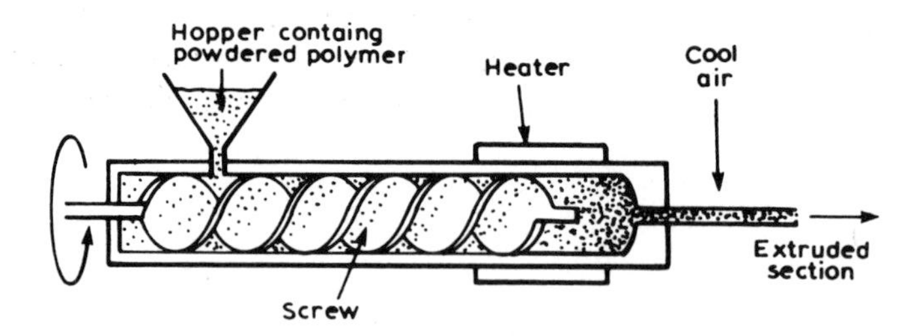

Figure 14.1 Extrusion

The extrusion process can be used with most thermoplastics and yields continuous lengths of product. Intricate shapes can be produced and a high output rate is possible. Curtain rails, household guttering, polythene bags and film are examples of typical products.

Extrusion blow moulding is a process used widely for the production of hollow articles such as plastic bottles. Containers from as small as 10^{-6} m^3 to 2 m^3 can be produced. The process involves the extrusion of a hollow thick-walled tube which is then clamped in a mould. Pressure is applied to the inside of the tube, which inflates to fill the mould.

A widely-used process for thermoplastics is *injection moulding*. With this process the polymer is melted and then forced into a mould (Figure 14.2). High production rates can be achieved and complex shapes with inserts, threads, holes, etc. can be produced. The process is particularly useful for small components. Typical products are beer or milk bottle crates, toys, control knobs for electronic equipment, tool handles, pipe fittings.

Foam plastic components can be produced by this method. Inert gases are dissolved in the molten polymer. When the hot polymer cools the gases come out of solution and expand to form a cellular structure. A solid skin is produced where the molten plastic comes in contact initially with the cold mould surface.

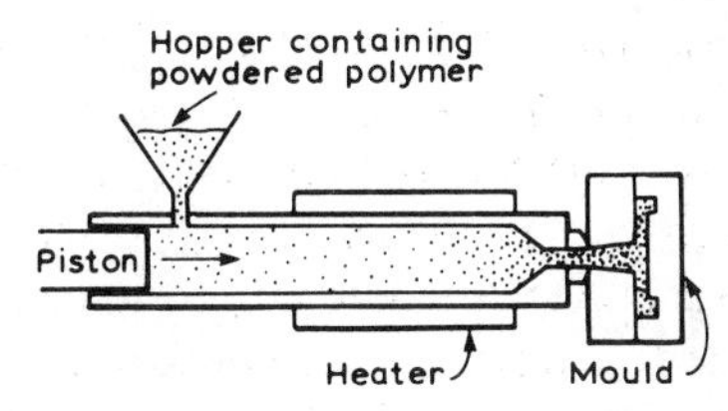

Figure 14.2 Injection moulding by a ram-fed machine. Screw-fed machines as in Figure 14.1 are also used

Widely-used processes for thermosetting plastics are *compression moulding* and *transfer moulding*. In compression moulding, the powdered polymer is compressed between the two parts of the mould and heated under this pressure (Figure 14.3). With transfer moulding, the powdered polymer is heated in a chamber before being transferred by a plunger into the mould (Figure 14.4).

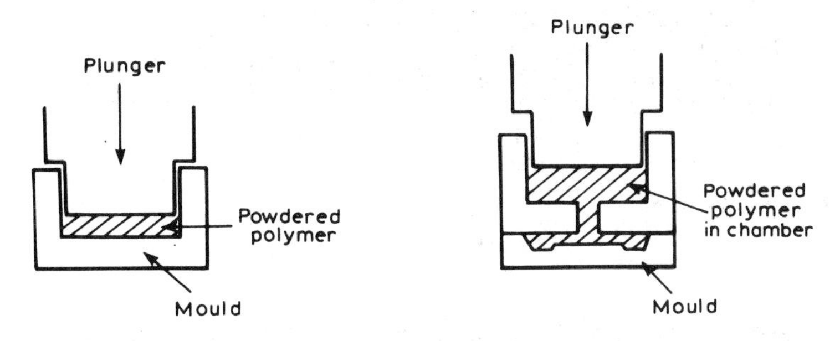

Figure 14.3 Compression moulding

Figure 14.4 Transfer moulding

One form of *casting* involves mixing substances of relatively short molecular chains, with any required additives, in a mould so that polymerization, i.e. the production of long-chain molecules, occurs during solidification. The term *cold-setting* is used generally for such polymers. Such methods are used for encapsulating small electrical components.

Powder casting involves the melting of powdered polymer inside a heated mould. The mould is often rotated during this operation and the term *rotational moulding* used to describe the process. It is a very useful process for the production of hollow articles. This process does however have a slow rate of production. Powder casting can be used to coat surfaces with films of polymers, e.g. non-stick surfaces of cooking pans.

Calendering is a process used for the production of continuous lengths of sheet thermoplastic, such as p.v.c. or polythene. The calender consists of essentially three or more heated rollers (Figure 14.5). The heated polymer is fed into the gap between the first pair of rollers, emerging as a sheet. The group of rollers determines the rate at which the sheet is produced, the thickness of the sheet and the surface finish.

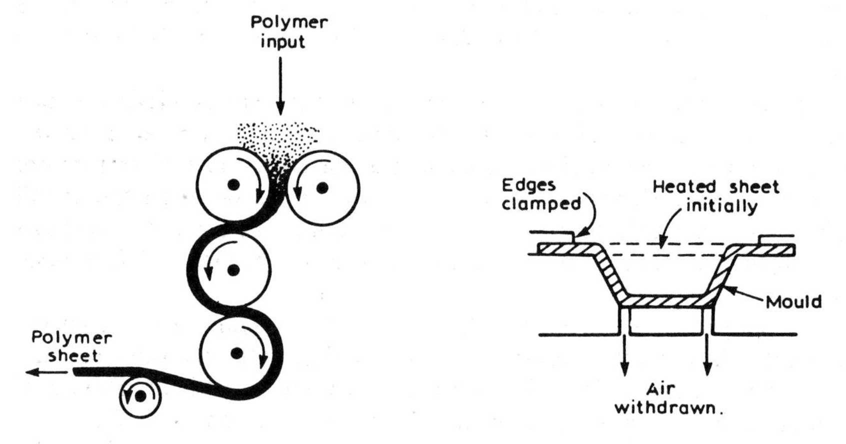

Figure 14.5 Calendering

Figure 14.6 Vacuum forming

Forming processes are used to mould articles from sheet polymer. The heated sheet is pressed into or around a mould. The term *thermoforming* is often used to describe this type of process. The sheet may be pressed against the mould by the application of air under pressure to one side of the sheet, *pressure forming*, or by the production of a drop in pressure, a vacuum, between the sheet and the mould, *vacuum forming* (Figure 14.6).

Thermoforming can have a high output rate, but dimensional accuracy is not too good and holes, threads, etc. cannot be produced. The method can be used for the production of large shaped objects but not very small items. Enclosed hollow shapes cannot be produced.

While polymers can be machined by most of the methods commonly used with metals, the process used to shape the polymer often produces the finished article with no further need of machining or any other process. Polymers tend to have low melting points and thus correct machining conditions, which do not result in high temperatues being produced, are vital if the material is not to soften and deform. Some polymer materials are brittle and so present problems in machining, shock loadings having to be avoided if cracking is not to occur.

Viscosity of polymers

Polymer-forming processes such as blow moulding involve the flow of the polymer, in this case through a die and during the extension of the material to form the hollow object. Polymer viscosity is thus an important property, determining, among other things, the speed with which the operation can be completed and hence the output possible from a machine. A high production speed would be facilitated by a low viscosity, however a high viscosity is required if the hot material extruded through the die is not to stretch too much under its own weight while cooling. A low viscosity aids the blowing of the material to form the hollow shape. A balance has thus to occur between these various demands.

A more convenient measure of viscosity for use with polymers is the *melt flow index*. This is the mass of polymer, in grammes, which is extruded in a given time through a standard-size nozzle under defined conditions of temperature and pressure. The higher the viscosity of a material the longer the time it will take to flow through the nozzle and so the smaller the amount of material flowing through in the time, hence a high viscosity means a low melt flow index value.

The viscosity, and hence melt flow index, of a polymer depends on the molecular chain length. The longer the chain the higher the viscosity and so the lower the melt flow index. The longer the chain the more chance there is of molecules becoming tangled, hence the increase in viscosity. Since the molecular weight of a polymer is related to its chain length, the higher the molecular weight the higher the viscosity and the lower the melt flow index.

Because the melt flow index is defined with different conditions for different polymers there is no simple way of comparing melt flow index values for different polymeric materials. However, as a rough guide, for blow moulding the value of the melt flow index should not be more than about seven.

Processing and properties

Polymers have high thermal expansivities which result in an amorphous polymer contracting in volume by about 6 per cent and a crystalline polymer by more than twice this when cooling from the processing temperature to room temperature. A volume change of 6 per cent means a change in linear dimensions of 2 per cent if the material is free to contract in all dimensions. Such changes have to be allowed for when the dimensions of moulds and dies are considered. However, the problem is rarely as simple as this, mainly because the ways by which a product can shrink are often restricted by the mould or die.

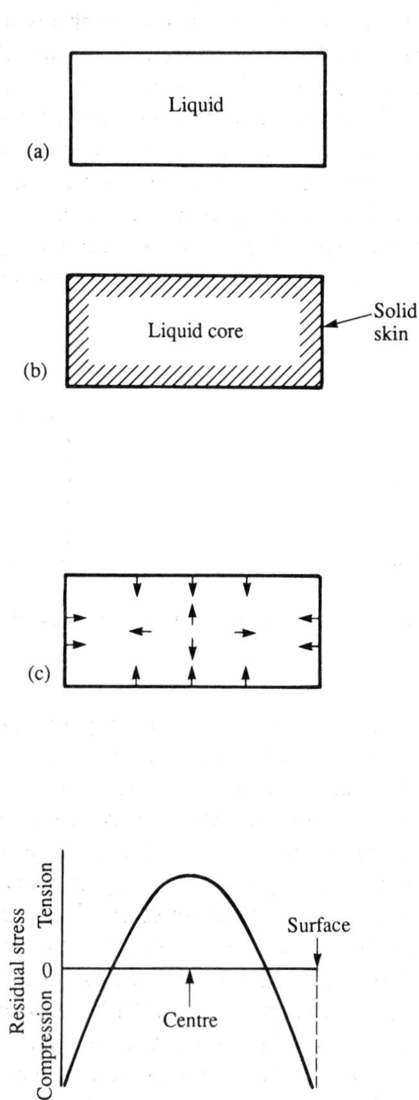

Figure 14.7 (a) Initial liquid polymer in mould, (b) Skin forms as outer layers cool faster than core, (c) Stresses develop, (d) The residual stresses produced

Problems can occur as a result of different parts of a polymer product cooling at different rates during processing. Thus, for example, if a rectangular block of polymer is cooling, the outer layers of the material will cool more rapidly than the inner core. This leads to a skin developing and as the skin is stiffer than the core it restrains the contraction of the core. This leads to the skin being compressed by the endeavours of the core to shrink and so compressive stresses develop in the skin. The core, since it is restrained by the skin from contracting, is in a state of tension. The net result is thus a system of *residual stresses* with the surface in compression and the core in tension (Figure 14.7).

These stresses can have the consequence of *sink marks* developing on the surfaces (Figure 14.8). The stresses might even be high enough to cause *voids* to develop in the core (Figure 14.9). Another consequence of the residual stresses is that if the product is machined and surface layers removed, the equilibrium between the surface and core stresses is disturbed and the product will distort.

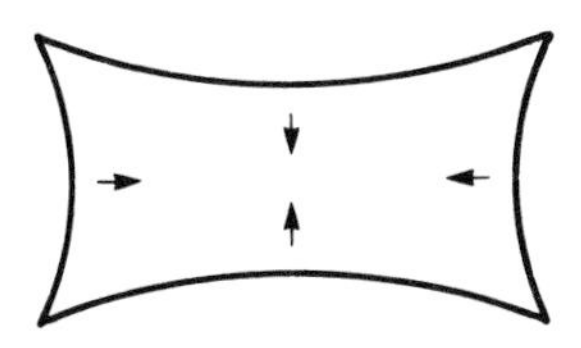

Figure 14.8 Sink marks as a result of internal stresses

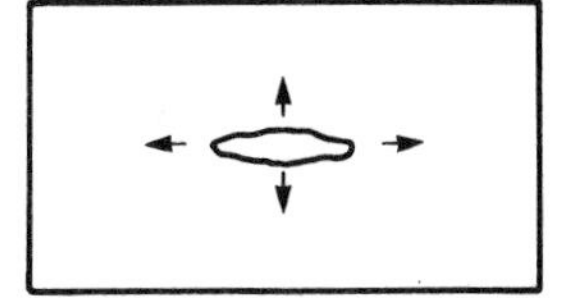

Figure 14.9 Void produced as a result of internal stresses

Another effect that can occur with polymers during their cooling from the liquid state is *orientation*. Polymer molecules can adopt an alignment in a particular direction if they are subject to a unidirectional stress during cooling from the liquid state. Thus in the extrusion process, when the extruded material, e.g. a rod, tube or sheet, is pulled away from the die by a slight tension applied by perhaps rollers, an orientated product is produced having a high tensile modulus and strength in the direction of the pull but a much lower modulus and strength in the transverse direction. Blown film is also likely to be strongly orientated. You might like to try stretching experiments on strips cut in different directions from a polythene bag. With injection moulding, as the melt flows into the cold-mould cavity, unidirectional shear forces develop and result in orientation, the maximum shear forces and hence orientation being at the surfaces of the moulding where the polymer melt is in contact with the cold surface.

If there are glass fibres present in the polymer melt, during the processing these, like the polymer molecules, can become aligned and give a directionality of properties.

A line of weakness, known as a *weld line*, can form in processing when two flowing streams of polymer melt meet and join together. This can occur if, within a mould, the polymer melt flows round an obstacle to join up again on the other side, e.g. round a pillar which is there to produce a hole in the resulting moulding. Another possibility is if in injection moulding the polymer melt enters through two or more gates. In extrusion blow moulding of containers an extruded, but still hot, thick-walled tube has its ends pressed together to form the base of the container. The weld line is a weakness because within a single melt flow the polymer molecules are orientated along the direction of flow at the flow boundary, so that when two flows meet and there is some outward flow of melt at right angles to the main flow direction an orientation is produced which is at right angles to the original direction (Figure 14.10). This gives a line of weakness. Also there can be inadequate fusion at the junction due to insufficient molecules diffusing across the junction and so bridging it.

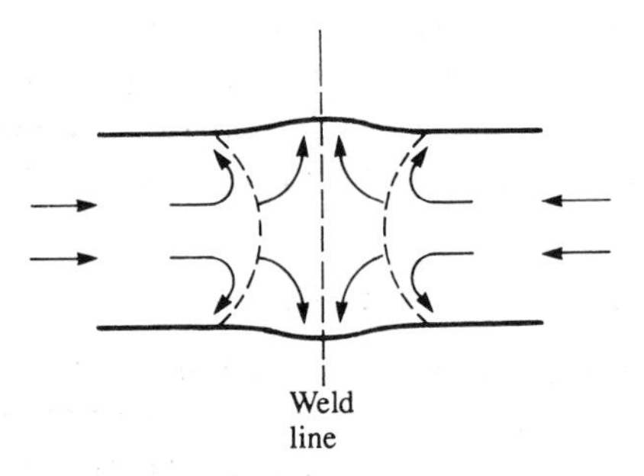

Figure 14.10 Production of a weld line

14.2 Foamed polymers

Foamed forms of many plastics and elastomers are produced and find wide application, e.g. all the cushions with foamed fillings. Examples of materials used in this way are polyurethanes, polyvinyl chloride, polystyrene, the phenolics, urea formaldehyde, natural rubber and SBR rubber. Three main types of processes are used to produce foamed polymers:

1. Bubbles are produced throughout a polymer emulsion or partially polymerized liquid by mechanical agitation.
2. Gas bubbles are produced within a liquid form of the polymer by heating or by a reduction in pressure, the process being said to involve a physical blowing agent.
3. Gas bubbles are generated within the polymer liquid by a chemical reaction, the process being said to involve a chemical blowing agent.

The foams so produced can have either a closed or open-cell structure. With a closed-cell structure each bubble is completely enclosed by polymer material and there is no access from one bubble to a neighbouring bubble. With an open-cell structure the bubbles have expanded to link one-with-another and so there is much more open and permeable structure. The mechanical agitation process usually produces an open-cell structure, the other processes being capable of producing either an open or closed-cell structure according to the conditions prevailing during the process.

Foamed polymers have a lower density and lower elastic modulus than the polymer from which they were formed. However, weight for weight foamed polymer lengths are less easily bent than the solid polymer. This is because although the elastic modulus is lower there is a much bigger second moment of area. Because of this, rigid foams are widely used as the filling in sandwich panel composites, with steel, aluminium or solid polymer sheets as the outer sheets. Such a composite can be used to give weight-saving for the same rigidity as that obtained with a solid material. Foamed elastomers are highly compressible, resilient materials and find many uses in packaging and for cushions. The presence of trapped air in a foamed polymer leads to it having a very low thermal conductivity and hence a use for heat insulation, the notable example of this being the use of foamed polystyrene in domestic buildings.

14.3 Manufacture of composites

Fibre-reinforced composites can be produced in a number of ways, depending on whether the fibres are to be continuous or discontinuous throughout the matrix and whether there is to be a specific orientation of the fibres in the product. The fibres used may thus be available as long continuous lengths, short lengths (referred to as chopped strand) or as a woven mat.

In the *hand-lay-up* process a mould defining the surface of the product is successively coated with layers of resin and the fibres, generally in the form of a woven cloth or chopped strand. The layers are built up until the required thickness is obtained, after which the resin is allowed to cure. The products produced by this method are generally of a plate or shell form, e.g. swimming pool linings, building panels, liquid storage tanks and boat hulls. The capital costs associated with such a process are low and the process is slow and labour intensive.

In the *spray-up process* bundles of chopped strand and resin are sprayed onto the mould surface and then compacted with rollers before curing. This enables a higher rate of production than the hand-lay-up process.

Matched-tool moulding involves the resin and fibres being placed between a pair of matched dies which are then closed to force the resin through the fibres and produce the required shape. Curing then follows at room temperature. Complex shapes can be produced with fine tolerances. Faster production, though with lower tolerances, is possible when a premixed moulding compound is used. This has the resin, fibres and other fillers already mixed to form a dough, referred to as a *dough-moulding compound* (DMC) or a *sheet-moulding compound* (SMC). With such compounds compressed between matched dies, heating is generally used to cure the material. Typical DMC products are power tool cases, small box-shaped containers and electrical distribution equipment, while typical SMC products are lorry cabs, fascias and car body parts.

Filament winding is used for axially symmetrical objects. e.g. pipes or large tanks, and involves winding a resin-coated continuous fibre onto a rotating mandrel. This gives a product with high strength in just one particular direction, the direction of the fibre orientation.

The above methods are characteristic of those used when the matrix is a polymer. When the matrix material is a metal there are essentially two forms of processes. In one, molten matrix material is infiltered into bundles of fibres or fibre mats. The material is then formed by rolling, extruding or machining. Examples of such composites are glass fibres in aluminium, alumina fibres in aluminium and tungsten fibres in copper. The other process involves the matrix material being vapour-deposited or electroplated onto fibres followed by hot pressing. The material can then be machined to its required form. Examples of such composites are alumina fibres in nickel, and tungsten fibres in nickel.

14.4 Forming processes with ceramics

In general, the method used to form ceramics is to mould the material to the required shape and then heat it in order to develop the bonding between the particles in the material. As most ceramics are both hard and brittle the shape produced has generally to be the final shape as machining or cold-working methods cannot be used.

A method used for the forming of clay shapes is *slip casting*. A suspension of clay in water is poured into a porous mould. Water is absorbed by the walls of the mould and so the suspension immediately adjacent to the mould walls turns into a soft solid. When a sufficient layer has built up, the remaining suspension is poured out, leaving a hollow clay object which is then removed from the mould and fired. This method is used for the production of wash basins and other sanitary ware.

The shaping of the block of wet clay on the potter's wheel is an example of *wet plastic forming*, the plastic clay mass being shaped by a tool before being fired. Another example of this type of forming is the extrusion of the plastic clay through dies, the extruded shape then being cut into appropriate lengths before being fired.

The *sintering* process, as described in Chapter 10 for metals, is used with ceramics. A version of this used with silicon involves compacting the silicon powder in an atmosphere of nitrogen at a temperatue of about 1400°C. During the sintering process the silicon is converted into silicon nitride. This type of process is known as *reaction sintering* as it involves a chemical reaction as well as the sintering process.

Problems

1. State a process that could be used for the production of the following products:
   (a) plastic guttering for house roofs,
   (b) plastic bags,
   (c) the plastic tubing for ball-point pens,
   (d) plastic rulers,
   (e) a small, lightweight, plastic toy train,
   (f) a plastic tea cup and saucer,
   (g) the plastic body for a camera,
   (h) a hollow plastic container for liquids,
   (i) coating a metal surface with a coloured layer of polymer for decorative purposes,
   (j) a plastic milk bottle.
2. Describe the types of product produced by the following polymer processes:
   (a) extrusion,
   (b) injection moulding,
   (c) calendering,
   (d) thermoforming,
   (e) casting.
3. What are the main types of processes used to produce foamed polymers?
4. Spectacle frames can be moulded from thermoplastic cellulose esters. Explain the process that would have been followed.
5. Why are plastic curtain rails made by extrusion rather than any other process?
6. What types of process are used with thermosetting materials?
7. Explain what is meant by sink marks and weld lines in polymers.
8. Explain how orientation can occur with a polymeric material and its effect on the properties.
9. What are dough-moulding compound and sheet-moulding compound?
10. What factors determine the strength of an adhesive bond?

15

Non-metals

15.1 Thermoplastics

The main properties and uses of thermoplastics commonly used in engineering are outlined below and in Tables 15.1 and 15.2. For more information about thermoplastics and their use the reader is referred to *The Selection and Use of Thermoplastics* by P. C. Powell (Oxford University Press).

Polyethylene

Polyethylene (PE), is referred to as high density (HDPE) when the molecular chain is linear and as low density (LDPE) when the molecular chain has branches and so the molecules cannot be so tightly packed and a less crystalline structure is feasible. LDPE has a density of about 918 to 935 kg/m^3 and HDPE about 935 to 965 kg/m^3. LDPE is flexible, tough and has good chemical resistance. Figure 15.1 shows the stress/strain graph and how it is affected by temperature. HDPE is much stronger and stiffer, though not as tough. It is likewise has good chemical resistance. The two forms of polyethylene can be blended to give plastics with properties intermediate between the two. Both forms can be extruded, blow moulded, rotationally moulded and injection moulded.

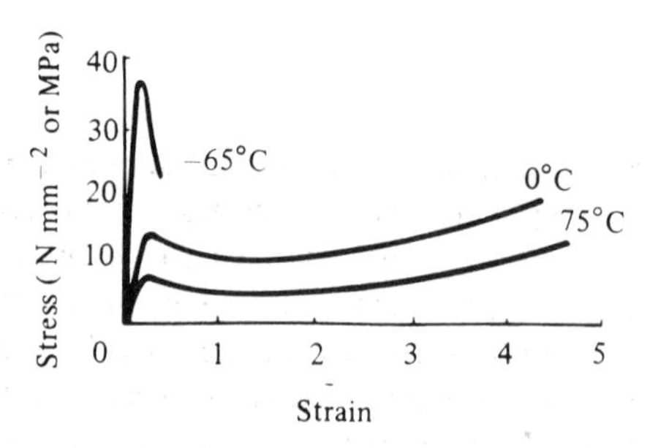

Figure 15.1 Stress/strain graph for low density polythene

Table 15.1 *Properties of typical thermoplastics used in engineering*

	State	T_g (°C)	Max. temp. (°C)	Tensile strength ($MN\ m^{-2}$ or MPa)	Tensile modulus ($GN\ m^{-2}$ or GPa)	Elongation (%)	Impact property
Polythene							
High density	SC	−120	125	22–38	0.4–1.3	50–800	2
Low density	SC	−90	85	8–16	0.1–0.3	100–600	3
Polypropylene	SC	−10	150	30–40	1.1–1.6	50–600	2
PVC							
With no plasticizer	G	87	70	52–58	2.4–4.1	2–40	2
Low plasticizer content	G	–	100	28–42	–	200–250	2
Polystyrene							
No additives	G	100	70	35–60	2.5–4.1	2–40	1
Toughened	G	–	70	17–24	1.8–3.1	8–50	2
ABS	G	100	70	17–58	1.4–3.1	10–140	2–3
Polycarbonate	G	150	120	55–65	2.1–2.4	60–100	2–3
Acrylic	G	100	100	50–70	2.7–3.5	5–8	1
Polyamides							
Nylon 6	SC	50	110	75	1.1–3.1	60–320	2–3
Nylon 6.6	SC	55	110	80	2.8–3.3	60–300	2–3
Nylon 6.10	SC	50	110	60	1.9–2.1	85–230	2–3
Nylon 11	SC	46	110	50	0.6–1.5	70–300	2–3
Polyethylene terephthalate	SC	69	120	50–70	2.1–4.4	60–100	2
Polyacetals							
Copolymer	SC	–	100	60	2.9	60–75	2
Homopolymer	SC	−76	100	70	3.6	15–75	2
PTFE	SC	−120	260	14–35	0.4	200–600	3
Cellulose acetate	G	120	70	24–65	1.0–2.0	5–55	2
Cellulose acetate butyrate	G	122	70	17–50	1.0–2.0	8–80	2

Notes: State SC is semi-crystalline; G is glass. Maximum temperature is the approximate maximum temperature at which the polymer can be in continuous use; higher temperatures are possible for intermittent use. Impact property – 1 is brittle, 2 is tough but not brittle, 3 is tough; all conditions referring to about 20°C.

Low-density polyethylene is used mainly in the form of films and sheeting, e.g. polythene bags, 'squeeze' bottles, ball-point pen tubing, wire and cable insulation. High-density polythene is used for piping, toys, filaments for fabrics, household ware. Both forms have excellent chemical resistance, low moisture absorption and high electrical resistance. The additives commonly used with polyethylene are carbon black as a stabilizer, pigments to give coloured forms, glass fibres to give increased strength and butyl rubber to prevent inservice cracking.

Table 15.2 *Chemical stability of thermoplastics at 20°C*

Polymer	*Water absorption*	*Acids*		*Alkalis*		*Organic solvents*
		Weak	*Strong*	*Weak*	*Strong*	
Polythene						
High density	L	R	AO	R	R	R
Low density	L	R	AO	R	R	R
Polypropylene	L	R	AO	R	R	R
PVC, unplasticized	M	R	R	R	R	A
Polystyrene	L	R	AO	R	R	A
ABS	M	R	AO	R	R	A
Polycarbonate	L	R	A	A	A	A
Acrylic	M	R	AO	R	R	A
Polyamides	H	A	A	R	R	R
Polyester	L	R	A	R	A	A
Polyacetal						
Copolymer	M	A	A	R	R	R
Homopolymer	M	R	A	R	A	R
PTFE	L	R	R	R	R	R
Cellulosics	H	R	A	R	A	A

Notes: R is resistant, A is attacked with AO being attacked by oxidizing acids. Water absorption – L is low, less than 0.1 per cent by weight in 24 h immersion; M is medium, between about 0.1 and 0.4 per cent by weight; H is high, greater than 0.4 per cent and often about 1 per cent.

Polypropylene

Polypropylene (PP) is used mainly in its crystalline form, being a linear polymer with side-groups regularly arranged along the chain. The presence of these side-groups gives a more rigid and stronger polymer than polyethylene in its linear form. The degree of crystallinity affects the physical properties. In addition, the properties are affected by the lengths of the molecular chains, i.e. the molecular weight. Low molecular weight tends to be associated with a higher degree of crystallinity. The greater the crystallinity, the stiffer the polypropylene. An increase in molecular weight increases the impact strength. It has good fatigue resistance, chemical resistance and electrical insulation properties. It can be extruded, blow moulded, injection moulded and sheet can be thermoformed.

Polypropylene is used for crates, containers, fans, car fascia panels, tops of washing machines, cabinets for radios and TV sets, toys and chair shells.

Polypropylene – ethylene copolymer

A copolymer of polypropylene with ethylene, polypropylene-ethylene copolymer (EPM), has properties determined by the proportion of ethylene present and how it is introduced into the chain structure. The effect of quite small percentages of ethylene, of the order of a few per cent, giving block copolymers with polypropylene, is to considerably improve the impact strength. Random copolymers of ethylene and propylene with higher percentages of ethylene produce an elastomer. Such an elastomer has good resistance to atmospheric degradation.

Polyvinyl chloride

Polyvinyl chloride (PVC) is a linear-chain polymer with bulky chlorine side-groups which prevent cystalline regions occurring. The polymer is mixed with a variety of additives to give a range of plastics. Normally PVC is a hard and rigid material but plasticizers can be added to give flexible forms. PVC is extruded to give sheet, film, pipe and cable covering, and calendered into sheet. It can be injection moulded, blow moulded, rotationally moulded and thermoformed.

The rigid form of PVC, i.e. the unplasticized form, is used for piping for waste and soil drainage systems, rainwater pipes, lighting fittings and curtain rails. Plasticized PVC is used for the fabric of 'plastic' raincoats, bottles, shoe soles, garden hose piping, gaskets and inflatable toys.

Vinyl chloride-vinyl acetate copolymer

A copolymer of vinyl chloride with vinyl acetate in a mass ratio of about 85 to 15 gives a rigid material, while a ratio of 95 to 5 gives a flexible material. The addition of plasticizers can also be used to modify the properties. It has properties similar to those of PVC but is easier to mould and thermoform. Like PVC it is non-crystalline.

Non-plasticized copolymer is used for gramophone records, while heavily plasticized copolymer is calendered to produce floor tiles. The material in this form has good abrasion and impact resistance.

Ethylene-vinyl acetate copolymer

Ethylene-vinyl acetate copolymer (EVA) is a linear chain polymer with short side branches which can show crystallinity. The properties depend on the relative proportions of the two constituents. Increasing the vinyl acetate component increases the flexibility, large amounts producing a polymer with properties more like those of an elastomer than a thermoplastic. EVA copolymers are flexible, resilient, tough and have good resistance to atmospheric degradation. They can be extruded, injection moulded, blow moulded and rotationally moulded.

EVA is used for road-marker cones, ice-cube trays, medical and surgical ware and as a major constituent in hot-melt adhesives.

Polystyrene

Polystyrene (PS) is a linear-chain polymer with bulky side-groups which prevent crystalline regions occurring. Polystyrene is available in many forms. General-purpose polystyrene is the term used to describe polystyrene which either has no additives or additives other than rubbers or copolymers. It is a rather brittle, transparent material with a smooth surface finish that can be printed on. It is used for injection-moulding containers for cosmetics, light fittings, boxes and ballpoint pen barrels.

Toughened or high-impact polystyrene is a blend of polystyrene with rubber particles. This blending improves the impact resistance but results in a decrease in tensile modulus, tensile strength and transparency. Injection moulding or thermoforming from extruded sheet is used to produce cups for vending machines and casings for cameras, projectors, radios, television sets and vacuum cleaners.

A widely used form of polystyrene is as expanded polystyrene, a foamed polymer. This is a rigid foam which is used for insulation and packaging.

Polystyrenes are attacked by many solvents, e.g. petrol, dry cleaning agents, greases, oxidizing acids and some oils. Detergents can lead to stress-cracking.

Acrylonitrile-butadiene-styrene terpolymer

Styrene-acrylonitrile copolymer (SAN) is a brittle, glassy, copolymer of

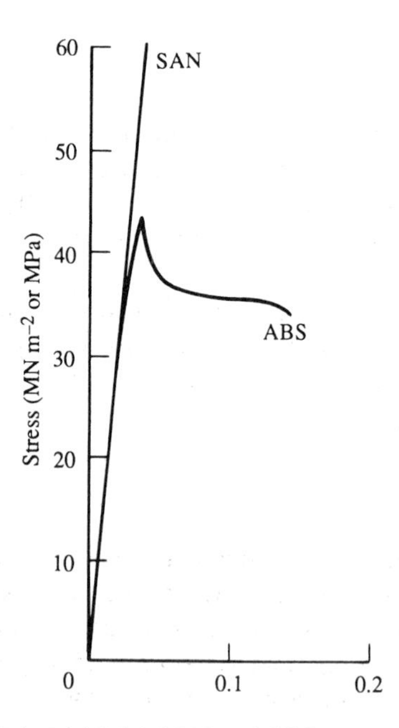

Figure 15.2 Stress/strain graph for SAN and ABS

styrene and acrylonitrile. It can, however, be toughened with polybutadiene. Some of the polybadiene forms a graft terpolymer with the styrene-acrylonitrile while some produces small rubber spheres which become dispersed through the terpolymer and SAN matrix. Figure 15.2 shows the stress/strain graphs for SAN and the acrylonitrile-butadiene-styrene (ABS). ABS is an amorphous material which is tough, stiff and abrasion resistant. It can be injection moulded, extruded, rotationally moulded and thermoformed.

ABS is widely used as the casing for telephones, vacuum cleaners, hair driers, radios, televison sets, typewriters, luggage, boat shells, and food containers.

Polycarbonate

Polycarbonate (PC) is a linear chain polymer which at room temperature is well below its glass transition temperature and hence amorphous. It is tough, stiff, strong and transparent. It also has reasonable outdoor weathering resistance and good electrical insulation properties. It may be injection moulded, blow moulded, extruded and vacuum formed.

Polycarbonate is used for applications where plastics are required to be resistant to impact abuse and, for plastics, relatively high temperatures. Typical applications are transparent streetlamp covers, infant-feeding bottles, machine housings, safety helmets, housings for car lights and tableware such as cups and saucers.

Acrylics

Acrylics are completely transparent thermoplastics, mostly based on polymethyl methacrylate (PMMA) which has linear chains with bulky side-groups and so gives an amorphous structure. They give a stiff, strong material with outstanding weather resistance. They can be cast, extruded, injection moulded and thermoformed.

Because of its transparency and weather resistance acrylic is used for windscreens, light fittings, canopies, lenses for car lights, signs and name-plates. Opaque acrylic sheet is used for domestic baths, shower cabinets, basins and lavatory cisterns.

Polyamides

Polyamides (PA), are commonly known as nylons and are linear non-ethnic polymers which give crystalline structures. There are a number of common polyamides: nylon 6, nylon 6.6, nylon 6.10 and nylon 11. The full stops separating the two numbers are sometimes omitted, e.g. nylon 66 is nylon 6.6. The numbers refer to the number of carbon atoms in each of the reacting substances used to give the polymer. The first number in fact indicates the number of carbon atoms in the resulting polymer chain before the chain repeats itself. The two most used nylons are nylon 6 and nylon 6.6. Nylon 6.6 has a higher melting point than nylon 6 and is also stronger and stiffer. Nylon 11 has a lower melting point and is more flexible.

In general, nylons are strong, tough materials with relatively high melting points. But they do tend to absorb moisture. The effect of this is to reduce their tensile strength and stiffness. It, however, increases toughness. Nylon 6.6 can absorb quite large amounts of moisture; nylon 11, however, absorbs considerably less. Nylons often contain additives, e.g. a stabilizer or flame retardent. Glass spheres or glass fibres are added to give improved strength and stiffness. Molybdenum disulphide is an additive to nylon 6 giving a material with very low frictional properties. Nylons are usually injection moulded, though they can be extruded and extrusion blow moulded.

Nylons are used for the manufacture of fibres for clothing, gears, bearings, bushes, housings for domestic and power tools, electric plugs and sockets.

Polyesters

Polyesters are available in both thermoset and thermoplastic form. The main thermoplastic form is polyethylene terephthalate (PETP). This is a linear chain polymer with side-groups. It gives crystalline structures and is below its glass transition temperature at room temperature. If it is rapidly quenched from a melt to below its glass transition point an amorphous structure is produced, the molecular chains not having sufficient time to become packed in an orderly manner. The polyester has properties similar to nylon and is usually processed by injection moulding.

It is used widely in fibre form for the production of clothes. Other uses are for electrical plugs and sockets, push-button switches, wire insulation, recording tapes, insulating tapes and gaskets.

Polyacetals

Polyacetals are linear-chain polymers with a backbone which consists of alternate carbon and oxygen atoms. One of the main forms of polyacetal is polyoxymethylene (POM), which is sometimes just referred to as acetal homopolymer. A copolymer, generally just referred to as acetal copolymer, has a modified backbone structure, the chain containing occasional ethylene units (Figure 15.3). The homopolymer is slightly stronger and stiffer than the copolymer but the copolymer has the advantage of better long-term strength at high temperatures. In general, acetals are strong, stiff and have a good impact resistance. They have low coefficients of friction and a good abrasion resistance. Glass-filled acetal is used where even higher stiffness is required. Processing is by injection moulding, extrusion or extrusion blow moulding.

Typical applications are pipe fittings, parts for water pumps and washing

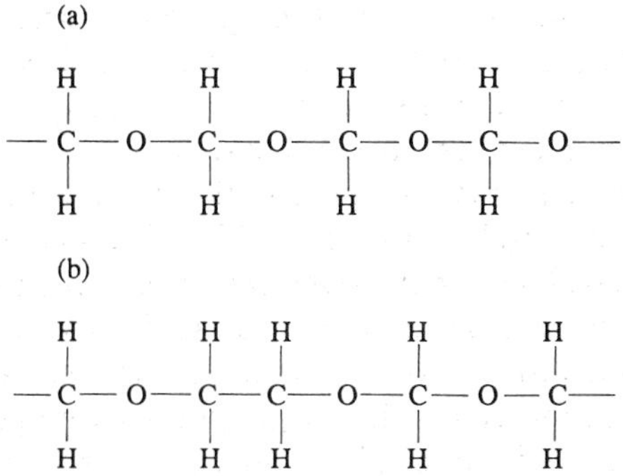

(a)

```
   H       H       H       H
   |       |       |       |
 — C — O — C — O — C — O — C — O —
   |       |       |       |
   H       H       H       H
```

(b)

```
   H       H   H       H       H
   |       |   |       |       |
 — C — O — C — C — O — C — O — C —
   |       |   |       |       |
   H       H   H       H       H
```

Figure 15.3 (a) Acetal homopolymer, (b) Acetal copolymer

machines, car instrument housing, bearings, gears, hinges and window catches, and seatbelt buckles.

Polytetrafluoroethylene

Polytetrafluoroethylene (PTFE) is a linear polymer like polyethylene, the only difference being that instead of hydrogen atoms there are fluorene atoms. It has a very high crystallinity as manufactured, about 90 per cent, though this degree of crystallinity can be reduced during processing to about 50 per cent if quench cooled, or 75 per cent with slow cooling. Tough and flexible, PTFE can be used over a wide range of temperature, 250°C down to almost absolute zero, and still retain the very important property of not being attacked by any reagent or solvent. It also has a very low coefficient of friction. No known material can be used to bond it satisfactorily to other materials.

PTFE is a relatively expensive material and is not processed as easily as other thermoplastics. It tends to be used where its special properties, i.e. resistance to chemical attack and very low coefficient of friction, are needed. Journal bearings with a PTFE surface can be used without lubrication because of the low coefficient of friction; they can even be used at temperatures up to about 250°C. Piping carrying corrosive chemicals at temperatures up to 250°C are made of PTFE. Other applications for PTFE are gaskets, diaphragms, valves, O-rings, bellows, couplings, dry and self-lubricating bearings, coatings for frying pans and other cooking utensils (known as 'non-stick'), coverings for rollers handling sticky materials, linings for hoppers and chutes, and electrical insulating tape.

Cellulosics

The most common cellulosic materials are cellulose acetate (CA), cellulose acetate butyrate (CAB), and cellulose acetate propionate (CAP). Cellulose acetate is hard, stiff and tough but has poor dimensional stability due to a high absorption of water. CAB is tougher and more resistant to water uptake and hence more dimensionally stable. Cellulose acetate propionate is slightly harder, stiffer and stronger. All three can be extruded and injection moulded.

Cellulose acetate is widely used for spectacle frames, having the advantage that it can be softened by a slight increase in temperature and hence adapated to facilitate adjustment to fit the wearer. Other applications are tool handles, typewriter keys and toys. CAB is used for internally-illuminated roadside signs, extruded piping, pens and containers. CAP is used for toothbrush handles, pens, knobs, steering wheels, toys and film for blister packaging.

15.2 Thermosetting polymers

Thermoplastic polymers soften if heated, and can be reshaped, the new shape being retained when the plastic cools. The process can be repeated. *Thermosetting polymers* cannot be softened and reshaped by heating. They are plastic in the initial stage of manufacture but once they have set they cannot be resoftened. The atoms in a thermosetting material form a three-dimensional structure of cross-linked chains. The bonds linking the chains are strong and not easily broken. Thus the chains cannot slide over one another but are essentially fixed in the position they occupied when the polymer was solidifying during its formation.

Thermosetting polymers are stronger and stiffer than thermoplastics and generally they can be used at higher temperatures than thermoplastics. As they cannot be shaped after the initial reaction in which the polymer chains are produced, the processes by which thermosetting polymers can be shaped are limited to those where the product is formed from the raw polymer materials. No further processing is possible (other than possibly some machining) and this limits the processes available to essentially just *moulding*. A number of different moulding methods are used but essentially all involve the combining together of the chemicals in a mould so that the cross-linked chains are produced while the material is in the mould. The result is a thermosetting polymer shaped to the form dictated by the mould.

The main properties and uses of common thermosets are outlined below and in Table 15.3.

Phenolics

Phenolics give highly cross-linked polymers. *Phenol formaldehyde* was the first synthetic plastic and is known as *Bakelite*. The polymer is opaque and initially light in colour. It does, however, darken with time and so is always mixed with dark pigments to give dark-coloured materials. It is supplied in the form of a moulding powder, including the resin, fillers and other additives such as pigments. When this moulding powder is heated in a mould the cross-linked

Table 15.3 *Properties of typical thermosets used in engineering*

	Density (10^3 *kg* m^{-3})	*Tensile strength* (*MN* m^{-2} *or MPa*)	*Tensile modulus* (*GN* M^{-2} *or GPa*)	*Elongation* (%)	*Max. service temp* (°C)
Phenol formaldehyde					
Unfilled	1.25–1.30	35–55	5.2–7.0	1–1.5	120
Wood flour filler	1.32–1.45	40–55	5.5–8.0	0.5–1	150
Asbestos filler	1.60–1.85	30–55	0.1–11.5	0.1–0.2	180
Urea formaldehyde					
Cellulose filler	1.5–1.6	50–80	7.0–13.5	0.5–1	80
Melamine formaldehyde					
Cellulose filler	1.5–1.6	55–85	7.0–10.5	0.5–1	95
Epoxy resin					
Cast	1.15	60–100	3.2	–	–
60% glass fabric	1.8	200–420	21–25	–	200
Polyester					
Unfilled	1.3	55	2.4	–	200
30% glass fibre	1.5	120	7.7	3	–
Polyurethane					
Foam	0.016–0.1	55	–	–	150

polymer chain structure is produced. The fillers account for some 50 to 80 per cent of the total weight of the moulding powder. Wood flour, a very fine soft wood sawdust, when used as a filler increases the impact strength of the plastic, asbestos fibres improve the heat properties, and mica the electrical resistance.

Phenol formaldehyde mouldings are used for electrical plugs and sockets, switches, door knobs and handles, camera bodies and ashtrays. Composite materials involving the phenolic resin being used with paper or an open-weave fabric, e.g. a glass fibre fabric, are used for gears, bearings, and electrical insulation parts.

Amino-formaldehydes

Amino-formaldehyde materials, generally *urea formaldehyde* and *melamine formaldehyde*, give highly cross-linked polymers. Both are used as moulding powders, like the phenolics. Cellulose and wood flour are widely used as fillers. Hard, rigid, high-strength materials are produced, with the melamine being harder and having better heat and stain resistance than the urea.

Both materials are used for tableware, e.g. cups and saucers, knobs, handles, light fittings and toys. Composites with open-weave fabrics are used as building panels and electrical equipment.

Epoxides

Epoxide materials are generally used in conjunction with glass, or other, fibres to give hard and strong composites. Epoxy resins are excellent adhesives giving very high adhesive strengths.

The unfilled epoxide has a tensile strength of 35 to 80 MN m^{-2}, considerably less than that of the composite. The composite is used for boat hulls and table tops.

Polyesters

Polyesters can be produced as either thermosets or thermoplastics. The thermoset form is mainly used with glass, or other, fibres to form hard and strong composites. Such composites are used for boat hulls, architectural panels, car bodies, panels in aircraft, and stackable chairs. They have a maximum service temperatue of the order of 200°C.

Polyurethanes

Polyurethanes can have properties ranging from elastomers to thermosets, depending on the degree of crosslinking produced between the molecular chains. As thermosets they are widely used to produce rigid foams. The foams have advantages over expanded polystyrene in having lower density, lower thermal conductivity and better oil, grease and heat resistance. The rigid foam can be formed in situ, e.g. in wall cavities for thermal insulation. Main uses are in refrigerators, structural sandwich panels in buildings and marine buoyancy applications.

15.3 Elastomers

Elastomers are polymers which show very large strains when subject to stress and which will return to their original dimensions when the stress is removed. Elastomers are essentially amorphous polymers with a glass transition temperature below their service temperature. The polymer structure is that of linear-chain molecules with some cross-linking between chains. Without the cross-linking, the molecular chains might slide past each other and give permanent deformations. The presence of these cross-links ensures that the material is elastic and returns to its original dimensions when the stress is removed. However, if there are too many cross-links the material becomes inflexible.

One way of classifying elastomers is in terms of the form of the polymer chains.

1 Only carbon in the backbone of the polymer chain. Natural rubber, butadiene-styrene, butadiene-acrylonitrile, butyl rubbers, polychloro-

Table 15.4 *Properties of common elastomers*

	Tensile strength ($MN\ m^{-2}$ or MPa)	Elongation (%)	T_g (°C)	Service temp. range (°C)	Resistance to oils and greases	Resilience
Natural rubber	20	800	−73	−50 to +80	Poor	Good
Butadiene-styrene	24	600	−58	−50 to +80	Poor	Good
Butadiene-acrilonitrile	28	700	−10	−50 to +100	Excellent	Fair
Butyl	20	900	−53	−50 to +100	Poor	Fair
Polychloroprene	25	1000	−48	−50 to +100	Good	Good
Ethyl-propylene	20	300	−58 to −52	−50 to +100	Poor	Good
Polypropylene oxide	14	300		−20 to +170	Poor	
Fluorosilicone	8	300	−123	−100 to +200	Good	Fair
Polysulphide	9	500	−50	−50 to +80	Good	Fair
Polyurethane	40	650	−35 to −70	−55 to +80	Poor	Poor
Styrene-butadiene-styrene	14	700		−60 to 80	Poor	

Notes: The service temperature range is for continuous use: To illustrate the term resilience, the more resilient a rubber ball the higher it will bounce up after being dropped from a fixed height onto the floor.

prene and ethylene-propylene are the examples considered in this chapter.

2 Polymer chains with oxygen in the backbone. Polypropylene oxide is an example of such an elastomer.
3 Polymer chains having silicon in the chain. Fluorosilicons are examples of such elastomers.
4 Polymer chains having sulphur in the chain. Polysulphide is an example of such an elastomer.
5 Thermoplastic elastomers. These are block copolymers with alternating hard and soft blocks. Examples are polyurethanes and styrene-butadiene-styrene.

The main properties and uses of the above elastomers are outlines below and in Table 15.4.

Natural rubber

Natural rubber is, in its crude form, just the sap from a particular tree. It consists of very long chain molecules, some 100 000 carbon atoms long. This is processed with sulphur to produce cross-links between chains, the amount of linkage being determined by the amount of sulphur added. This is called *vulcanization*. In addition, anti-oxidants, plasticizers and reinforcing fillers are added (see Section 16.3 for a discussion of particle reinforcement of rubber by carbon black). Temperature has a marked effect on the properties of natural rubber; at low temperatures the rubber can lose its elasticity and become brittle and at high temperatures it can lose all its stiffness. Natural rubber is inferior to synthetic rubbers in oil and solvent resistance, oils causing it to swell and deteriorate.

Natural rubber (NR), is used for balloons and rubber bands and as a high percentage of materials used for tyre treads, inner tubes, conveyor belts and seats. Rubber foams are manufactured from natural rubber and are used as carpet backing and car seats.

Butadiene-styrene

Butadiene-styrene rubbers (SBR) are copolymers produced from butadiene and styrene. Like natural rubber the molecular chains are cross-linked by vulcanization with sulphur and reinforcement with carbon black is used. SBR is cheaper than natural rubber and is widely used as a synthetic substitute for it. SBR has good low-temperature properties, good wear and weather resistance and good tensile properties. It has, however, poor resistance to fuels and oils and poor fatigue resistance.

SBR is used in the manufacture of tyres, hosepipes, conveyor belts, footwear, and cable insulation. It has poorer resilience than natural rubber and so has been unable to replace it for large tyres where such properties are essential.

Butadiene acrylonitrile

Butadiene acrylonitrile rubbers (NBR), are commonly referred to as nitrile rubbers. Like SBR they are copolymers produced from butadiene with, this time, acrylonitrile. The acrilonitrile component is varied between about 20 to 50 per cent by weight of copolymer. Such a rubber has excellent resistance to fuels and oils, even at comparatively high temperatures. The greater the acrylonitrile content the greater the resistance to fuels and oils. NBR is vulcanized by either sulphur or peroxides and reinforced by carbon black.

NBR is a high-cost rubber and thus tends to be used where its excellent resistance to fuels and oils is a vital requirement. It thus finds uses as hoses, gaskets, seals, tank linings, rollers and valves.

Butyl rubbers

Butyl rubbers are copolymers of isobutylene and isoprene with vulcanization by sulphur with reinforcement by a suitable form of carbon black. The rubber has the important property of extreme impermeability to gases, with low resilience. Because of these properties, it is used for inner linings of tubeless tyres, steam hoses and diaphragms.

Polychloroprene

Polychloroprene (CR) is commonly known as neoprene. It is a homopolymer with zinc and magnesium oxides used to give vulcanization. The cross-links are probably oxygen atoms. It has good resistance to oils and a variety of other chemicals, and also has good weathering characteristics. It is widely used in the chemical industry, because of these properties, for oil, and petrol hoses, gaskets, seals, diaphragms and chemical tank linings.

Ethylene-propylene

Ethylene-propylene rubber (EPM) is a random copolymer formed from ethylene and propylene. It is cross-lined by various materials to give a rubber which is very highly resistant to heat and attack by oxygen and ozone. This high resistance, coupled with good electrical insulation properties, gives it wide application for cable covering.

A terpolymer of propylene, ethylene and a diene, after vulcanization by sulphur, gives a rubber (EDPM) with very high resistance to heat and attack by oxygen and ozone. It is widely used in the car industry.

Polypropylene oxide

Polypropylene oxide rubbers (PPO) have polymer chains containing oxygen. Such rubbers have good low-temperature properties and good weathering properties but are not resistant to acids and hydrocarbons. The mechanical properties and resilience are good. The main uses are for electrical insulation and moulded mechanical products.

Fluorosilicones

Fluorosilicones (FVMQ) have polymer chains containing silicone instead of carbon. Because of this they have, for elastomers, exceptional high temperature properties, being usable up to temperatures of the order of 200°C. They have good resistance to oils, fuels and organic solvents, good electrical insulation properties and good weathering properties. Because such elastomers are expensive compared with others, uses are restricted to parts where resistance to oils and solvents is required at high temperatures, e.g. O-rings, seals, gaskets and hoses.

Polysulphides

Most rubbers have polymer chains with backbones which are entirely carbon. Polysulphide differs from this in having a backbone containing both oxygen and sulphur atoms (Figure 15.4). Without fillers the strength of the rubber is low. It has poor abrasion resistance, swells in hot water, has low resistance to high temperatures and will burn rapidly. The rubber also has a strong odour. It has good resistance to oils and solvents and low permeability to gases, however it can be attacked by microorganisms.

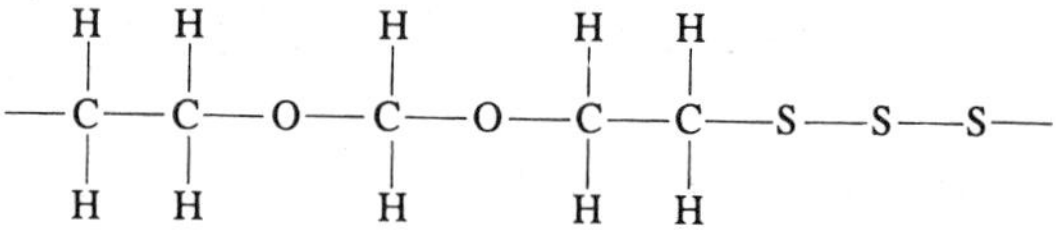

```
   H   H       H       H   H
   |   |       |       |   |
 —C—C—O—C—O—C—C—S—S—S—
   |   |       |       |   |
   H   H       H       H   H
```

Figure 15.4 Polysulphide

It has a limited number of uses – linings of oil and paint hoses, printing rolls and a coating for fabrics. In liquid form it is used as a sealant in building work and for paints and adhesives.

Polyurethanes

Polyurethanes (PUR) can form *thermoplastic elastomers*. Such elastomers have the characteristic form of being block polymers with hard and soft blocks alternating down the polymer chain. With polyurethane rubbers the hard blocks are polyurethane and the soft blocks polyether or polyester. The hard blocks are glassy polymers and act as the physical links in forming the network links between chains. The effectiveness of the links diminishes rapidly above the glass transition temperature of the polyurethane. The hard blocks also act as a fine particle reinforcing filler for the material.

Polyurethane elastomers are widely used for flexible foams, having properties similar to those of natural rubber but with better resistance to ageing. Hence they are widely used as cushioning or for packaging. Polyurethanes have higher tensile strengths than other rubbers.

Styrene-butadiene-styrene

Styrene-butadiene-styrene (SBS) is another example of a thermoplastic elastomer. It is a block copolymer with essentially a soft block, polybutadiene, terminated at each end by a hard block, polystyrene. The polystyrene blocks form the links between polymer chains (Figure 15.5). The hard blocks are in the glassy state and the effectiveness of the links and hence the elastomer properties of the material diminishes above the glass transition temperature of the polystyrene. Above such a temperature the material becomes completely thermoplastic and can be processed by normal thermoplastic processing methods.

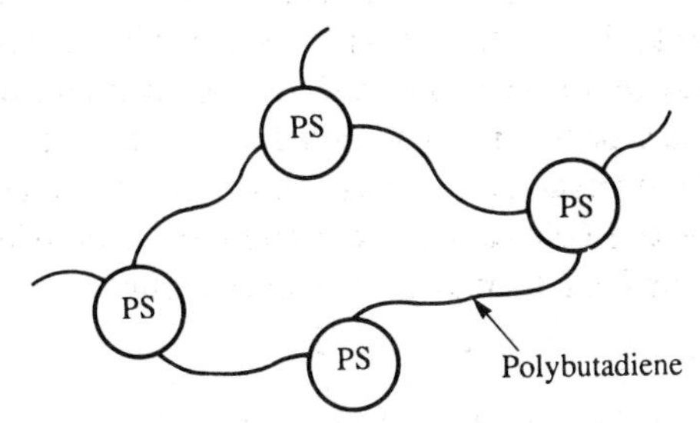

Figure 15.5 Styrene-butadiene-styrene elastomer

The properties of the material are controlled by the ratio of styrene to butadiene, the percentage of styrene being in the range 15 to 30 per cent. The properties are similar to those of natural rubber, with reasonable abrasion resistance and weathering properties, poor resilience and flame resistance. The main uses are in adhesives, carpet backing and as an additive to other polymers in the production of plastics.

15.4 Thermal, electrical, optical and chemical properties of polymers

Solid polymers have very low thermal conductivities when compared with metals, and cellular polymers even lower. For this reason they can be used for thermal insulation. Thus polyurethane foam has a thermal conductivity of about 0.4 W m^{-1} K^{-1}, compared with values of hundreds for metals.

Polymers have higher thermal expansivities than metals, in general about 2 to 10 times greater. Polypropylene has a linear thermal expansivity of about 9 $\times 10^{-5}$ K^{-1} while copper has a value of 1.7×10^{-5} K^{-1}.

Solid polymers have specific heat capacities two to five times those of metals. For example, low density polythene has a specific heat capacity of 1.9 kJ kg^{-1} K^{-1}, copper 0.38 kJ kg^{-1} K^{-1}.

Polymers are electrical insulators, having resistivities which are about 10^{20} times greater than those of metals. If a piece of polymer is rubbed with a piece of cloth it will become electrically charged, the high resistivity of the polymer allowing large electrostatic charges to accumulate. Such charges can give rise to problems. For example, in the production of polymer sheets, charges on the sheets can result in their sticking together or in dust sticking to them.

Polymer transparency ranges from highly transparent to completely opaque to visible light. Polypropylene, for example, is translucent and a 1 mm thickness allows about 10 per cent of the incident light to be transmitted through it. Acetal plastics are virtually opaque. PVC in a 1 mm thickness will allow over 90 per cent of incident light to be transmitted through it.

In general, polymers are resistant to weak acids, weak alkalis and salt solutions but not necessarily resistant to organic solvents, oils and fuels. The effect of solvents on polymers is usually that the polymer dissolves. The degree of resistance depends on the polymer concerned. Thus, for example, polythene is almost completely insoluble in all organic solvents at room temperature, but above 70°C it dissolves in many of them. Acrylics, however, have a poor resistance to organic solvents at room temperature.

Polymers are generally affected by exposure to the atmosphere and to sunlight. The effect can show as a slow ageing process with the material becoming more brittle. This is due to bonds being formed between neighbouring molecular chains. Such effects may be reduced by the inclusion of suitable additives with the polymer in the plastic or rubber.

Some polymers when stressed and in contact with certain environments can develop brittle cracking. This effect is known as *environmental stress cracking*. Polythene, for example, can show this effect in the presence of detergents.

Elastomers can be particularly affected by atmospheric ozone. If the elastomer is under stress in such conditions, cracking can occur; this effect is known as ozone cracking. Stabilizers are usually included with the polymer in the rubber to inhibit such cracking.

15.5 Ceramics

The term *ceramics* covers a wide range of materials, e.g. brick, concrete, stone, clay, glasses and refractory materials. Ceramics are formed from combinations of one or more metals with a non-metallic element such as oxygen, nitrogen or carbon. Ceramics are usually hard and brittle, good electrical and thermal insulators, and have good resistance to chemical attack. They tend to have a

Table 15.5 *Properties of typical engineering ceramics*

	Density (10^3 *kg* m^{-3})	*Coefficient of expansion* (10^{-6} K^{-1})	*Melting point* (°C)	*Thermal conductivity* (*W* m^{-1} K^{-1})	*Tensile strength* (*MN* m^{-2} *or MPa*)	*Compressive strenth* (*MN* m^{-2} *or MPa*)	*Tensile modulus* (*GN* m^{-2} *or GPa*)
Alumina	3.9	8	2040	12–30	400	3800	380
Silicon carbide	3.1	4.5	Decomposes at 2300	40–100	200–800	1400	200–400
Silicon nitride	3.2	2.9	Sublimes at 1900	4–16	200–900		150–300

low thermal shock resistance, because of their low thermal conductivity and a low thermal expansivity: think of the effect of pouring a very hot liquid into a drinking glass.

Ceramics are generally crystalline, though amorphous states are possible. Thus if silica in the molten state is cooled very slowly it crystallizes at the freezing point. However, if the molten silica is cooled more rapidly it is unable to get all its atoms into the orderly arrangements required of a crystal and the resulting solid is a disorderly arrangement which is called a glass.

Some common engineering ceramics with their properties and typical applications are described below and in Table 15.5.

Alumina

Alumina is an oxide of aluminium and is widely used for electrical insulators. It is used for sparking plug insulators where the material withstands rapid fluctuations of temperature and pressure, high voltages and also maintains gas-tight joints with the metal conductor and base. Its high melting point means that it can be used as a refractory material for the lining of high temperature furnaces. It has high compressive strength and resistance to wear and so can be used for tool tips and grinding tools (corundum used for emery paper and grinding wheels is a mixture of alumina and iron oxide).

Silicon nitride

Silicon nitride is a compound of silicon and nitrogen. Nitrides, in general, are fairly brittle and oxidize readily. Silicon nitride has high thermal conductivity with a low thermal expansion and so has good thermal shock resistance. It also has high strength and so finds uses in heat exchangers, furnace components, crucibles and high temperature bearings.

Silicon carbide

Silicon carbide is a compound of silicon with carbon. Carbides in general have very high melting points, however most of them cannot be used unprotected at high temperatures because they oxidize; the exception is silicon carbide. It has a high thermal conductivity combined with a low thermal expansion and so is resistant to thermal shock. It has very good abrasion resistance and chemical inertness. It is used for ball and roll bearings, combustion tubes, rocket nozzles and high temperature furnaces.

Refractories

These are special materials used in construction which are capable of withstanding high temperatures, e.g. furnace linings. The term *refractoriness* is used to describe the ability of a material to withstand high temperatures without appreciable deformation or softening under service conditions.

One of the most widely used refractories consists of silica and aluminium oxide. Figure 15.6 shows the thermal equilibrium diagram. As the diagram indicates, mixtures of silica and aluminium oxide with between about 3 and 8 per cent aluminium oxide should be avoided since they are close to the eutectic point and so the melting point is relatively low. Refractoriness increases with an increase in aluminium oxide content above the eutectic point. With about 20 to 40 per cent aluminium oxide, the product finds a use as fireclay refractory bricks. This increase in aluminium oxide increases the temperature range over which the material softens, the amount of liquid being produced at, say, 1600°C decreasing. This improves the performance of the material in service. For more severe conditions the amount of aluminium oxide can be increased still further. With more than 71.8 per cent the material can be used up to 1800°C.

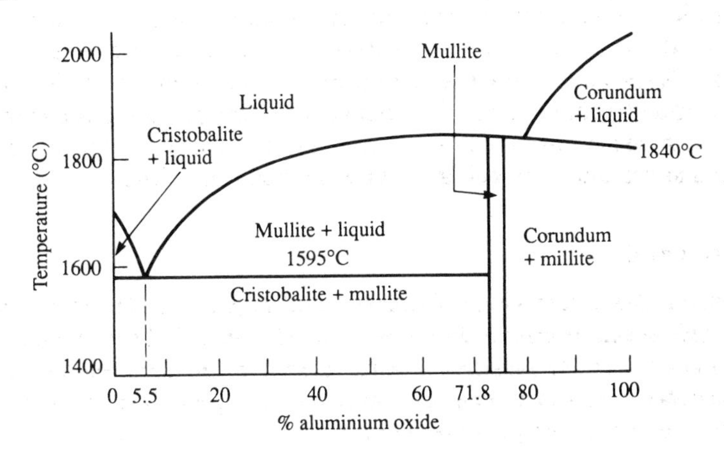

Figure 15.6 Thermal equilibrium diagram for silica-aluminium oxide

Glasses

The basic ingredient of most glasses is sand, i.e. silica – silicon dioxide. Ordinary window glass is made from a mixture of sand, limestone (calcium carbonate) and soda ash (sodium carbonate). Heat-resistant glasses such as Pyrex are made by replacing the soda ash by boric oxide. Many of the mechanical properties of glasses are almost independent of their chemical composition. They thus tend to have a tensile modulus of about 70 GN m^{-2} (GPa). The tensile strength in practice is markedly affected by microscopic defects and surface scratches and for design purposes a value of about 50 MN m^{-2} (MPa) is generally used. They have low ductility, being brittle. They have low thermal expansivity and low thermal conductivity. They are electrical insulators, with resistivities of the order of 10^{14} Ω m or higher. They are resistant to many acids, solvents and other chemicals. The maximum service temperature tends to be about 500°C to 1300°C, depending on the composition of the glass. Table 15.6 gives typical properties.

Table 15.6 *Properties of typical glasses*

	Density $(10^3\ kg\ m^{-3})$	*Coefficient of expansion* $(10^{-6}\ K^{-1})$	*Max. service temperature (°C)*	*Tensile modulus* $(GN\ m^{-2}$ *or GPa)*
Soda–lime–silica glass	2.5	9.2	460	70
Pyrex	2.2	3.2	490	67

Problems

1 How do the properties of high and low density polythene differ?
2 How do the properties of polystyrene with no additives and toughened polystyrene differ?
3 To what temperature would high density polythene have to be heated to be hot formed?
4 Which is stiffer at room temperature, PVC with no plasticizer or polypropylene?
5 The casing of a telephone is made from ABS. How would the casing behave if someone left a burning match or cigarette against it?
6 What are the special properties of PTFE which render it useful despite its high price and processing problems?
7 Figure 15.7 shows the effect on the Charpy impact strength for nylon 6 of percentage of water absorbed. As the percentage of water absorbed increases, is the material becoming more or less brittle?

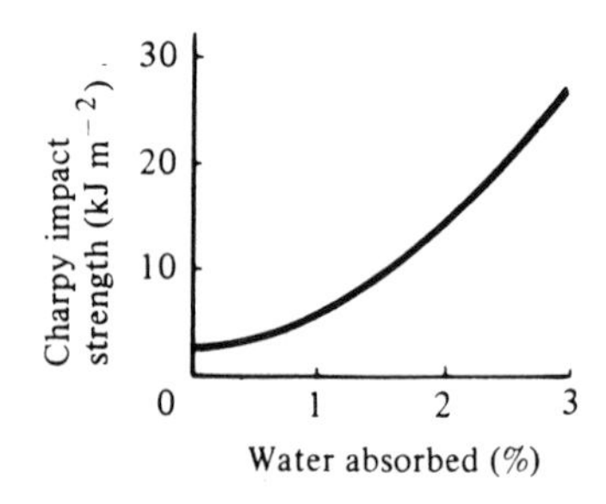

Figure 15.7 The effect of water absorption on the impact strength of nylon 6

8 How do the mechanical properties of thermosets differ, in general, from those of thermoplastics?
9 What is Bakelite and what are its mechanical properties?
10 Describe how cups made of melamine formaldehyde with a cellulose filler might be expected to behave in service.
11 What is meant by vulcanization?
12 Describe the basic structure of a thermoplastic elastomer.
13 Which polymers would be suitable for the following applications? (a) an ashtray, (b) a garden hose pipe, (c) a steam hose, (d) insulation for electric wires, (e) the transparent top of an electrical meter, (f) a plastic raincoat, (g) a toothbrush, (h) a camera body, (i) a road marker cone, (j) cups and saucers, (k) a door knob, (l) a conveyor belt, (m) a fuel hose, (n) cushioning.
14 Explain the reasons for the use of ceramic materials as (a) sparking plug insulators, (b) tool tips, (c) rocket nozzles, (d) furnace components.
15 Describe the basic properties of glasses.

16

Composites

16.1 Composites

The term *composite* is used for a material composed of two different materials bonded together with one serving as the matrix surrounding fibres or particles of the other. A common example of a composite is reinforced concrete. This has steel rods embedded in the concrete (Figure 16.1). The composite enables loads to be carried that otherwise could not have been carried by the concrete alone. Concrete itself is a composite, without the presence of steel reinforcement. It is made by mixing cement, sand, aggregate and water. Stone chips or gravel are often used as the aggregate. The resulting concrete consists of the aggregate in a matrix (Figure 16.2).

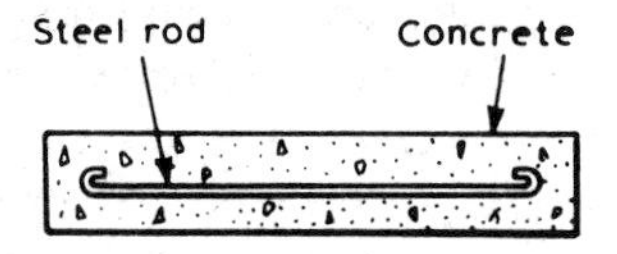

Figure 16.1 Reinforced concrete, steel rods in a matrix

Figure 16.2 Concrete aggregate in a matrix

There are many examples of composite materials encountered in everyday components. Many plastics are glass fibre or glass particle reinforced. Vehicle tyres are rubber reinforced with woven cords. Wood is a natural composite material with tubes of cellulose bonded by a natural plastic called lignin (Figure 16.3). Cermets, widely used for cutting tool tips, are composites involving ceramic particles in a metal matrix.

Composites can be classified into three categories:

1 Fibre reinforced, e.g. vehicle tyres.
2 Particle reinforced, e.g. cermets.

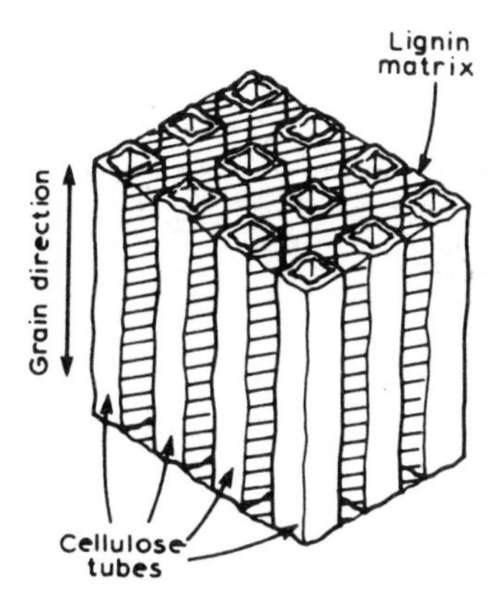

Figure 16.3 Wood, cellulose fibres in a lignin matrix

3 Dispersion strengthened, e.g. precipitation hardened aluminium–copper alloy.

16.2 Fibre-reinforced materials

The main functions of the fibres in a composite are to carry most of the load applied to the composite and provide stiffness. For this reason fibre materials have high tensile strength and a high tensile modulus. Ceramics are frequently used for the fibres in composites. Ceramics have high values of tensile strength and tensile modulus, and also a useful asset of low density. However, ceramics are brittle and the presence of quite small surface flaws can markedly reduce the tensile strength. By incorporating such fibres in a ductile matrix it is possible to form a composite which makes use of the high strength/high modulus properties of the fibres and the protective properties of the matrix material to give a composite with properties considerably better than is possible with just the matrix material or the properties of damaged fibre material. The properties required of a suitable matrix material are that it adheres to the fibre surfaces so that forces applied to the composite are transmitted to the fibres since they are primarily responsible for the strength of the composite; that it protects the fibre surfaces from damage and that it keeps the fibres apart to hinder crack propagation.

The fibres used may be continuous, i.e. in lengths running the full length of the composite, or discontinuous, i.e. in short lengths. They may be aligned so that they are all in the same direction or randomly orientated. Aligning them all in the same direction gives a directionality to the properties of the composite. An aligned continuous fibre composite will have a higher strength than an aligned discontinuous fibre composite made with the same materials.

Table 16.1 gives the properties of some commonly-used reinforcing

Table 16.1 *Properties of fibres and whiskers*

Fibre/whisker	*Density ($10^3 kg/m^3$)*	*Tensile modulus (GN/m^2 or GPa)*	*Tensile strength (GN/m^2 or GPa)*
E-glass	2.5	70	3.5
Silica	2.2	75	6.0
Alumina	3.2	170	2.1
Alumina whisker	3.9	1550	20.8
Carbon	1.8	544	2.6
Graphite whisker	2.2	704	20.7

materials in composites. Whiskers differ from fibres in that they are grown as single crystals rather than polycrystalline.

Examples of composites formed using fibres or whiskers are: glass-fibre reinforced plastics, alumina fibres in nickel, carbon fibres in epoxy resins or aluminium.

Reinforced plastics consist of a stiff, strong, material combined with the plastic. Glass fibres are probably the most used additive. The fibres may be long lengths, running through the length of the composite, or discontinuous short lengths randomly orientated within the composite. Another form of composite uses glass fibre mats or cloth in the plastic. The effect of the additives is to increase both the tensile strength and the tensile modulus of the plastic, the amount of change depending on both the form the additive takes and the amount of it. The continuous fibres give the highest tensile modulus and tensile strength composite but with a high directionality of properties. The strength along the direction of the fibres could be perhaps 800 MPa while that at right-angles to the fibre direction may be as low as 30 MPa, i.e. just about the strength of the plastic alone. Random orientated short fibres do not lead to this directionality of properties but do not give such high strength and tensile modulus. The composites with glass fibre mats or cloth tend to give tensile strength and modulus values intermediate between those of the continuous and short length fibres. Table 16.2 gives examples of the strength and modulus values obtained with reinforced polyester.

Table 16.2 *Properties of reinforced polyester*

Material	*Percentage weight of glass*	*Tensile modulus (GPa)*	*Tensile strength (MPa)*
Polyester	0	2 to 4	20 to 70
With short fibres	10 to 45	5 to 14	40 to 180
With plain weave cloth	45 to 65	10 to 20	250 to 350
With long fibres	50 to 80	20 to 50	400 to 1200

Table 16.3 gives some examples of fibre-reinforced metals and the tensile strengths achieved.

Table 16.3 *Properties of reinforced metals*

Composite	*Tensile strength ($MN\ m^{-2}$ or MPa)*
Aluminium + 50% silica fibres	900
Aluminium + 50% boron fibres	1100
Nickel + 8% boron fibres	2700
Nickel + 40% tungsten fibres	1100
Copper + 50% tungsten fibres	1200
Copper + 80% tungsten fibres	1800

See Section 14.3 for details of the processes used to manufacture fibre-reinforced polymers and metals.

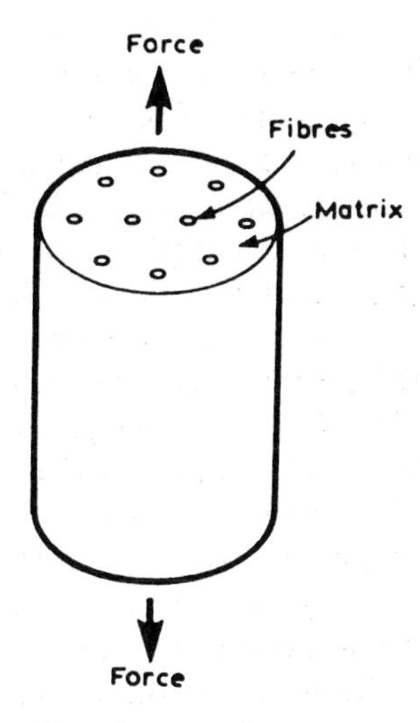

Figure 16.4 Continuous fibres in a matrix

Continuous fibres

Consider a composite rod made up of continuous fibres, all parallel to the rod axis, in a matrix (Figure 16.4). These could be glass fibres in a plastic or steel reinforcement rods in concrete. Each element in the composite has a share of the applied force, thus:

Total force = force on fibres + force on matrix

But the stress on the fibres is equal to the force on them divided by their cross-sectional area. Similarly the stress on the matrix is equal to the force on the matrix divided by its area. Hence:

Total force = stress on fibres × area of fibres + stress on matrix × area of matrix

Dividing both sides of the equation by the total area of the composite gives

$$\frac{\text{Total force}}{\text{total area}} = \text{stress on fibres} \times \frac{\text{area of fibres}}{\text{total area}} + \text{stress on matrix} \times \frac{\text{area of matrix}}{\text{total area}}$$

The fraction of the cross-sectional area that is fibre f_f is given by the area of the fibres divided by the total area, similarly the fraction of the cross-section that is matrix f_m is the area of the matrix divided by the total area. The total area divided by the total force is the stress applied to the composite. Thus:

Stress on composite = $\sigma_f f_f + \sigma_m f_m$
where σ_f = stress on fibres and σ_m = stress on matrix.

If the fibres are firmly bonded to the matrix then the elongation or contraction of the fibres and matrix must be the same and equal to that of the composite as a whole. Thus:

Strain on composite = strain on fibres = strain on matrix

Dividing both sides of the stress equation by the strain gives an equation in terms of the tensile moduli (stress/strain = tensile modulus). Thus:

Modulus of composite = $E_f f_f + E_m f_m$
where E_f = tensile modulus of fibres and E_m = modulus of matrix.

Suppose we have glass fibres with a tensile modulus of 76 GPa in a matrix of polyester having a tensile modulus of 3 GPa. Then, if the fibres occupy 60 per cent of the cross-sectional area, the tensile modulus of the composite will be given by:

$$\begin{aligned}\text{Modulus of composite} &= 76 \times 0.6 + 3 \times 0.4 \\ &= 46.8 \text{ GPa}\end{aligned}$$

The composite has a tensile modulus considerably greater than that of the polyester.

Not only has the composite a higher tensile modulus but also a higher strength than that of the matrix material. Thus for the 60 per cent glass fibres in polyester, the tensile strength of the polyester may be 50 MPa and that of the glass fibres 1500 MPa. Thus since:

Stress on composite $= \sigma_f f_f + \sigma_m f_m$
Strength of composite $= 1500 \times 0.6 + 50 \times 0.4$
$= 920$ MPa

Consider another example, that of a column of reinforced concrete with steel reinforcing rods running through the entire length of the column and parallel to the column axis. If the concrete has a modulus of elasticity of 15 GPa and a tensile strength of 2.8 MPa, the steel with a tensile modulus of elasticity of 210 GPa and a tensile strength of 400 MPa, and the steel rods occupy 10 per cent of the cross-sectional area of the column, then the modulus of the composite is given by:

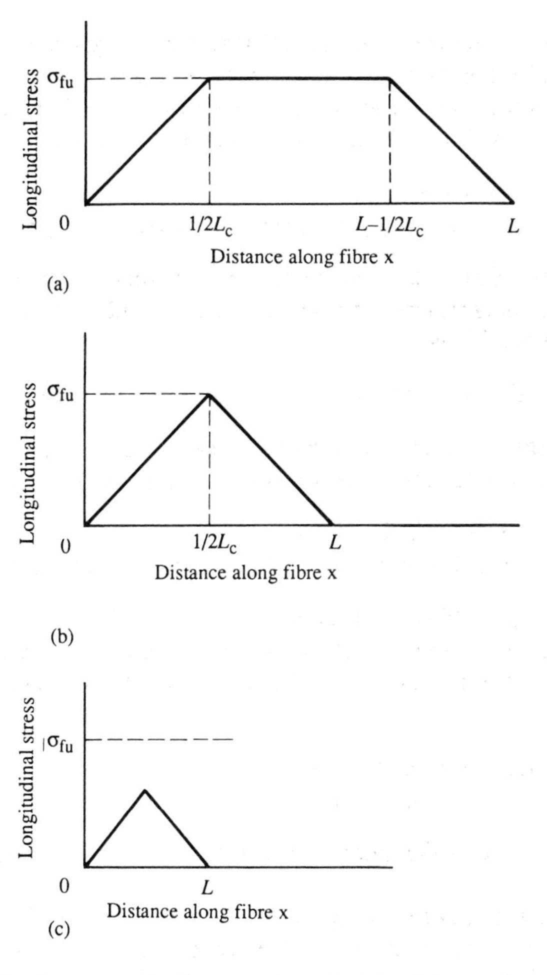

Figure 16.5 Stress transfer from matrix to fibre for different fibres lengths L (a) $L > L_c$ (b) $L = L_c$ (c) $L < L_c$

$$\text{modulus of composite} = E_f f_f + E_m f_m = 210 \times 0.1 + 15 \times 0.9 = 34.5 \text{ GPa}$$

The tensile strength of the composite is:

$$\text{strength of composites} = \sigma_f f_f + \sigma_m f_m = 400 \times 0.1 + 2.8 \times 0.9 = 42.5 \text{ MPa}$$

Discontinuous fibres

In a fibre-reinforced composite the matrix has the task of transferring the externally-applied load to the fibres. This is done by shear forces at the fibre-matrix interface. The extent to which this can be done depends on the surface area of a fibre and hence this is a vital factor in determining the strength of a composite. Figure 16.5 shows the stress transfer from a matrix to a fibre with fibres of different lengths. At the ends of a fibre the stress transferred is zero, rising to a maximum value in the central region of the fibre. Provided the fibre is equal to, or greater than, some critical length L_c the stress reaches the maximum value possible, the tensile strength σ_{fu}. If, however, the length of the fibre is less than L_c the stress does not reach such a value and so the matrix cannot transfer as much of the applied load to the fibre. Thus, provided the fibre lengths in a composite are greater than the critical length L_c the fibre-reinforced composite can realize its full potential strength.

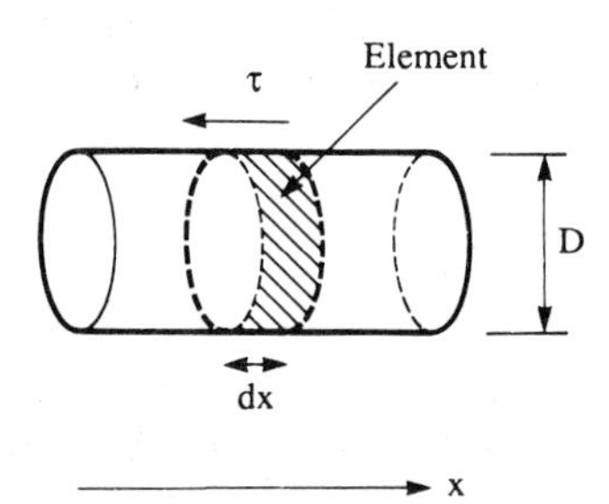

Figure 16.6

If τ is the interfacial shear stress then the shear force acting on a section of the fibre, length dx and of uniform cross-sectional diameter D (Figure 16.1a), is the product of the shear stress and the surface area of the element, i.e.

$$\text{Shear force} = \tau \pi D dx$$

This shear force results in a longitudinal stress in the fibre, $d\sigma_f$. Hence the force balancing the shear force is the product of $d\sigma_f$ and the cross-sectional area of the fibre. Thus:

$$\tau\pi D dx = d\sigma_f \tfrac{1}{4}\pi D^2$$

Hence

$$\frac{d\sigma_f}{dx} = \frac{4\tau}{D}$$

Integrating this gives:

$$\sigma_f = \frac{4\tau x}{D}$$

The stress increases from zero at the end of a fibre, i.e. when $x=0$, to its maximum possible value, its tensile strength σ_{fu}, when $x=\tfrac{1}{2}L_c$ (as in Figure 16.5a). Thus the maximum value is given by:

$$\text{Tensile strength } \sigma_{fu} = \frac{4\,\tau\,(\tfrac{1}{2}L_c)}{D}$$

$$\sigma_{fu} = \frac{2\,\tau\,L_c}{D}$$

The critical length to diameter ratio L_c/D must not be less than $\sigma_{fu}/2\tau$ if the composite is to fully realize the potential of the fibre.

If for a glass-fibre-polyester composite σ_{fu} is 1500 MPa and the shear strength τ is 25 MPa then L_c/D is 30. For fibres of diameter 5 μm then L_c is 150 μm or 0.15 mm. If the fibres used are of greater diameter then the critical length is increased.

The average stress in a fibre of length L, where L is greater than L_c and the stress in the mid-region of the fibre is equal to the fibre tensile strength, is given by:

$$\text{Average stress } \bar{\sigma}_f = \sigma_{fu}\left(1 - \frac{L_c}{2L}\right)$$

This average stress value can be used with the equation developed earlier for the strength of a composite with continuous fibres, the σ_f in that equation being replaced by $\bar{\sigma}_f$. Hence:

$$\text{Strength of composite} = \bar{\sigma}_f f_f + \sigma_m f_m$$

Thus for the glass-fibre-polyester composite referred to above, if the tensile strength of the polyester is 50 MPa,the fraction of the composite which is fibre is 60 per cent and the fibres are of length 3 mm:

$$\text{Strength of composite} = 1500\left(1 - \frac{0.15}{2 \times 3}\right) 0.6 + 50 \times 0.4$$

$$= 897.5 \text{ MPa}$$

This compares with a strength of 920 MPa that occurs with the fibres continuous. If the fibres had been equal to the critical length then the strength of the composite would have been 475 MPa.

16.3 Particle-reinforced materials

Particle-reinforced materials have particles of the order of 1 μm or more in diameter dispersed throughout the matrix, the particles often accounting for a quarter to a half, or more, of the volume of the composite. Particle-reinforced materials include many combinations of metals, polymers and ceramics.

Cermets

Cermets, or cemented carbides, are examples of particle-reinforced composites in which hard ceramic particles are in a metal matrix. The ceramics used have high strengths, high values of tensile modulus and high hardness, but are by themselves brittle substances. By comparison, the metals are weaker and less stiff, but ductile. By incorporating ceramic particles, often about 80% by volume, in a metal matrix, a composite can be produced which is strong, hard and tough and can be used as a tool material.

For example, tungsten carbide is a very hard (about 2000 HV) ceramic material, with a high tensile modulus, but also very brittle. Tools made from this material are thus extremely brittle. A cermet involving tungsten carbide in a metal matrix, cobalt, can be made by mixing tungsten carbide powder with cobalt powder and heating the compacted powders to a temperature above the melting point of the cobalt. The liquid cobalt then melts and flows round each tungsten carbide particle. After solidification the cobalt acts as a binder for the tungsten carbide. The composite has a better toughness than the tungsten carbide alone, since crack propagation though the material is hindered. When in use, the tungsten carbide particles in the surface of the material provide the tool with its cutting ability. As the tungsten carbide particles at the cutting surface become blunted, they either fracture or pull out of the cobalt matrix and expose fresh tungsten carbide particles which can continue to provide cutting ability. For a fine cutting tool the amount of cobalt in the composite is low, and the tungsten carbide particles fine, so that the tungsten carbide particles pull out easily and the tool remains sharp. For a rough cutting tool the amount of cobalt is increased to improve toughness and coarser tungsten carbide particles are used. Table 16.4 shows the composition and application of some cobalt–tungsten carbide composites used as tool materials.

Table 16.4 *Cobalt–tungsten carbide tool materials*

Cobalt (%)	*Tungsten carbide* (%)	*Grain size*	*Typical applications*
3	97	Medium	Machining of cast iron, non-ferrous metals and non-metallic materials
6	94	Fine	Machining of non-ferrous and high temperature alloys
6	94	Medium	General purpose machining for metals other than steels, small and medium size compacting dies and nozzles
6	94	Coarse	Machining of cast iron, non-ferrous metals and non-metallic materials, compacting dies
10	90	Fine	Machining steel, milling, form tools
10	90	Coarse	Percussive drilling bits
16	84	Fine	Mining and metal forming tools
16	84	Coarse	Mining and metal forming tools, medium and large size dies where high toughness is required
25	75	Medium	Heavy impact metal forming tools such as heading dies, cold extrusion dies

Particle-reinforced polymers

Many polymeric materials incorporate fillers, these being particulate. Examples of such fillers are glass beads, silica flour and rubber particles (see section 4.3). Thus, for example, the toughness of some polymers is increased by incorporating tiny rubber particles in the polymer matrix. Polystyrene is toughened this way by polybutadiene to give a product referred to as high impact polystyrene (HIPS). The rubber increases the plane strain fracture toughness K_{Ic} (see section 6.4) of the polystyrene from about 1 to 1.7 $MN\,m^{-3/2}$. The rubber particles block the transmission of cracks and, since they deform readily, absorb energy. Styrene–acrylonitrile–copolymer is toughened with polybutadiene or styrene–butadiene–copolymer to give acrynitrile–butadiene–styrene terpolymer (ABS). With such materials the rubber particles account for about 30% of the volume of the composite. The rubber has a lower tensile modulus than the matrix material and the net result is a lowering of the tensile modulus and tensile strength, but much greater elongations before breaking and a tougher material.

Carbon black, which consists of very fine particles of carbon, is widely used as a filler with vulcanized rubber. The carbon black enhances the strength, stiffness, hardness, wear resistance and heat resistance of the rubber.

Foams

Foams are a form of particulate composite in which the component bound by the matrix is not a solid but bubbles of a gas. Such foams are used as cushioning in furniture, energy-absorbent packaging and padding, for thermal insulation, for buoyancy, and as the filling in sandwich panels (see section 16.5). The parameters determining the characteristics of foams are the ratio of the bulk density of the foam to that of the unfoamed matrix material, and the cellular structure of the foam. The foam can be open-cell, closed-cell, or a mixture of the two. With a closed-cell structure the gas bubbles in the foam are discrete and not interconnected, whereas in an open-cell structure the bubbles have coalesced and are interconnected. Structural forms and sandwich foams have solid skins covering a foamed core.

Figure 16.7 shows the typical forms of compressive stress–strain graphs for polymeric foams. Over the initial straight line part of the graph the cell walls just bend under the action of the applied stress. The next stage is when the walls elastically buckle, often giving a plateau of deformation at almost constant stress. This deformation is still elastic and so recoverable. Finally the cell walls suffer irrecoverable buckling collapse. With a material used for, say, a cushion the foam is required to give continual increasing resistance to increasing load and so a plateau is not required and the stress–strain graph is required to be of the form shown as A in the figure. Foams used for packaging do, however, need to absorb the energy involved when packages are dropped and so a plateau is highly desirable, like that indicated for B in the figure, since it indicates a high energy absorption. A low density packaging foam might

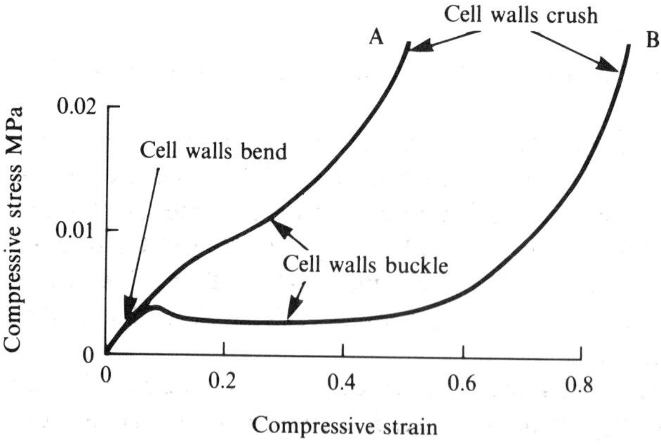

Figure 16.7 Stress–strain graphs for foamed polymers

have a density ratio bulk foam to unfoamed plastic of 0.01 and be used for packaging small delicate instruments, heavier density foams being used for packaging heavier components.

For structural and sandwich foams, to a reasonable approximation, the elastic modulus E_f of the foam is related to the modulus E_s of the matrix material when solid by

$$E_f = V_s E_s$$

where V_s is the fraction of the bulk volume of the composite that is matrix.

$$V_s = \frac{v_s}{v_s + v_g}$$

where v_s is the volume of the foam that is solid and v_g the volume that is gas. The bulk density of the foam $\rho_f = (m_s + m_g)/(v_s + v_g)$, with m_s being the mass of the solid and m_g that of the gas in the foam. Thus

$$V_s = \frac{v_s \rho_f}{m_s + m_g}$$

Since $m_g \approx 0$ and the density of the solid material $\rho_s = m_s/v_s$, then

$$V_s \approx \frac{\rho_f}{\rho_s}$$

Thus we can write

$$E_f \approx \frac{\rho_f}{\rho_s} E_s$$

For foams with uniform densities but without the structural skins, a better relationship has been found experimentally to be

$$E_f = \left(\frac{\rho_f}{\rho_s}\right)^n E_s$$

where n has the approximate value of 1.5 for such foams in tension and 2 in compression.

Thus, for example, foamed polystyrene with a volume fraction of polymer of 0.5 would have a tensile modulus which is about 0.35 times that of the unfoamed polystyrene. The compression modulus would be about 0.25 times that of the unfoamed polystyrene. Expanded polystyrene used for thermal insulation and packaging has a volume fraction of about 0.05. The tensile modulus is then about 0.011 and the compression modulus about 0.0025 that of the unfoamed polystyrene.

The lower modulus means that foamed plastics are much more flexible than when unfoamed. However, in bending this reduction in modulus can be more than offset by the ability to increase the second moment of area. For the same mass of polymer, the foamed plastic has material much further away from the neutral axis than the unfoamed plastic and so a greater second moment of area. The important parameter in determining the stiffness of a beam is the product of the modulus E and the second moment of area I. For example, for a cantilever of length L subject to a force F at its free end, the deflection y at the free end is given by

$$y = \frac{FL^3}{3EI}$$

A stiff cantilever is one where the force per deflection F/y is large. This means that a large value of EI is required. Because with a foamed plastic the fractional increase in I is much greater than the fractional decrease in E, a foamed version of a plastics is stiffer per unit mass than when unfoamed.

16.4 Dispersion-strengthened metals

The strength of a metal can be increased by small particles dispersed throughout it. Thus solution treatment followed by precipitation hardening for an aluminium–copper alloy can lead to a fine dispersion of an aluminium–copper compound throughout the alloy. The result is a higher tensile strength material because the movement of dislocations is hindered, see Table 16.5.

Another way of introducing a dispersion of small particles throughout a metal involves *sintering*. This process involves compacting a powdered metal powder in a die and then heating it to a temperature high enough to knit together the particles in the powder.If this is done with aluminium the result is a fine dispersion of aluminium oxide (about 10 per cent) throughout an aluminium matrix. The aluminium oxide occurs because aluminium in the presence of oxygen is coated with aluminium oxide. When the aluminium powder is compacted, much of the surface oxide film becomes separated from the aluminium and becomes a fine powder dispersed throughout the metal. The aluminium oxide powder, a ceramic, dispersed throughout the aluminium matrix gives a stronger material than that which would have been given by the aluminium alone. At room temperature the tensile strength of the sintered

Table 16.5 *The effect of precipitation hardening on an aluminium alloy*

Aluminium alloy	*Condition*	*Tensile strength (MPa)*
4.0% Cu, 0.8% Mg, 0.5% Si, 0.7% Mn.	Annealed	190
	Solution treated, precipitation hardened	430

aluminium powder is about 400 MPa, compared with that of about 90 MPa for commercial, annealed, aluminium. The sintered aluminium has an advantage over the precipitation-hardened aluminium alloy in that it retains its strength better at high temperatures (Figure 16.8). This is because, at the higher temperatures, the precipitate particles tend to coalesce or go into solution in the metal.

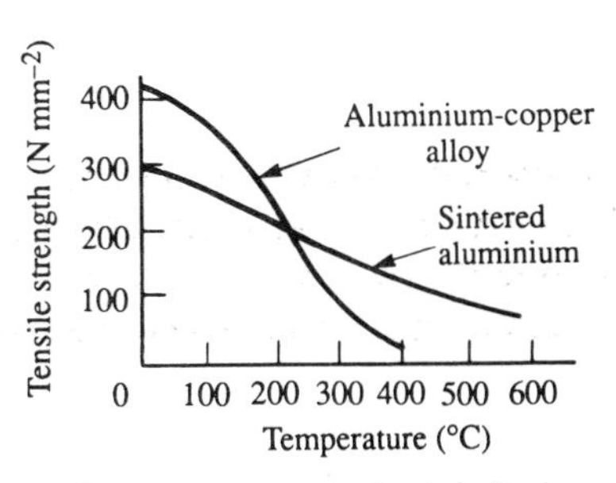

Figure 16.8 The effect of temperature on the tensile strength of an aluminium–copper alloy and sintered aluminium

16.5 Laminates

Plywood is an example of a laminated material. It is made by gluing together thin sheets of wood with their grain directions at right angles to each other (Figure 16.9). The grain directions are the directions of the cellulose fibres in the wood and thus the resulting structure, the plywood, has fibres in mutually perpendicular directions. Thus, whereas the thin sheet had properties that were directional the resulting laminate has no such directionality.

The term *laminated wood* is generally used to describe the product obtained by sticking together thin sheets of wood but with the grain of each layer parallel to the grain of the others. Large wooden arches and beams in modern buildings are likely to be laminated rather than a solid piece of wood. By carefully choosing the wood used to build up the beam a better quality beam can be produced than would otherwise be produced by nature.

It is not only wood that is laminated, metals are too. The *cladding* of aluminium–copper alloy with aluminium to give a material with a better corrosion resistance than that of the alloy alone is an obvious example. Galvanized steel can be considered another example, a layer of zinc on the steel in order to give better corrosion resistance. Steel for use in food containers is often plated with tin to improve corrosion resistance. Many of the metals clad or plated with other metals are in this form for improved corrosion resistance.

Corrugated cardboard is another form of laminated structure (Figure 16.10), consisting of paper corrugations sandwiched between layers of paper. The resulting structure is much stiffer, in the direction parallel to the corrugations,

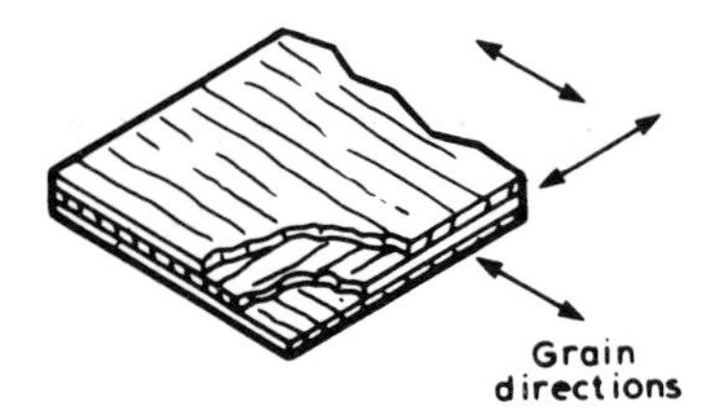

Figure 16.9 Plywood

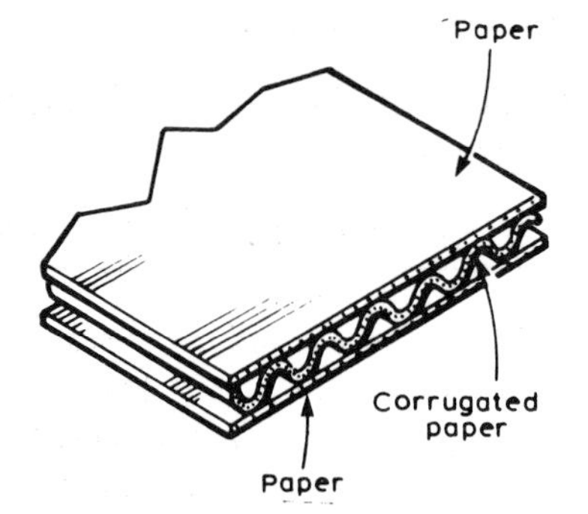

Figure 16.10 Corrugated cardboard

than the paper alone. A similar type of material is produced with metals, a metal *honeycomb structure* sandwiched between thin sheets of metal. Such a structure has good stiffness and is very light. Aluminium is often used for both the honeycomb and the sheets.

A similar laminated structure to the corrugated cardboard or the honeycomb between two plates is that of a plastic foam sandwiched between two skins or plates. With these forms of structure, the bending stiffness of a panel depends on the value of the product of the modulus of elasticity and the second moment of area (see section 16.3). The effect of the corrugations or honeycomb or foam is to give a very low density core and so enable the main mass of a panel to be pushed further away from the neutral axis than would otherwise occur with a similar mass but solid panel. This increase in second moment in area more than outweighs the lower modulus of core.

16.6 Properties of composites

The following are the properties characteristic of many composites.

1 Specific strength

Composite materials often have an advantage over many other materials of a high specific strength:

$$\text{Specific strength} = \frac{\text{tensile strength}}{\text{density}}$$

Thus, for example, a glass fibre reinforced epoxy might have a tensile strength of 0.8 GPa and a density of 2200 kg/m^3, hence a specific strength of 0.36 MPa/kg m^{-3}. This is a higher specific strength than mild steel, tensile strength 0.46 GPa and density 7800 kg/m^3 and hence specific strength of 0.06 MPa/kg m^{-3}.

2 Specific modulus

Composites often have a high specific modulus when compared with other materials:

$$\text{Specific modulus} = \frac{\text{tensile modulus}}{\text{density}}$$

Thus, for example, a glass fibre reinforced epoxy might have a tensile modulus of 56 GPa and a density of 2200 kg/m^3, hence a specific modulus of 25 MPa/kg m^{-3}. This is a specific modulus comparable with that of mild steel, tensile modulus 210 GPa and density 7800 kg/m^3 and hence specific modulus of 27 MPa/kg m^{-3}.

3 Anisotropy

For those composites with fibres all aligned in the same direction, the tensile properties can differ markedly with the angle of the applied forces to the fibre direction. Thus the tensile strength at right angles to the fibre direction can be about one-sixth of that in the direction of the fibres. The tensile strength at just 10° from the fibre direction can be only about half that in the direction of the fibres. The material is thus said to show anisotropy. Composites made with randomly orientated fibres do not show this effect, however their tensile strengths are much less.

4 Fatigue strength

The fatigue strength for plain polymeric materials is generally not very good, reinforced polymeric materials can, however, have much better properties. A glass fibre reinforced polyester can have an endurance limit of about 80 MPa, comparable with that of many aluminium alloys. Fatigue failure in composites with a metal matrix is primarily determined by the properties of the matrix. Thus, for example, a fibre reinforced aluminium composite has fatigue properties almost identical to those of the aluminium alone.

5 Creep resistance

The reinforcement of materials can improve their creep resistance, and in doing so increase the temperature at which the materials can be used. Figure 16.11 shows the type of effect on creep that incorporating glass fibres in nylon can have.

6 Impact properties

In some composites the increases in tensile strength, tensile modulus and hardness produced by, say, the inclusion of fibres can result in a reduction in impact strength since the material becomes less ductile. However, ductile

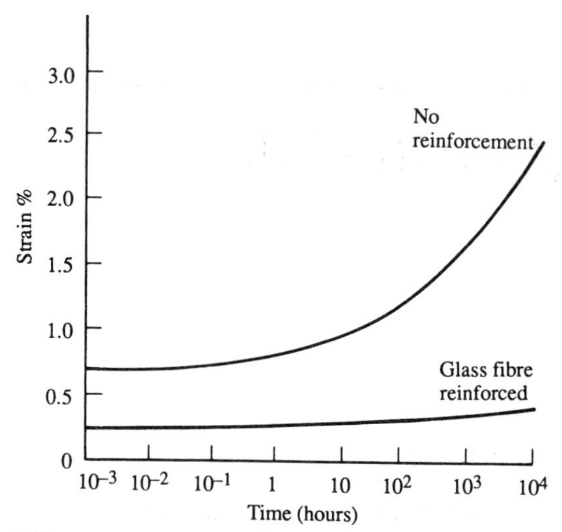

Figure 16.11 Effect on creep properties of nylon of glass fibre reinforcement at 20°C and a stress of 20 MPa

fibres in a brittle matrix can lead to an increase in impact strength. High impact polystyrene has rubbery particles incorporated in the polymer and this results in a significant increase in impact strength.

Problems

1 Explain the term composite.
2 Give examples of composites involving (a) plastics and (b) metals.
3 Describe how the mechanical properties of a fibre composite depend on the form of the fibres and their orientation.
4 Calculate the tensile modulus of a composite consisting of 45 per cent by volume of long glass fibres, tensile modulus 76 GPa, in a polyester matrix, tensile modulus 4 GPa. In what direction does your answer give the modulus?
5 In place of the glass fibres referred to in Question 4, carbon fibres are used. What would be the tensile modulus of the composite if the carbon fibres had a tensile modulus of 400 GPa?
6 Explain the significance of the 'critical length' when discontinuous fibres are used for a composite.
7 Calculate (a) the critical length and (b) the tensile strength of a composite made with discontinuous fibres of diameter 6 μm, length 2 mm and, tensile strength 2000 MPa in a matrix which gives a shear strength of 22 MPa.
8 What is a cermet?
9 Explain how carbon black particles improve the tensile modulus of rubbers.
10 Explain what is meant by the term 'dispersion strengthened metals' and give an example of one.
11 Explain how plywood and corrugated cardboard obtain their stiffness.

17

Joining materials

17.1 Joining methods

The main joining processes can be summarized as:

1 Adhesive bonding
2 Soldering and brazing
3 Welding
4 Fastening systems

The factors that determine the joining process to be chosen are:

1 The materials involved.
2 The shape of the components being joined.
3 Whether a permanent or temporary joint is required.
4 Limitations imposed by the environment.
5 Cost.

17.2 Adhesives

An adhesive can be defined as any substance that is placed between two surfaces in order to hold the two surfaces together. For an adhesive to work it must wet the surfaces being joined. This means that it should flow over the surface and not roll up into globules. The surfaces can be considered to be held together as a result of mechanical bonding between a surface and the adhesive and intermolecular forces between molecules in the adhesive and in the surface. The mechanical bonding is because the adhesive flows into crevices in the surface and, following solidification of the adhesive, is held to the surface by the interlocking adhesive protrusions into the surface. For this reason, surfaces are generally roughened before the adhesive is applied. The intermolecular forces between the adhesive molecules and the surface molecules can be as a result of van der Waals forces.

The bond strength of an adhesive joint depends on the way the joint is

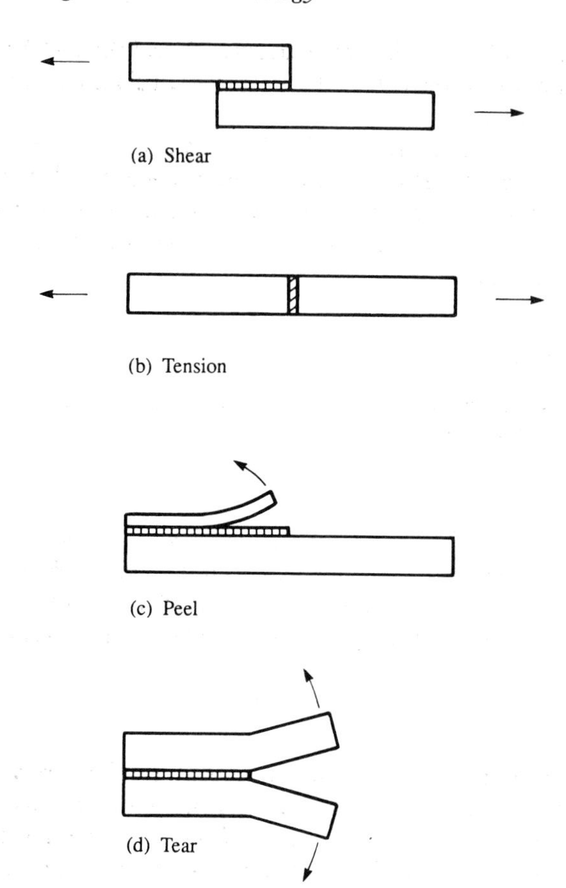

Figure 17.1 Modes of loading adhesive joints

loaded. The maximum strength tends to be when the joint is loaded in shear, with tension, peel and tear modes being generally much weaker (Figure 17.1). Typically, a room temperature cured epoxy resin has a shear strength at room temperature of about 17 MPa, the strengths in other modes being lower. Thus, if a joint is required to withstand a load of, say, 500 N then with the joint loading mode being in shear the bond area must be:

$$\text{Area} = \frac{500}{17 \times 10^6}$$
$$= 2.9 \times 10^{-5}\ \text{m}^2 = 29\ \text{mm}^2$$

For the maximum strength joint to be realized the maximum area of bonding should be used in shear.

Failure of an adhesive bonded joint can occur due to either the bonds between the metal and the adhesive failing or bonds within the adhesive layer failing. For many adhesives the bonds between it and the metal are stronger than those within the adhesive itself. For this reason thin layers of adhesive are preferable to thick layers.

The strength of an adhesive bonded joint thus depends on the type of adhesive used, the mode of loading, the area that is bonded and the thickness of the adhesive layer. In addition, the maximum strength of an adhesive depends on the curing time and temperature, and in some instances the pressure applied during the curing.

Types of adhesives

Adhesives can be classified according to the type of chemical involved. The main types are as follows:

Natural adhesives

Vegetable glues made from plant starches are typical examples of natural adhesives. These types are used on postage stamps and envelopes. However, such adhesives give bonds with poor strength which are susceptible to fungal attack and are also weakened by moisture. They set as a result of solvent evaporation.

Elastomers

Elastomeric adhesives are based on synthetic rubbers; they also set as a result of solvent evaporation. Strong joints are not produced as they have low shear strength. The adhesive is inclined to creep. These adhesives are mainly used for unstressed joints and flexible bonds with plastics and rubbers.

Thermoplastics

These include a number of different setting types. An important group are those, such as polyamides, which are applied hot, solidify and bond on cooling. They are widely used with metals, plastics, wood, etc. and have a wide application in rapid assembly work such as furniture assembly and the production of plastic film laminates.

Another group are the acrylic acid diesters which set when air is excluded, the reaction being one of a build-up of molecular chain length. Cyanoacrylates, the 'super-glues', set in the presence of moisture, in a similar way, with the reaction taking place in seconds. This makes them very useful for rapid assembly of small components. Other forms of thermoplastic adhesives set by solvent evaporation, e.g. polyvinyl acetate.

In general, thermoplastic adhesives have a low shear strength and under high loads are subject to creep, so they are generally used in assemblies subject to low stresses. They have poor to good resistance to water but good resistance to oil.

Thermosets

These set as a result of a build-up of molecular chains to give a rigid cross-linked. Epoxy resins, such as Araldite, are one of the most widely used thermoset adhesives. These are two-part adhesives, in that setting only starts to occur when the two components of the adhesive are brought together. They will bond almost anything and give strong bonds which are resistant to water, oil and solvents.

Phenolic resins are another example of thermoset adhesives. Heat and pressure are necessary for setting. They have good strength and resistance to water, oil and solvents and are widely used for bonding plywood.

Two-polymer types

Thermosets by themselves give brittle joints, but combined with a thermoplastic or elastomer a more flexible joint can be produced. Phenolic resins with nitrile or Neoprene rubbers have high shear strength, excellent peel strength, good resistance to water, oils and solvent and good creep properties. Phenolic resins with polyvinyl acetate, a thermoplastic, give similar bond strengths but with even better resistance to water, oils and solvents. These adhesives are used for bonding laminates and metals. Joints using them can be subjected to high stresses and can often operate satisfactorily up to temperatures around 200°C.

Advantages of adhesives

The use of adhesives to bond materials together can have advantages over other joining methods, i.e.

(a) Dissimilar materials can be joined, e.g. metals to polymers.
(b) Jointing can take place over large areas.
(c) A uniform distribution of stress over the entire bonded area is produced – with a minimum of stress concentration.
(d) The bond is generally permanent.
(e) Joining can be carried out at room temperature or temperatures close to it.
(f) A smooth finish is obtained.

Disadvantages are that optimum bond strength is usually not produced immediately: a curing time has to be allowed. The bond can be affected by environmental factors such as heat, cold and humidity. Many adhesives generally cannot be used at temperatures above about 200°C.

17.3 Soldering and brazing

With *soldering*, the joining agent is different from the two materials being joined but alloys locally with them. The joining agent, the solder, is heated together with the materials being joined until it melts and alloys with their surfaces. On cooling, the alloy solidifies forming a bond between the two materials. The joining process requires temperatures below 425°C and often below 300°C.

Solders are only weak structural materials when compared with the metals they are used to join; there is thus a need to ensure that the strength of the soldered joint does not rely on solder strength and is designed so that the materials interlock in some way (Figure 17.2).

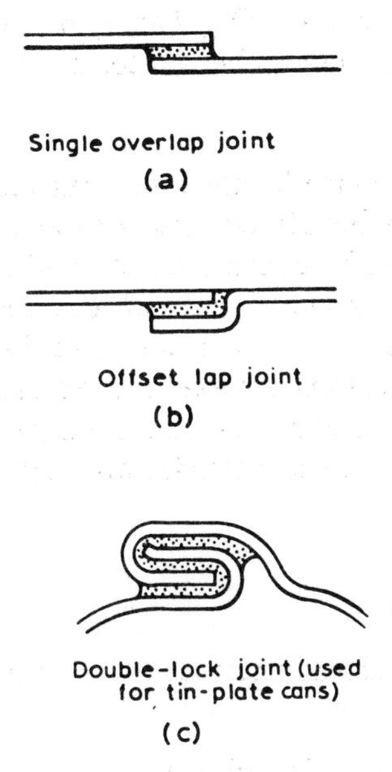

Figure 17.2 Solder joints

The hot solder must wet the metal surfaces being joined, which requires not only a suitable choice of solder material but clean surfaces, as unclean surfaces may make soldering impossible. Cleaning may involve abrasion of the surfaces as well as degreasing. Soldering flux is then applied to the surfaces. Fluxes, when heated, promote or accelerate the wetting of the surfaces by the solder. They remove oxide layers from both metal and solder and prevent them reforming during soldering. Fluxes are grouped in three categories: corrosive, intermediate and non-corrosive, and the least corrosive flux which will give a good joint should be used. After soldering the residues should be removed; if left, they can result in corrosion of the metal surrounding the joint.

Solders are alloys of tin and other metals such as lead or antimony. The solder composition used depends on the metals being joined and the type of joint concerned. A 50 per cent tin/50 per cent lead solder could be used for joining sheet metal. For use at temperatures above 100°C a 95 per cent tin/5 per cent antimony solder may be used.

Brazing is a process similar to that of soldering but involves temperatures above 425°C. Brazing can be used with aluminium and its alloys, nickel and copper alloys, cast iron, steels, and many other less-used metals. Dissimilar metals can be joined. The term 'braze' comes from the use of brass as the substance used to make the joint. A 50 per cent copper/50 per cent zinc is used for general work, with a melting point of about 870°C. Other alloys, e.g. a copper/silver alloy, are used.

The procedure for brazing is similar to that for soldering, but the end result is a stronger joint than that given with solder. Brazing may be achieved by techniques which just involve heating a band of metal on either side of the joint, e.g. by means of a torch, or by heating the whole component in a furnace. The temperatures attained during brazing can affect the mechanical properties of hardened and tempered steels or precipitation-hardened alloys. This can mean, in the case of a torch-brazing operation, that a band of metal on either side of the joint is effectively re-tempered by the brazing operations and thus will have different properties to the rest of the metal. With furnace welding the entire component will effectively be re-tempered. It is possible to combine the tempering and brazing operations. Likewise, similar changes in properties can occur with precipitation-hardened alloys and again the operations can be combined.

Strength of brazed joints

The strength of a brazed joint is markedly affected by the joint clearance, i.e. the separation of the two joint material faces. The maximum strength tends to occur with joint clearances of the order of 0.1 mm, greater clearances resulting in a considerable reduction in strength. The reason for this is that a degree of alloying occurs between the brazing material and the metal or metals being joined and this new alloy has a higher strength than the original brazing material. If the clearance is too large not all the material changes to the new alloy and so the strength is not increased. With joint clearances less than 0.1 mm it is difficult to obtain sound joints and so the strength is not as high.

In addition to the alloying of the brazing material improving the strength, even higher strengths can occur in certain circumstances, e.g. the copper brazing of low carbon steels in a reducing atmosphere. With such a joint when tensile stresses are applied, the joint, being a ductile material, exhibits necking. The effect of this is to put the joint material in shear and since the shear strength is greater than the tensile strength the joint strength is improved.

17.4 Welding

With brazing and soldering, the joint is effected by inserting a metal between the two metal surfaces being joined, the inserted metal having a lower melting point than that of the materials being joined. With *welding*, the joint is effected directly between the parts being joined by the application of heat or pressure. In *fusion welding* an external heat source is used to melt the interfaces of the joint materials and so cause the materials to fuse together. With *solid-state welding*, pressure is used to bring the two interfaces of the joint materials into intimate contact and so fuse the two materials together. Welding processes are capable of producing high-strength joints. The temperatures involved in producing the welds may, however, cause detrimental changes in the materials being joined. These may be local distortions due to uneven thermal expansion, residual stresses, or microstructural changes.

There are four main types of process used for fusion welding:

1 Electric arcs
2 Electrical resistance

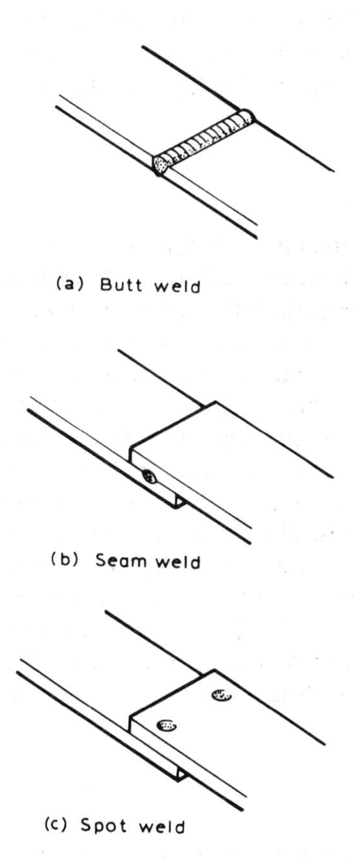

Figure 17.3 Examples of resistance-welded points (a) Butt weld, (b) Seam weld, (c) Spot welds

3 Radiation
4 Thermochemical

With *electric arc welding* an arc is produced between the workpiece and an electrode. Temperatures of the order of 20 000 K are produced with currents between the electrode and workpiece of the order of 200 A. With *electrical resistance welding* the high temperatures are produced by passing an electric current across the interface of the joint and result from the passage of current through the electrical resistance of the joint. With *radiation welding* the high temperatures occur as the result of focusing a beam of electrons, in a vacuum or low pressure, on to the joint area. An alternative is to use a laser to focus a beam of radiant energy on to the joint. *Thermochemical welding* uses chemical reactions to generate the heat. One form of this uses a thermit reaction to produce liquid steel. Another form has oxygen and some fuel gas combining in a flame, oxygen and acetylene being very common.

Arc welding gives high quality welds, is a flexible method, and is low in cost. It is used for joints on bridges, piping, ships, etc. Electrical resistance welding can be used to give butt welds between two surfaces which butt up to each other, seam welds (a line of welded material between two sheets), or spot welds in which the weld occurs only in a small region, spots (Figure 17.3). Spot welding is used in the car industry for bodywork, seam welding is used in sheet metal fabrication. Thermit welding is used for the repair of iron and steel castings, railway lines, shafts, etc.

There are a number of forms of solid-state welding. *Pressure welding* involves a ductile material being pressed against a similar or dissimilar metal; aluminium and copper can be welded, cold, by this method. *Friction welding* involves sliding one material surface, under pressure, over the other. The friction breaks up any surface films and softens the surfaces by virtue of the rise in temperature produced by the friction. With *explosive welding* the two surfaces are impacted together by an explosive charge.

Solid-state welding is widely used for cladding sheets with a thin layer of some other metal. Aluminium alloy sheets are often clad with aluminium.

Welds in steel

Figure 17.4 shows a cross-section of a weld between two plates. Molten steel is produced between the two plates in the welding process and during this melting and then solidification heat is conducted into the plates on either side of the weld. The term *heat-affected zone* is used to describe those parts of the plates which have their temperatures raised to above the critical point A_1.

In the weld pool the metal shows the typical cast-metal structure of columnar grains (see Figure 2.1). In the case of a multi-pass weld deposit each pass forms a heat-affected zone in the weld metal immediately below it. The result is a more complex microstructure.

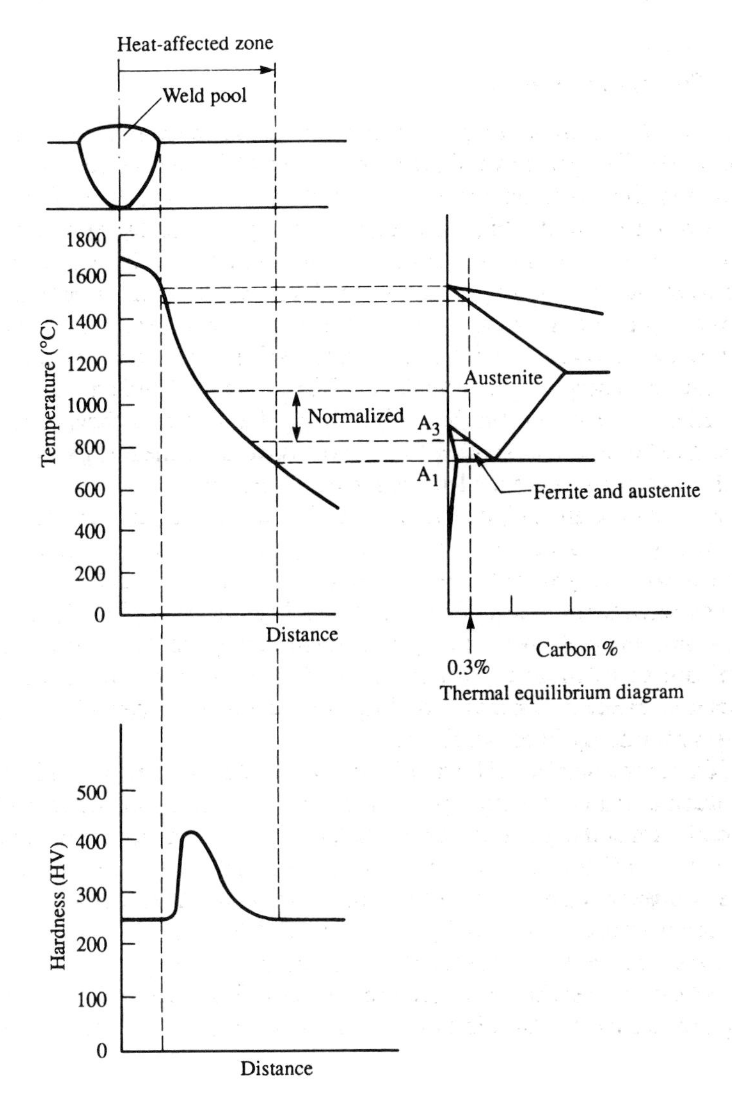

Figure 17.4 The effect of welding heat on the hardness of a 0.3 per cent carbon steel

In the heat-affected zone there will be a wide variation in temperature as illustrated in Figure 17.4. For a 0.3 per cent carbon steel, close to the weld pool the temperature will rise to well above the A_3 temperature, while at the edge of the heat-affected zone it will be just A_1. Thus, close to the weld pool the metal will have been heated into the austenitic state while near the edge of the heat-affected zone it will be ferrite plus austenite. With the low-carbon steel where the temperature rises to above A_3 grain growth occurs – the closer to the weld pool the larger the grain size produced. For the region just above the A_3 temperature the cooling may result in the metal becoming normalized. Where the temperature had risen to between A_1 and A_3 the original ferrite and pearlite is likely to reform. These various effects can be indicated by a hardness traverse across the welded and heat-affected zone. A peak in hardness occurs within that part of the heat-affected zone that was heated to the higher temperature regions of the austenitic state.

For higher-carbon and low-alloy steels the rate of cooling is likely to be such that martensite forms, particularly in that part of the heat-affected zone that reaches the highest temperatures. The result of such changes is an even greater rise in hardness. The hardness produced depends on the hardenability of the steel, the cooling rate and to some extent on the prior austenite grain size. The likelihood of martensite forming can be reduced by reducing the cooling rate. This can be done by preheating the whole structure in the vicinity of the joint before welding.

The hardness of the heat-affected zone is a measure of the tensile strength and, for a given alloy, an indication of the degree of embrittlement. The degree of embrittlement that can occur can be sufficient to cause cracking either during the welding operation or when the welded material is in service.

The hardenability of a steel depends on the elements present in the alloy, different elements having different effects. An overall measure is the *carbon equivalent*. The greater the carbon equivalent the greater will be the hardenability and hence the greater the hardness in the weld heat-affected zone. The carbon equivalent thus becomes a measure of the likelihood of cracking occurring and hence weldability. A number of different carbon equivalent formulae exist, a commonly used one being:

$$\text{Carbon equivalent} = \%\text{C} + \frac{\%\text{Mn}}{6} + \frac{\%\text{Cr}+\%\text{Mo}+\%\text{V}}{5} + \frac{\%\text{Ni}+\%\text{Cu}}{15}$$

Thus, for example, a steel with 0.23 per cent carbon and 0.60 per cent manganese will have a carbon equivalent of 0.33 per cent. Materials with carbon equivalents of more than about 0.50 per cent are very susceptible to cracking; below about 0.40 per cent it is relatively unlikely. Thus the steels with a high risk of cracking are alloy steels, tool steels and plain carbon steels with carbon contents greater than about 0.3 to 0.4 per cent.

The hardness of the heat-affected zone may be reduced by a post-welding heat treatment. This can temper the material and give stress relief.

Welds in aluminium

Fusion welding of aluminium is mainly used with pure aluminium, the non-heat-treatable aluminium-manganese and aluminium-magnesium alloys, and the aluminium-magnesium-silicon and aluminium-zinc-magnesium heat-treatable alloys. The higher strength aluminium-copper-magnesium and duraluminium alloys cannot be effectively fusion-welded and such alloys, which are mainly used for aircraft structures, are normally joined by riveting.

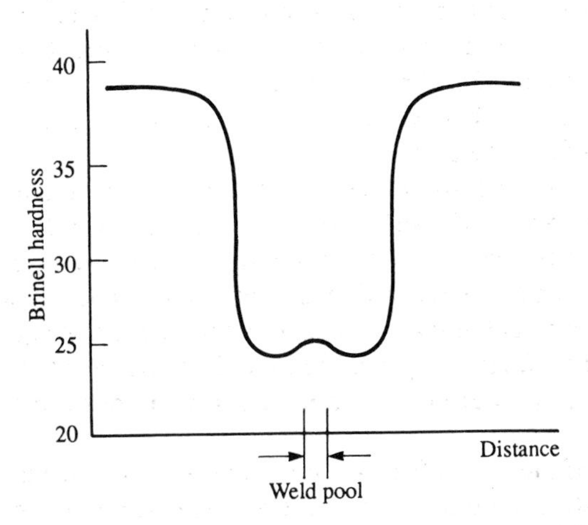

Figure 17.5 Hardness of weld zone in a non-heat-treatable aluminium alloy

Figure 17.5 shows the results of a hardness traverse of a weld between two plates of a non-heat-treatable aluminium alloy. Within the heat-affected zone the alloy is fully or partially annealed by the temperatures produced during the welding. The result is that the work-hardened material is much softer in the weld region than in the unaffected material. Since the tensile strength is related to the hardness this means there is a drop in strength. This effect is in the main irreversible, although the strength of the weld pool area can be improved by rolling or hammering.

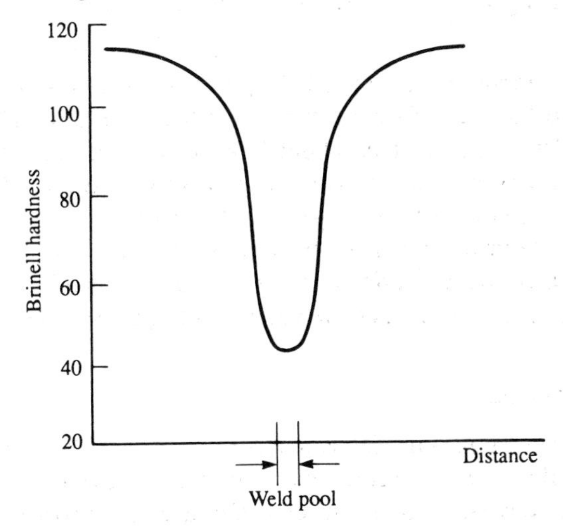

Figure 17.6 Hardness of weld zone in a heat-treatable aluminium alloy

Figure 17.6 shows the results of a hardness traverse of a weld between two plates of a heat-treatable aluminium alloy. Softening and a reduction in tensile strength occur within the heat-affected zone. However, the strength may be almost completely recoverable by solution treatment and ageing of the component.

Welds in copper

Copper oxide reacts with hydrogen, at temperatures above about 500°C, to give copper and steam. A consequence of this is that the steam gives rise to porosity in the weld pool, as well as causing fissures to develop and embrittlement to occur in the heat-affected zone close to the weld pool. To reduce this effect the copper has to be deoxidized prior to welding. This can be achieved by the presence of small amounts of phosphorus in the copper. Hence, welds of acceptable strength are possible in phosphorus-deoxidized copper.

Brasses are difficult to weld because of the volatilization of zinc, this leading to porosity. Tin bronzes, aluminium bronzes and silicon bronzes can be welded without the porosity problem occurring. Aluminium bronzes, however, are susceptible to cracking during welding. Cupronickels can be welded without porosity if a deoxidized alloy or filler rods containing deoxidant are used.

The strength of work hardened or age-hardened copper alloys is reduced by welding. After welding the strength of a fusion welded joint in a hard temper copper is probably about that of the annealed metal.

Corrosion of welds

Because of differences in the microstructure and composition of welded areas compared with the parent metal, selective corrosion of welds is likely in many corrosive environments. Thus welded carbon steels exposed to a marine environment may show corrosion more markedly in the weld metal or heat-affected zone than the parent metal. Heating a stainless steel to about 500 to 700°C can lead to the precipitation of carbides at grain boundaries. This results in the removal of chromium from grains to the boundary and hence a reduction in corrosion resistance. The effect is known as weld decay since such effects occur during the welding of stainless steels. The defect can be overcome by heat treating the steel or using a stabilized stainless steel, i.e. one which includes niobium or titanium.

Weldability

The term *weldability* is used to describe the ease with which a sound weld can be made between materials. Sound welds between dissimilar metals are only feasible if they are soluble in each other. With both similar and dissimilar materials weldability problems can occur if the metals have a high sulphur content. This is because such metals crack in the weld as a result of the sulphur causing low strength in the solidifying metal. A high carbon, or equivalent carbon content, can cause problems as a result of embrittlement. Embrittlement may also occur as a result of the presence of hydrogen.

17.5 Fastening systems

The choice of fastener will depend on a number of factors:

1 Environmental, e.g. temperature, corrosive conditions.
2 Nature of the external loading on the fastener, e.g. tension, compression, shear, cyclic, impact, and its magnitude.
3 Life and service requirements, e.g. frequent assembly and disassembly.
4 Design of the components being joined and types of material involved.
5 Quantity of fasteners required and their cost.

Fasteners provide a clamping force between two pieces of material. A wide variation exists in types of fastener and the materials used for making them. The types of fastener available can be classified as threaded, non-threaded and special-purpose. Steel is probably the most common material used, although aluminium alloys, brass and nickel are among other metals used. Aluminium alloy fasteners have the advantage over steel of being much lighter, non-magnetic and more corrosion resistant. Nickel has the particular advantage of strength at high temperatures.

With a *threaded fastener*, the clamping force holding the two pieces of material together is produced by a torque being applied to the fastener and being maintained during the service life of the fastener. Bolts mated with nuts, and screws with threads in the material, are examples of threaded fasteners.

Nails, rivets and pins are examples of *non-threaded fasteners*. Nails are used extensively for making joints between pieces of wood. Rivets, however, are used for joining dissimilar or similar materials, both metallic and non-metallic. Both nails and rivets are low-cost fasteners designed for making joints which are intended to be permanent and non demounted.

Fatigue properties of fastened joints

The use, for example, of bolts or rivets as fasteners for joints can introduce fretting damage. *Fretting* is the wear process that occurs at the areas of contact of two metals undergoing small amplitude cyclic slip. On steels this damage may be visible as the red oxide of iron, on aluminium as black oxides. The damage is referred to as *galling* or *scuffing*. Such damage can lower the fatigue strength by factors as high as three for aluminium alloys. One form of anti-fret treatment is to separate the metal surfaces by using PTFE shims or a coating of a paint containing the solid lubricant molybdenum disulphide.

17.6 Joining methods for plastics

The joining methods that can be used with plastics can be considered to fall in four main groups:

1 *Welding*
Thermoplastics can be welded by a number of methods, all involving the melting of the interface between the plastic surfaces being joined. With hot-gas welding, hot-plate welding and hot-wire welding the interface is melted by direct heating. For example, hot-gas welding is similar to oxyacetylene welding with heat being applied using a welding torch to blow hot gas onto the interface. With spin welding, vibration welding and ultrasonic welding the interface is melted by frictional heat being developed by the two components rubbing together. Ultrasonic welding involves the application to the joint area of vibrations with frequencies of the order of 20 kHz. This method results in consistently high weld strengths.

2 *Adhesive bonding*
The types of adhesive used may be elastomers, thermoplastics, thermoset or two-polymer types. Some plastics may be bonded by the use of a solvent to soften the interfaces of the joint, then light pressure is applied to bring the surfaces into close contact.

3 *Riveting*
Both metal and thermoplastic rivets are used. With this method, joints can be made between plastics and metals.

4 *Press and snap fits*
This is an important way of making both permanent and recoverable assemblies. One of the advantages of plastics is that they can be subject to quite severe elastic distortion and still return to their original shape when

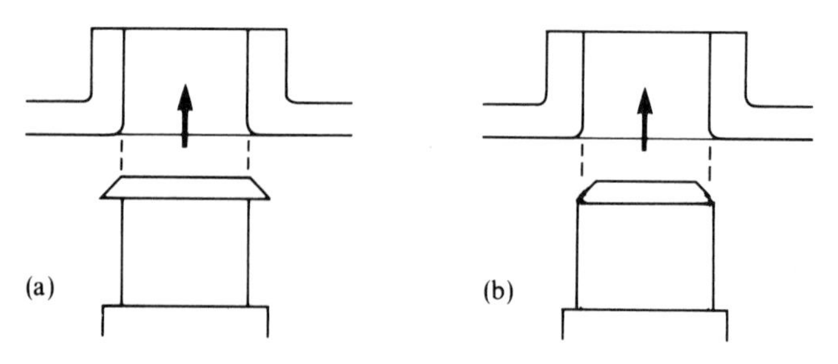

Figure 17.7 (a) A snap-fit (b) A press-fit

the load is removed. Press and snap fits rely on this. A common form of snap fit is the hook joint (Figure 17.7(a)). When the component is pushed into the hole, the end is deformed so that it can slide through the hole until it emerges from the other end. Then it expands and locks the component in position. Figure 17.7(b) shows a press-fit. The component is just a tight fit in the hole.

5 *Thread systems*
Screw threads and self-tapping screws are widely used.

Problems

1 Discuss the merits of joining by means of adhesives and the limitations of the method.
2 List the factors that determine the strength of an adhesive-bonded joint.
3 Distinguish between soldering, brazing, fusion welding and cold welding.
4 State and explain the factors that determine the strength of a brazed joint.
5 Explain the effect of brazing on the structure and properties in a brazed joint.
6 Describe and explain the hardness and structural changes that can occur with the welding of steel plates.
7 Why are there problems in welding high-carbon steels?
8 Explain the reasons for preheating and post-welding heat treatment.
9 Describe and explain the hardness and structural changes that can occur with the welding of non-heat-treatable and heat-treatable aluminium alloys.
10 State the factors which have to be considered in making a choice of fastener.
11 Explain what fretting is and its significance for fastened joints.

PART FIVE

Selection

18

Selection of materials

18.1 General considerations

A number of vital questions need to be posed before a decision can be made as to the specification required of a material and hence its selection. The questions can be grouped under four headings: properties, availability, processing parameters and cost.

Properties

1 *What mechanical properties are required?* For example: hardness, tensile strength, ductility, stiffness, behaviour in compression, impact strength, creep resistance, fatigue resistance, wear properties, shear properties. Will the properties be required at low temperatures, room temperature or high temperatures?

2 *What physical properties will be required?* For example: density, melting point, electrical resistance, thermal expansivity, heat capacity, flammability, thermal conductivity.

3 *What chemical properties will be required?* For example: chemical structure, chemical bonding, composition, resistance to environmental degradation.

4 *What dimensional conditions are required?* For example: flatness, surface finish, dimensional stability, size.

Availability

1 Is the material readily available? Is it, for example, already in store, or perhaps quickly obtainable from the normal suppliers?

2 *Are there any ordering problems?* Is the material only available from special suppliers? Is there a minimum order quantity?

3 *What forms is the material usually supplied in?* Is the material, for example, usually supplied in bars or perhaps sheets?

Processing parameters

1 *Are there any special processing requirements?* For example, does the material have to be cast? or perhaps extruded?

2 *Are there any material treatment requirements?* For example, does the material have to be annealed or perhaps solution hardened? Are any surface treatments required?
3 *Are there any special tooling requirements?* Does, for instance,the hardness of the material mean that special cutting tools are required?

Cost

1 *What is the cost of the raw material?* Could a cheaper material be specified?
2 *What quantity is required?* What quantity of product is to be produced per week, per month, per year? What stocking policy should be adopted for the material?
3 *What are the cost implications of the process requirements?* Does the process require high initial expenditure? Are the running costs high or low? Will expensive skilled labour be required?
4 *What are the cost penalties from overspecification?* If the material is, for example, stronger than is required will this significantly increase the cost? If the product is manufactured to a higher quality than is required what will be the cost implications?

This chapter is a consideration of the questions raised above concerning the selection in relation to the properties required of materials. Chapter 19 is a consideration in relation to the processing requirements and Chapter 20 in relation to costs.

18.2 Selection for static strength

Static strength can be defined as the ability to resist a short-term steady load at moderate temperatures without breaking or crushing or suffering excessive deformations. If a component is subject to a uniaxial stress the yield stress is commonly taken as a measure of the strength if the material is ductile and the tensile strength if it is brittle. Measures of static strength are thus yield stress, proof stress, tensile strength, compressive strength and hardness, the hardness of a material being related to the tensile strength of a material.

If the component is subject to biaxial or triaxial stresses, e.g. a shell subject to internal pressure, then there are a number of theories which can be used to predict material failure. The maximum principal stress theory, which tends to be used with brittle materials, predicts failure as occurring when the maximum principal stress reaches the tensile strength value, or the elastic limit stress value, that occurs for the material when subject to simple tension. The maximum shear stress theory, used with ductile materials, considers failure to occur when the maximum shear stress in the biaxial or triaxial stress situation reaches the value of the maximum shear stress that occurs for the material at the elastic limit in simple tension. With biaxial stress, this occurs when the difference between the two principal stresses is equal to the elastic limit stress. Another theory that is used with ductile materials is that failure occurs when the strain energy per unit volume is equal to the strain energy at the elastic limit in simple tension.

It should be recognized that a requirement for strength in a component requires not only a consideration of the static strength of the material but also the design. Thus for bending, an I-beam is more efficient than a rectangular cross-section beam because the material in the beam is concentrated at the top and bottom surfaces where the stresses are high and is not 'wasted' in regions where the stresses are low. A thin shell or skin can be strengthened by adding ribs or corrugations.

For most ductile wrought materials, the mechanical properties in compression are sufficiently close to those in tension for the more readily available tensile properties to be used as an indicator of strength in both tension and compression. Metals in the cast condition, however, may be stronger in compression than in tension. Brittle materials, such as ceramics, are generally stronger in compression than in tension. There are some materials where there is significant anisotropy, i.e. the properties depend on the direction in which it is measured. This can occur with, for example, wrought materials where there are elongated inclusions and the processing results in their becoming orientated in the same direction (see section 10.3) or in composite materials containing unidirectional fibres.

The mechanical properties of metals are very much affected by the treatment they undergo, whether it be heat treatment or working. Thus it is not possible to give anything other than a crude comparison of alloys in terms

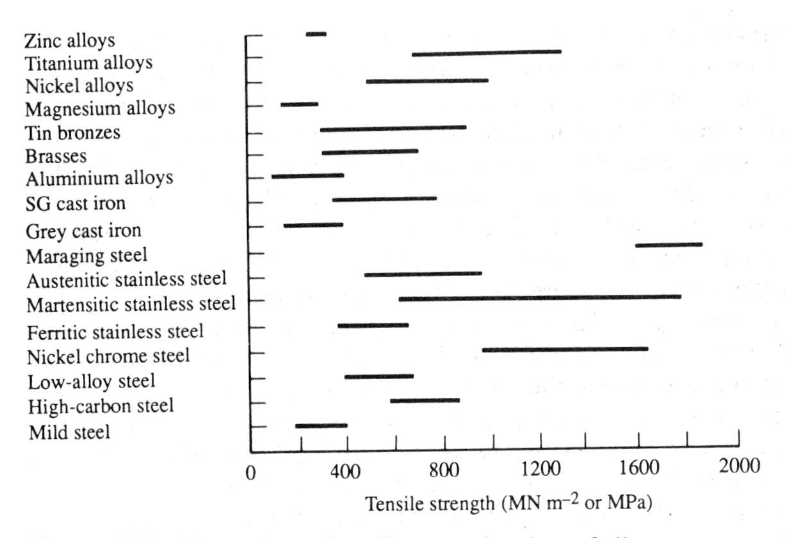

Figure 18.1 Comparison of tensile strength values of alloys

of tensile strengths. Figure 18.1 shows such a comparison. The properties of polymeric materials are very much affected by the additives mixed in with them in their formulation and thus only a crude comparison of mechanical properties of different polymers is possible. Figure 18.2 shows such a comparison for tensile strengths for thermoplastics and Figure 18.3 for thermosets. Figure 18.4 shows the range of hardness values that occurs with different types of materials. All the data in Figures 18.1, 18.2, 18.3 and 18.4 refer to temperatures around about 20° C.

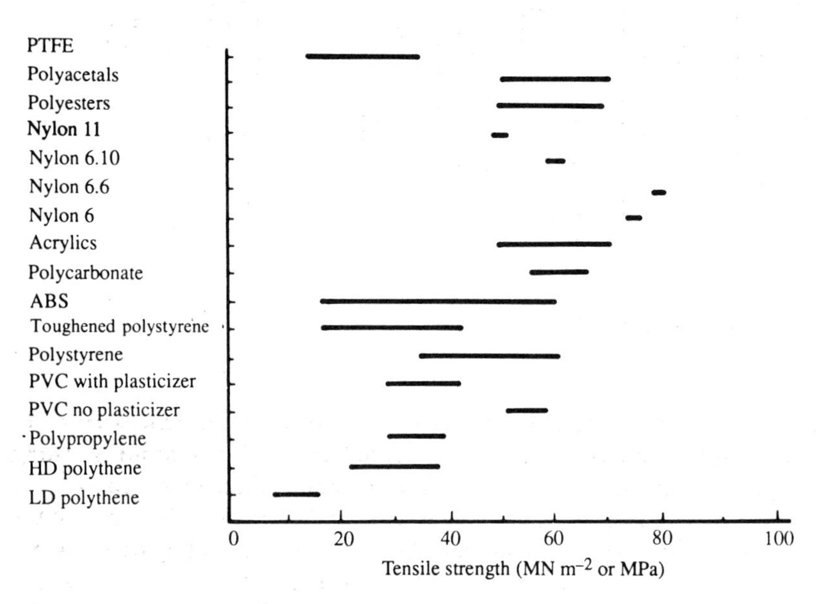

Figure 18.2 Comparison of tensile strength values of thermoplastics

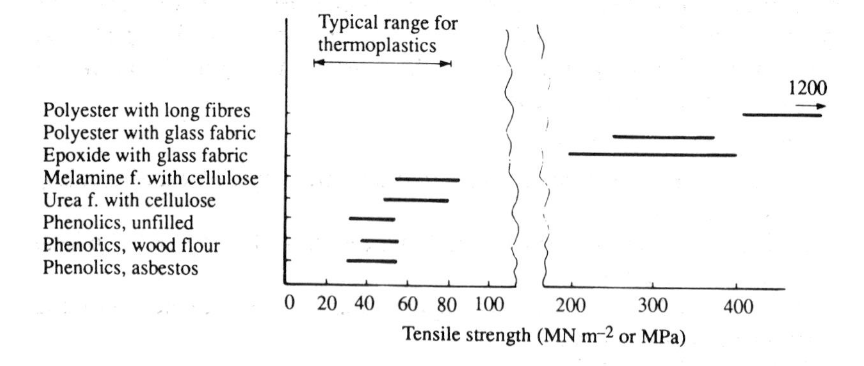

Figure 18.3 Comparison of tensile strength values of thermosets

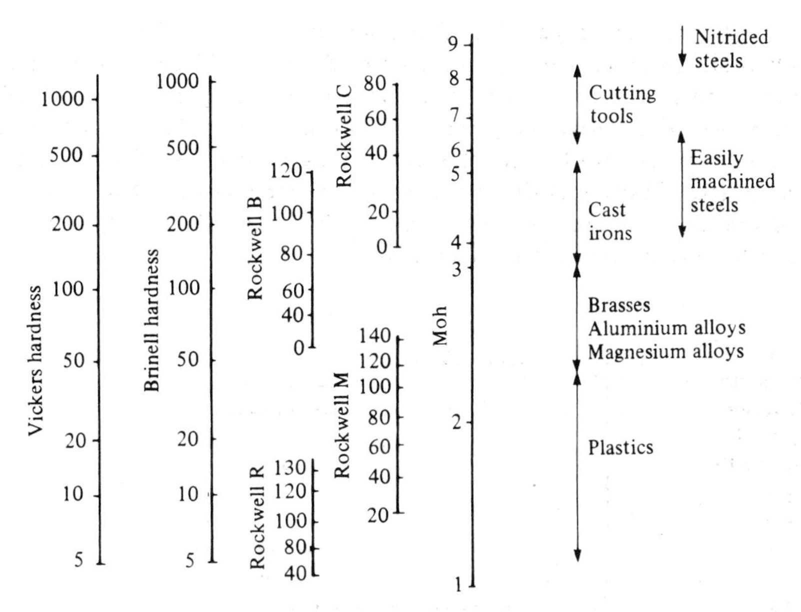

Figure 18.4 Hardness values

When mechanical properties of a steel are quoted by manufacturers or in standard tables, different values are quoted for different limiting ruling sections. The limiting ruling section is the maximum diameter of round bar at the centre of which the specified properties may be obtained. The reason for this is that during the heat treatment different rates of cooling occur at the centres of bars, or indeed any cross-section, due purely to differences in sizes. Thus in selecting a material to give specific properties, the size of the component being made has to be taken into account.

There are many different steel compositions and each can often be heat treated in a number of ways to give different properties. There is thus often a very large number of possibilities in selecting steels to give particular properties. The following are, however, the most commonly made choices for different ranges of tensile strength.

1 *620–770 MN m^{-2} (MPa)*
080M40 Medium carbon steel, hardened and tempered, limiting ruling section 63 mm.
150M36 Carbon-manganese steel, hardened and tempered, limiting ruling section 150 mm.
503M40 A 1 per cent nickel steel, hardened and tempered, limiting ruling section 250 mm.

2 *700–850 MN m^{-2} (MPa)*
150M36 A 1.5 per cent manganese steel, hardened and tempered,

limiting ruling section 63 mm.
708M40 A 1 per cent Cr-Mo steel, hardened and tempered, limiting ruling section 150 mm.
605M36 A 1.5 per cent Mn-Mo steel, hardened and tempered, limiting ruling section 250 mm.

3 *770–930 MN m^{-2} (MPa)*
708M40 A 1 per cent Cr-Mo steel, hardened and tempered, limiting ruling section 100 mm.
817M40 A 1½ per cent Ni-Cr-Mo steel, hardened and tempered, limiting ruling section 250 mm.

4 *850–1000 MN m^{-2} (MPa)*
630M40 A 1 per cent Cr steel, hardened and tempered, limiting ruling section 63 mm.
709M40 A 1 per cent Cr-Mo steel, hardened and tempered, limiting ruling section 100 mm.
817M40 A 1½ per cent Ni-Cr-Mo steel, hardened and tempered, limiting ruling section 250 mm.

5 *930–1080 MN m^{-2} (MPa)*
709M40 A 1½ per cent Cr-Mo steel, hardened and tempered, limiting ruling section 63 mm.
817M40 A 1½ per cent Ni-Cr-Mo steel, hardened and tempered, limiting ruling section 100 mm.
826M31 A 2½ per cent Ni-Cr-Mo steel, hardened and tempered, limiting ruling section 250 mm.

6 *1000–1150 MN m^{-2} (MPa)*
817M40 A 1 per cent Ni-Cr-Mo steel, hardened and tempered, limiting ruling section 63 mm.
826M31 A 2½ per cent Ni-Cr-Mo steel, hardened and tempered, limiting ruling section 150 mm.

7 *1080–1240 MN m^{-2} (MPa)*
826M31 A 2½ per cent Ni-Cr-Mo steel, hardened and tempered, limiting ruling section 100 mm.
826M40 A 2½ per cent Ni-Cr-Mo steel, hardened and tempered, limiting ruling section 250 mm.

8 *1150–1300 MN m^{-2} (MPa)*
826M40 A 2½ per cent Ni-Cr-Mo steel, hardened and tempered, limiting ruling section 150 mm.

9 *1240–1400 MN m^{-2} (MPa)*
826M40 A 2½ per cent Ni-Cr-Mo steel, hardened and tempered, limiting ruling section 150 mm.

10 *1540 MN m^{-2} (MPa) minimum*
835M30 A 4 per cent Ni-Cr-Mo steel, hardened and tempered, limiting ruling section 150 mm.

18.3 Selection for stiffness

Stiffness can be considered to be the ability of a material to resist deflection when loaded. Thus if we consider a cantilever of length L subject to a point load F at its free end, then the deflection y at the free end is given by

$$y = \frac{FL^3}{3EI}$$

with E being the tensile modulus and I the second moment of area of the beam cross-section with respect to the neutral axis. Thus for a given shape and length cantilever, the greater the tensile modulus the smaller the deflection. Similar relationships exist for other forms of beam. Hence we can state that the greater the tensile modulus the greater the stiffness.

The tensile modulus of a metal is little affected by changes in its composition or heat treatment. However, the tensile modulus of composite materials is very much affected by changes in the orientation of the fillers and the relative amounts. Figure 18.5 shows typical tensile modulus values for metal alloys, Figure 18.6 for thermoplastics and Figure 18.7 for thermosets.

The deflection of a beam is a function of both E and I. Thus, for a given material, a beam can be made stiffer by increasing its second moment of area. The second moment of area of a section is increased by placing as much as possible of the material as far as possible from the axis of bending. Thus an I-section is a particularly efficient way of achieving stiffness. Similarly a tube is more efficient than a solid rod.

Another situation which is related to the value of EI is the *buckling* of columns when subject to compressive loads. The standard equation used for buckling has it occurring for a column of length L when the load F reaches the value

$$F = \frac{\pi^2 EI}{L^2}$$

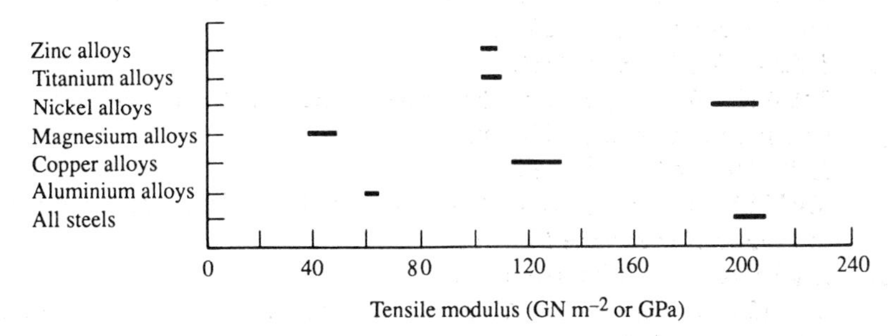

Figure 18.5 Comparison of tensile modulus values of alloys

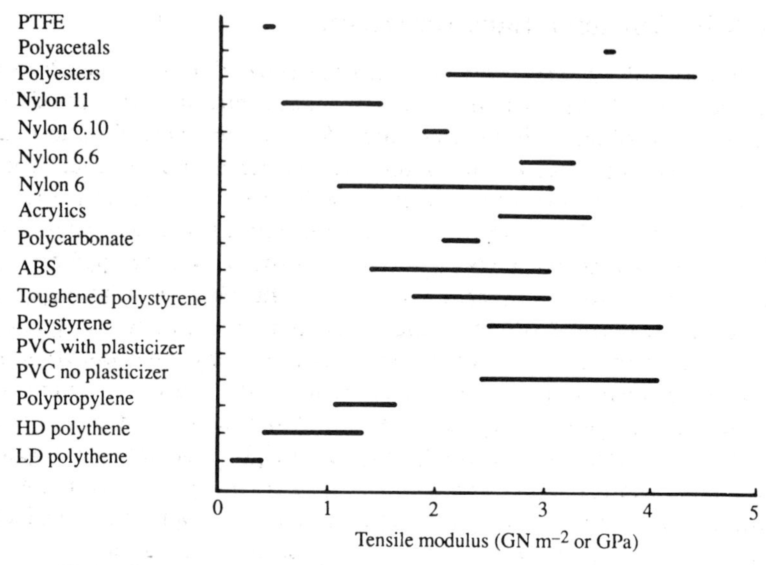

Figure 18.6 Comparison of tensile modulus values of thermoplastics

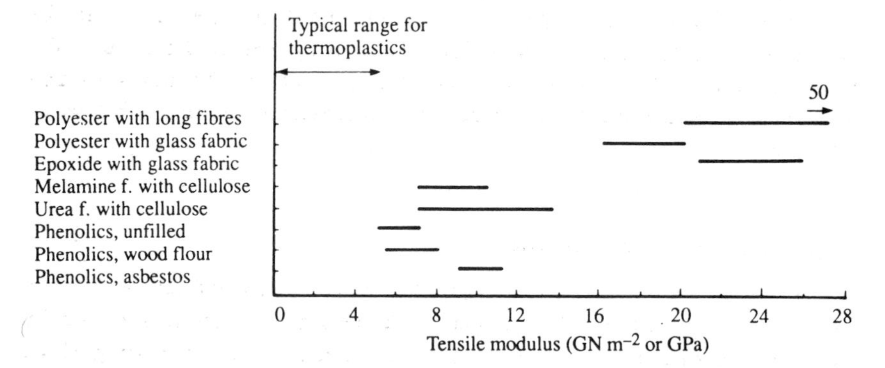

Figure 18.7 Comparison of tensile modulus values of thermosets

This equation is known as Euler's equation. Thus the bigger the value of *EI* the higher the load required to cause buckling. Hence we can say that the column is stiffer the higher the value of *EI*. Note that a short and stubby column is more likely to fail by crushing when the yield stress is exceeded rather than buckling. Buckling is however more likely to be the failure mode if the column is slender.

18.4 Selection for fatigue resistance

In section 18.2 the concern was with the static strength of materials; here we are concerned with the *dynamic strength*, i.e. the fatigue property. The failure of a component when subject to fluctuating loads is the result of cracks which tend to start at some discontinuity in the material and grow until failure occurs (see Chapter 7). The main factors affecting fatigue properties of metals are discussed in section 7.3. They are stress concentrations caused by component design, corrosion, residual stresses, surface finish/treatment, temperature, the microstructure of the alloy and its heat treatment. Only to a limited extent does the choice of material determine the fatigue resistance of a component.

In general, for metals the fatigue limit or endurance limit at about 10^7 to 10^8 cycles lies between about a third and a half of the static tensile strength. For steels the fatigue limit is typically between 0.4 and 0.5 that of the static strength. Inclusions in the steel, such as sulphur or lead to improve machinability, can however reduce the fatigue limit. For grey cast iron the fatigue limit is about 0.4 that of the static strength, for nodular and malleable irons in the range 0.5 for ferritic grades to 0.3 for the higher strength pearlitic irons, for blackheart, whiteheart and the lower strength pearlitic malleable irons about 0.4. With aluminium alloys the endurance limit is about 0.3 to 0.4 that of the static strength, for copper alloys about 0.4 to 0.5.

Fatigue effects with polymers are complicated by the fact that the alternating loading results in the polymer becoming heated. This causes the elastic modulus to decrease and at high enough frequencies this may be to such an extent that failure occurs. Thus fatigue in polymers is very much frequency dependent.

18.5 Selection for toughness

Toughness (see section 6.4) can be defined as the resistance to fracture. A tough material is resistant to crack propagation. A measure of toughness is given by two main measurements: the resistance of a material to impact loading which is measured in the Charpy or Izod tests by the amount of energy needed to fracture a test piece (see Chapter 5) and the resistance of a material to the propagation of an existing crack in a fracture toughness test (see section 6.4).

Within a given type of metal alloy there is an inverse relationship between yield stress and toughness, the higher the yield stress the lower the toughness. Figure 18.8 illustrates this. Thus if, for instance, the yield strength of low alloy, quenched and tempered steels is pushed up by metallurgical means then the toughness declines. Steels become less tough with increasing carbon content and larger grain size.

The toughness of plastics is improved by incorporating rubber or another tougher polymer, copolymerization, or incorporating tough fibres. For

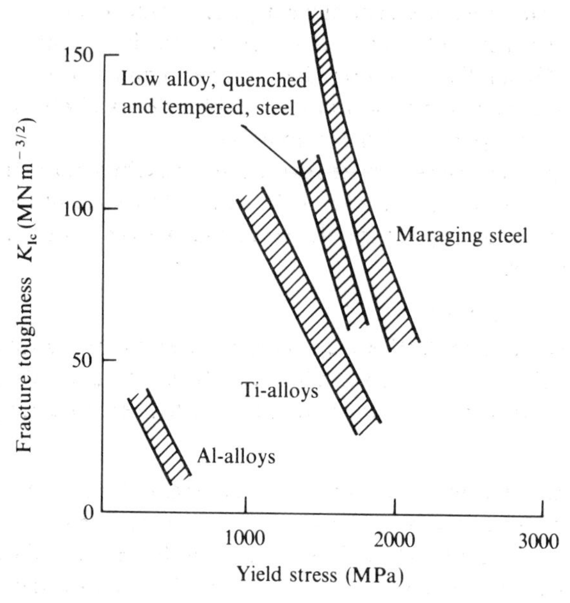

Figure 18.8 Variation of fracture toughness with yield stress

example, styrene–acrylonitrile (SAN) is brittle and far from tough. It can however be toughened with the rubber polybutadiene to give the tougher acrylonitrile–butadiene–styrene (ABS) (see section 15.1 and Figure 15.2).

18.6 Selection for creep and temperature resistance

The creep resistance of a metal can be improved by incorporating a fine dispersion of particles to impede the movement of dislocations (see section 8.2). The Nimonic series of alloys, based on an 80/20 nickel–chromium alloy, have good creep resistance as a consequence of fine precipitates formed by the inclusion of small amounts of titanium, aluminium, carbon or other elements. Creep increases as the temperature increases (see section 8.3) and is thus a major factor in determining the temperature at which materials can be used. Another factor is due to the effect on the material of the surrounding atmosphere. This can result in surface attack and scaling which gradually reduces the cross-sectional area of the component and so its ability to carry loads. Such effects increase as the temperature increases. The Nimonic series of alloys have good resistance to such attack and thus this, combined with their good creep resistance, makes them materials which can be used at elevated temperatures. Typically they can be used up to temperatures of the order of 900°C (see Table 8.2).

For most metals creep is essentially a high temperature effect, however this is not the case with plastics. Here creep can be significant at room temperatures. Generally thermosets have higher temperature resistance than thermoplastics, however the addition of suitable fillers and fibres can improve the temperature properties of thermoplastics.

Figure 18.9 shows typical high temperature limits for some commonly used engineering materials. The following is an amplification of the figure and considers the types of engineering materials that are typically used in various temperature ranges.

Room temperature to 150°C

In general, few thermoplastics are recommended for prolonged use above about 100°C. Glass-filled nylon can however be used up to 150°C. The only engineering metal which has limits within this temperature range is lead.

150° to 400°C

Magnesium and aluminium alloys can in general only be used up to about 200°C, though some specific alloys can be used to higher temperatures. For example, the aluminium alloy LM13 (AA336.0), an aluminium–silicon–copper–magnesium alloy, is used for pistons in engines and experiences temperatures of the order of 200° to 250°C while some cast aluminium bronzes can be used up to about 400°C with wrought aluminium bronzes up to about 300°C. Plain carbon and manganese–carbon steels are widely used for temperatures in this range.

400° to 600°C

Plain carbon and manganese–carbon steels cannot be used above about 400° to 450°C. For such temperatures low-alloy steels are used. These commonly

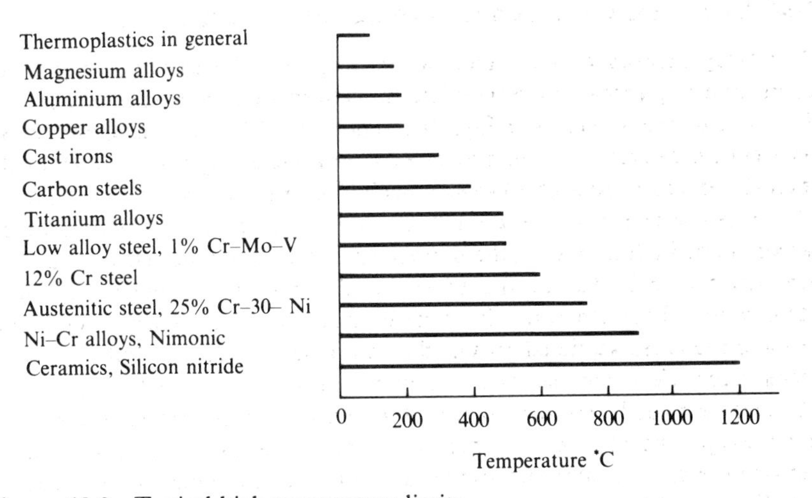

Figure 18.9 Typical high temperature limits

include such elements as molybdenum, chromium and vanadium. For temperatures up to about 500°C a carbon–0.5% molybdenum steel might be used, up to about 525°C a 1% chromium–0.5% molybdenum steel, up to about 550°C a 0.5% chromium–molybdenum–vanadium steel, and up to about 600°C a steel with 5 to 12% chromium. Titanium alloys are also widely used in this temperature range. The alpha-beta alloy 6% aluminium–4% vanadium (IMI318) (see Chapter 13) is used up to about 450°C. Near alpha alloys can be used to higher temperatures, for example the alloy with 5.5% aluminium–3.5% tin, 3% zirconium–1% niobium–0.25% molybdenum–0.25% tin (IMI 829) is used up to about 600°C.

*600° to 1000°*C

The metals most widely used in this temperature range are the austenitic stainless steels, nickel–chromium and nickel–chromium–iron alloys, and cobalt base alloys. Austenitic stainless steels with 18% chromium–8% nickel can be used at temperatures up to about 750°C. There is a range of high temperature alloys based on the nickel–chromium base (see section 13.5), the term superalloy often being used for such alloys that are able to maintain their strength, resistance to creep and oxidation resistance at high temperatures. The Nimonic series contains alloys such as the precipitation hardening Nimonic 90 which can be used for temperatures up to about 900°C, Nimonic 901 to about 1000°C. Another series of high temperature alloys are the nickel–chromium–iron alloys, such as the Inconel and Incoloy series. For example, Inconel 600 can be used up to virtually 1000°C and Incoloy 800H to 700°C. Cobalt-base alloys are not used as extensively at the nickel-base alloys, tending to be used only where a prime requirement is high corrosion resistance.

Above 1000°C

The materials which can be used at temperatures in excess of 1000°C are the refractory metals, i.e. molybdenum, niobium, tantalum and tungsten, and ceramics. The refractory metals, and their alloys, can be used at temperatures in excess of 1500°C. Surface protection is one of the main problems facing the use of these alloys at high temperatures. Ceramics can also be used at such high temperatures but tend to suffer from the problems of being hard, brittle and vulnerable to thermal shock. Alumina is used in furnaces up to about 1600°C, silicon nitride to about 1200°C and silicon carbide to about 1500°C.

18.7 Selection for corrosion resistance

For metals subject to atmospheric corrosion the most significant factor in determining the chance of corrosive attack is whether there is an aqueous electrolyte present (see Chapter 9). This could be provided by condensation of moisture occurring as a result of the climatic conditions. The amount of pollution in the atmosphere can also affect the corrosion rate. Corrosion can often be much reduced by the selection of appropriate materials (see section 9.3 for other ways of reducing corrosion). For metals immersed in water, the corrosion depends on the substances that are dissolved or suspended in the water.

Carbon steels and low alloy steels are not particularly corrosion resistant, rust being the evidence of such corrosion. In an industrial atmosphere, in fresh and sea water, plain carbon steels and low alloy steels have poor resistance. Painting, by providing a protective coating of the surface, can reduce such corrosion. The addition of chromium to steel can markedly improve its corrosion resistance. Steels with 4–6% chromium have good resistance in an industrial atmosphere, in fresh and sea water, while stainless steels have an excellent resistance in an industrial atmosphere and fresh water but can suffer some corrosion in sea water. The corrosion resistance of grey cast iron is good in an industrial atmosphere but not so good in fresh or sea water, though still better than that of plain carbon steels.

Aluminium when exposed to air develops an oxide layer on its surface which then protects the substrate from further attack. Wrought alloys are often clad with thin sheets of pure aluminium or an aluminium alloy to enhance the corrosion resistance of such alloys. Thus in air, aluminium and its alloys have good corrosion resistance. When immersed in fresh or sea water, most aluminium alloys offer good corrosion resistance, though there are some exceptions which must be clad in order to have good corrosion resistance.

Copper in air forms a protective green layer which protects it from further attack and thus gives good corrosion resistance. Copper has also good corrosion resistance in fresh and sea water, hence the widespread use of copper piping for water distribution systems and central heating systems. Copper alloys likewise have good corrosion resistance in industrial atmospheres, fresh and sea water, though demetallification can occur with some alloys, e.g. dezincification of brass with more than 15% zinc.

Nickel and its alloys have excellent resistance to corrosion in industrial air, fresh and sea water. Titanium and its alloys have excellent resistance, probably the best resistance of all metals, in industrial air, fresh and sea water, and are thus widely used where corrosion could be a problem.

Plastics do not corrode in the same way as metals and thus, in general, have excellent corrosion resistance: hence, for example, the increasing use of plastic pipes for the transmission of water and other chemicals. Polymers can deteriorate as a result of exposure to ultraviolet radiation, e.g. that in the rays from the sun, heat and mechanical stress. To reduce such effects, specific additives are used as fillers in the formulation of a plastic (see section 4.3).

Most ceramic materials show excellent corrosion resistance. Glasses are exceedingly stable and resistant to attack, hence the widespread use of glass containers. Enamels, made of silicate and borosilicate glasses, are widely used as coatings to protect steels and cast irons from corrosive attack.

18.8 Selection for wear resistance

Wear is the progressive loss of material from surfaces as a result of contact with other surfaces. It can occur as a result of sliding or rolling contact between surfaces or from the movement of fluids containing particles over surfaces. Because wear is a surface effect, surface treatments and coatings play an important role in improving wear resistance. See, for example, section 12.7 and Table 12.5 concerning surface hardening and the wear resistant treatments. Lubrication can be considered to be a way of keeping surfaces apart and so reducing wear.

Mild steels have poor wear resistance. However, increasing the carbon content increases the wear resistance. Surface hardenable carbon or low alloy steels enable wear resistance to be improved as a result of surface treatments such as carburizing, cynaniding or carbonitriding. Even better wear resistance is provided by nitriding medium carbon chromium or chromium–aluminium steels, or by surface hardening high carbon high chromium steels. Grey cast iron has good wear resistance for many applications. Better wear resistance is, however, provided by white irons. Among non-ferrous alloys, beryllium coppers and cobalt-base alloys, such as Stellite, offer particularly good wear resistance.

Metallic materials for use as bearing surfaces need to be hard and wear resistant, with a low coefficient of friction, but at the same time sufficiently tough. Generally these requirements are met by the use of a soft, but tough, alloy in which hard particles are embedded. The soft alloy is capable of yielding to accommodate any localized high pressures resulting from slight misalignments and starting up, while the hard particles provide the wear resistance. Such materials include the white bearing metals, copper base bearing metals and aluminium base bearing metals. The white bearing metals are tin or lead based materials. The tin base materials, known as Babbit metals, are tin–antimony–copper alloys with possibly some lead. The hard particles are provided by antimony–tin and copper–tin compounds. The lead based materials are lead–antimony–tin alloys with antimony–tin compounds providing the hard particles. The main copper based metals used are phosphor bronzes with 10-15% tin and copper–lead alloys. The main aluminium bearing materials are aluminium–tin alloys.

Self-lubricating plastics, e.g. nylon 6.6 with the lubricating additive of 18% PTFE-2% silicone, offer very good wear resistance and are widely used for low wear applications such as bearings and gears.

18.9 Selection for physical characteristics

As indicated in section 18.1, there are many physical properties which need to be considered in relation to the selection of a material. In this section just density, electrical conductivity, thermal conductivity and thermal expansion are discussed.

Density

The density of a material is an important factor when the weight of a component is critical, as for example in aircraft or where a low weight makes a product a more attractive proposition, e.g. domestic appliances which need to be light and easy to move. What is often required is a combination of strength or stiffness with low weight and thus the important factor is tensile strength/density or tensile modulus/density.

Table 18.1 shows some values of densities and these ratios for typical engineering materials. Magnesium is the lightest metal and, though magnesium alloys have relatively low strengths, the strength to density ratio is better than that of steels and makes them an attractive proposition in situations where low weight is required. Titanium alloys have lower densities than steels and strengths which compare with the best of steels. They thus have an exceptionally high strength to density ratio. Composites can also be designed to have very high strength to density ratios.

Electrical resistivity/conductivity

The electrical resistivity ρ of a length L of material with a cross-sectional area A and electrical resistance R is given by

$$\rho=\frac{RA}{L}$$

Table 18.1 *Strength/density and modulus/density values*

Material	*Density* Mg/m^3	*Strength/ density* $MPa/Mg\,m^{-3}$	*Modulus/ density* $GPa/Mg\,m^{-3}$
Low carbon steels	7.8	51	26
High carbon steel	7.8	109	26
Cr steel	7.8	128	26
Austenitic stainless	7.8	128	26
Aluminium–copper alloys	2.7	148	26
Aluminium–magnesium alloys	2.7	74	26
Brasses	8.7	23	13
Magnesium alloys	1.8	139	22
Ti alpha-beta alloys	4.5	244	24
ABS polymer	1.1	36	2
Nylon	1.1	68	2
Rigid PVC	1.4	39	2
Epoxy + 72% E glass composite	2.2	745	25
Epoxy + 58% carbon composite	1.7	894	97

The electrical conductivity σ is the reciprocal of the resistivity, i.e.

$$\sigma = \frac{L}{RA}$$

In general, metals are good electrical conductors, having low resistivities/high conductivities. Typical resistivities are of the order of 10^{-7} to $10^{-8}\,\Omega$m. Polymers and ceramics are electrical insulators, having very high resistivities/very low conductivities. Typical resistivities are of the order of 10^{-10} to $10^{-18}\,\Omega$m.

The metals in common use in engineering which have the highest electrical conductivities are silver, copper and aluminium. In each case the conductivity is highest when the material is of the highest purity and in the fully annealed condition. Often, however, a compromise has to be reached in that the high purity, fully annealed, metals do not have sufficient strength to enable them to be, for example, strung between posts. Electrical conductivities for copper and aluminium are usually expressed on an IACS (International Annealed Copper Standard) scale. The value of 100% corresponds to annealed copper at 20°C, a conductivity of $5.800 \times 10^{7}\,\Omega^{-1}\,\mathrm{m}^{-1}$. However it needs to be pointed out that some high conductivity coppers have values greater than 100%. Table 18.2 gives some values for materials commonly used for their electrical conductivity.

Some metals are selected for their relatively high resistivities, e.g. for electrical resistors and electric heating elements. One of the main groups of alloys used for such purposes are the nickel–chromium alloys, the nickel–chromium–iron alloys and the iron–chromium–aluminium alloys. Table 18.3 gives some values for metals used for heating elements.

Thermal conductivity

In general, metals have high thermal conductivities while polymers and ceramics have low conductivities. Good thermal conductivity and good electrical conductivity go together.

Carbon steels tend to have thermal conductivities, at about 20°C of the order of $50\,\mathrm{W\,m^{-1}\,K^{-1}}$, low alloy steels about 30–$50\,\mathrm{W\,m^{-1}\,K^{-1}}$, high alloy and stainless steels about 11–$15\,\mathrm{W\,m^{-1}\,K^{-1}}$. Aluminium typically has values of the order of $240\,\mathrm{W\,m^{-1}\,K^{-1}}$, aluminium alloys 160–$190\,\mathrm{W\,m^{-1}\,K^{-1}}$, copper $400\,\mathrm{W\,m^{-1}\,K^{-1}}$, copper alloys 20–$100\,\mathrm{W\,m^{-1}\,K^{-1}}$. Polymers have low thermal conductivities, e.g. nylon 66 with $0.025\,\mathrm{W\,m^{-1}\,K^{-1}}$ and PVC with $0.0019\,\mathrm{W\,m^{-1}\,K^{-1}}$. Glass has a thermal conductivity of about $1\,\mathrm{W\,m^{-1}\,K^{-1}}$, alumina about $2\,\mathrm{W\,m^{-1}\,K^{-1}}$.

Thermal expansion

Metals expand when heated and the problems of differential expansion between different materials in a component can often be an important

Table 18.2 *Electrical conductivities*

Materials	*Condition*	*Electrical conductivity (% IACS)*
Al > 99.50% pure	Annealed	61
Al–5.5% Cu–0.4% Bi–0.4% Pb	Solution heated, treated and aged	39
Al–0.7% Mg–0.4% Si	Annealed	58
	Solution heat treated and aged	53
Electrolytic tough-pitch h.c. Cu	Soft to hard rolled or drawn	101.5–100
Oxygen free h.c. Cu	Soft to hard rolled or drawn	101.5–100
Fire refined tough-pitch h.c. Cu	Soft to hard rolled or drawn	95–89
Phosphorus deoxidized Cu	Soft to hard rolled or drawn	90–70
Phosphorus deoxidized arsenical Cu	Soft to hard rolled or drawn	50–35
Cadmium–copper, Cu–1% Cd	Soft strip or bars	80–92
	Hard strip or bars	75–87
	Soft sheet	27
Fine silver		106

Table 18.3 *Electrical heating element metals*

Material	*Conductivity (% ALCS)*	*Resistivity (10^{-8} Ωm)*
78.5% Ni–20% Cr–1.5% Si	1.6	108
73.5% Ni–20% Cr–5% Al–1.5% Si	1.2	138
68% Ni–20% Cr–8.5% Fe–2% Si	1.5	116
60% Ni–16% Cr–22.5% Fe–1.5% Si	1.5	112
35% Ni–20% Cr–43.5% Fe–1.5% Si	1.7	101
72% Fe–23% Cr–5% Al	1.3	139
55% Fe–37.5% Cr–7.5% Al	1.2	166
Molybdenum	34	5.2
Platinum	16	10.6
Tantalum	14	12.5
Tungsten	30	5.7

concern. The coefficient of expansion α is defined as being the change in length per unit length per degree change in temperature, i.e.

$$\alpha = \frac{\text{change in length}}{\text{original length} \times \text{change in temperature}}$$

Typical values for the coefficient of linear expansion in the temperature region 0° to 100°C are: aluminium and its alloys $22–24 \times 10^{-6}\,K^{-1}$, copper and its alloys $16–20 \times 10^{-6}\,K^{-1}$, steels $10–12 \times 10^{-6}\,K^{-1}$, magnesium and its alloys $25–27 \times 10^{-}\,K^{-1}$, nickel and its alloys $11–13 \times 10^{-6}\,K^{-1}$, titanium and its alloys $8–9 \times 10^{-6}\,K^{-1}$.

18.10 Selection for dimensional conditions

The requirement for a particular dimensional condition, such as surface texture, dictates the type of process that can be used for the production of a component. This in turn dictates the types of materials that can be used. For example, if a fine surface is required without further work after the initial processing then die casting might be used. This has implications for the types of materials that can be used. This issue is discussed in more detail in Chapter 19.

18.11 Available forms of materials

A major factor affecting the choice of material is the form and size it can be supplied in. Thus if the design indicates, for example, an I-girder with specified dimensions in a particular steel and that is not the size in which that material is normally supplied then there may well have to be a change in the material used. The 'as supplied' form and size can determine whether further processing is required and, if so, to what extent. Coupled with such consideration has to be a recognition of the tolerances with which a material is supplied – tolerances on dimensions, angles, twist, etc. The surface conditions of the supplied material may also be important, particularly if the material is to be used without further processing. Thus, for example, a hot-rolled steel could have a loose, flaky scale on its surface and have to be machined.

To give some idea of the range of forms materials can be supplied in, the following is the range mild steel is supplied in by a typical supplier: rounds, squares, flats, tees, channels, circular hollow sections, square hollow sections, rectangular hollow sections, rolled steel joists, universal beams and columns, sheets, galvanized sheets, plates etc. To give an indication of the range of sizes, mild steel rounds are available in the following diameters: 5.5, 6.0, 6.5, 8.0, 9.5, 10.0, 12.0, 12.5, 13.0, 16.0, 20.0 . . . 70.0, 75.0, 80.0 . . . 270.0, 280.0, 300.0, 305.0 mm. Coupled with this are the specifications of the materials for which the sizes are available, e.g. Table 18.4.

Table 18.4 *Specifications for black steel round and square bars to BS 4360: 1972*

Grade	*Tensile strength* *(MN/m^2)*	*Yield stress* *Size range*	*(MN/m^2)*	*El.% min on* $5.65\sqrt{S_0}$
43A	430–510	Up to and including 25 mm	255	22
		Over 25 mm to 50 mm	245	
		Over 50 mm to 63 mm	240	
		Over 63 mm to 100 mm	230	
50B	490–620	Up to and including 25 mm	355	20
		Over 25 mm to 50 mm	355	
		Over 50 mm to 63 mm	345	
		etc.		

In some cases materials can be supplied in a variety of surface finishes. The following is an extract from a manufacturer's catalogue for cold-rolled steel strip (Courtesy of Arthur Lee and Sons Ltd).

Finishes

Bright hard rolled: General standard of finish as produced by cold rolling, which gives a smooth bright surface suitable for most requirements.

Bright annealed: Finally heat treated, maintaining a finish similar to bright hard rolled. Achieved by specialized and rigid atmospheric control in the annealing furnaces in which the heat treatment is carried out.

Plating or mirror finish: A superfine finish, provided in several plating grades for specific applications and requirements.

Matt finish: Produced by passing trip through specially prepared finishing rolls to give a dull but smooth surface. Very suitable for making parts which are to be lacquered, enamelled or painted. This strip has the advantage of retaining lubricants in the deep drawing process, thus facilitating this operation.

Coppered: Obtained by a process of electrolytic copper plating. Apart from its main application, it is also beneficial as a deep drawing lubricant, and as a base for articles to be plated, thereby saving production costs by the use of this partially prepared surface.

Patterned strip: Normally supplied in the annealed and pinch passed condition.

Polymeric materials are generally supplied as granules ready for processing. However, the composition of the granules, for what is the same polymeric material, can vary. This variation is often due to the other materials, such as fibres, which have been added to the polymeric in order to constitute the plastic. Thus, for example, Bayer supply ABS, under the trade name Novodur, in nineteen different types. The general-purpose moulding types are graft copolymers with grafted polybutadiene in a coherent styrene-acrylonitrile copolymer matrix. Another form has a butadiene-acrylonitrile

Table 18.5 *Properties of Novodur*

Properties	*Graft polymers, general-purpose moulding types*				
	Yellow PX	*White PH-AT*	*Orange PHGV*	*White PHGV-AT*	*Black PHGV-7*
Yield stress (MPa)	54	50	57	48	55
Elongation (%)	~3	~3	~2	~2	~2.5
Flexural yield stress (MPa)	76	70	86	80	90
Elastic modulus (GPa)	2.30	2.38	5.00	4.77	3.59
Flexural modulus (GPa)	2.19	2.13	3.66	3.15	2.87
Impact strength (kJ/m^2) + 23°C	87	90	14	15	17
Coefficient of linear expansion ($10^{-6}K^{-1}$)	93	88	47	39	
Density (g/cm^3)	1.05	1.09	1.15	1.22	1.08
Water absorption from dry state (%) etc.	0.5	0.6	0.4	0.4	0.4

Note: PX is listed as a general-purpose type, very hard grade; PH-AT as a general purpose type; PHGV as containing 20 per cent by weight glass fibre; PHGV-AT as containing 20 per cent by weight glass fibre; PHGV-7 as containing 7 per cent by weight glass fibre.

copolymer mixed in with a styrene-acrylonitrile copolymer. Some forms include glass fibres and some are different colours. To give some indication of the type of specifications given by manufacturers, Table 18.5 is a brief extract from the Bayer UK Ltd data sheet for ABS.

18.12 The costs of materials

Table 18.6 gives some idea of the relative costs of metals in various forms of supply. In general, the cost of plate and sheet material is about 1.4 times that of bar material, the cost of castings about 2.4 times that of bar material. Table 18.7 gives some idea of the costs of polymeric materials. Compared with metals, the cost per kilogramme of polymeric material is such that polyethylene is about three times the cost of mild steel. However, the much lower densities of polymer materials mean that the cost per cubic metre of polyethylene is about one-fifth that of mild steel.

Problems

1 Suggest the main property or properties required of a material and the types of materials that might be used in the following situations:

(a) for pipes to be used in the distribution of water,

(b) for a component where strength is required at temperatures in the region of 700°C,

(c) for a component which is required to be stiff,

Table 18.6 *Relative costs of metals*

Bar materials	*Relative cost/kg*	*Relative cost/m³*
Mild steel	1.0	1.0
Medium-carbon steel	1.6	1.6
High-carbon steel	2.3	2.3
Aluminium	8.5	2.9
Nickel-chrome steel	4.6	4.6
Brass	6.6	7.4
Stainless steel	9.6	9.6
Phosphor bronze	16.0	18.3
Plate materials		
Mild steel	1.0	1.0
Aluminium	4.9	1.7
Brass	3.9	4.4
Copper	5.1	5.9
Stainless steel	7.3	7.3
Sheet material		
Mild steel	1.0	1.0
Duralumin	6.5	2.2
Brass	3.7	4.1
Stainless steel	6.1	6.1
Copper	5.9	6.8
Castings		
Cast iron	1.0	1.0
Aluminium	4.1	1.1
Brass	2.5	3.1
Manganese bronze	7.6	9.1
Phosphor bronze	6.1	7.6

(d) for a thin spherical shell to be subject to a high internal pressure,
(e) for a component where strength is required in a marine environment,
(f) for a clutch plate,
(g) for a container to hold acids,
(h) for a heating element to be used at temperatures in excess of 1000°C,
(i) for a component which can be used with direct stresses of the order of 1000 MPa,
(j) for a component subject to impact loading,
(k) for a component subject to cyclic loading.

2 List the factors affecting the usefulness of the following materials:
(a) mild steel,
(b) grey cast iron,

Table 18.7 *Relative costs of polymeric materials*

Material	*Relative cost/kg or /m³*
Polyethylene	1
Polypropylene	1
Polystyrene	1
PVC	2
ABS	4
Phenolics	4
Acrylics	4
Cellulose acetate	5
Acetals	5
Polycarbonate	12
Nylons	15
Polyurethane	20
PTFE	30
Fluorosilicons	80

(c) brasses,
(d) aluminium alloys,
(e) ceramics,
(f) thermoplastics.

19

Selection of processes

19.1 General considerations

A number of vital questions need to be posed before a decision is made about the manufacturing process to be used for a product.

1. *What is the material?* The type of material to be used affects the choice of processing method. Thus, for example, if casting is to be used and the material has a high melting point then the process must be either sand casting or investment casting.
2. *What is the shape?* The shape of the product is generally a vital factor in determining which type of process can be used. Thus, for example, a product in the form of a tube could be produced by centrifugal casting, drawing or extrusion but not generally by other methods.
3. *What type of detail is involved?* Is the product to have holes, threads, inserts, hollow sections, fine detail etc? Thus, for example, forging could not be used if there was a requirement for hollow sections.
4. *What dimensional accuracy and tolerances are required?* High accuracy would rule out sand casting, though investment casting might well be suitable.
5. *Are any finishing processes to be used?* Is the process to be used to give the product its final finished state or will there have to be an extra finishing process. Thus, for example, planing will not produce as smooth a surface as grinding.
6. *What quantities are involved?* Is the product a one-off, a small batch, a large batch or continuous production? While some processes are economic for small quantities, others do not become economic until large quantities are involved. Thus open die forging could be economic for small numbers but closed die forging would not be economic unless large numbers were produced.

In considering surface finish a measure used is the roughness. *Roughness* is defined as the irregularities in the surface texture which are inherent in the production process but excluding waviness and errors of form. Waviness may arise from such factors as machine or work deflections, vibrations, heat treatment of warping strains.

One measure of roughness is the arithmetical mean deviation, denoted by the symbol R_a. This is the arithmetical average value of the variation of the profile above and below a reference line throughout the prescribed sampling length. The reference line may be the centre line, this being a line chosen so that the sums of the areas contained between it and those parts of the surface profile which lie on either side of it are equal (Figure 19.1). Table 19.1 indicates the significance of R_a values and Table 19.2 shows what is achievable with different processes.

The degree of roughness that can be tolerated for a component depends on its use. Thus, for example, precision sliding surfaces will require R_a values of the order 0.2 to 0.8 μm with more general sliding surfaces 0.8 to 3 μm. Gear teeth are likely to require R_a to be 0.4 to 1.5 μm, friction surfaces such as clutch plates 0.4 to 1.5 μm, mating surfaces 1.5 to 3 μm.

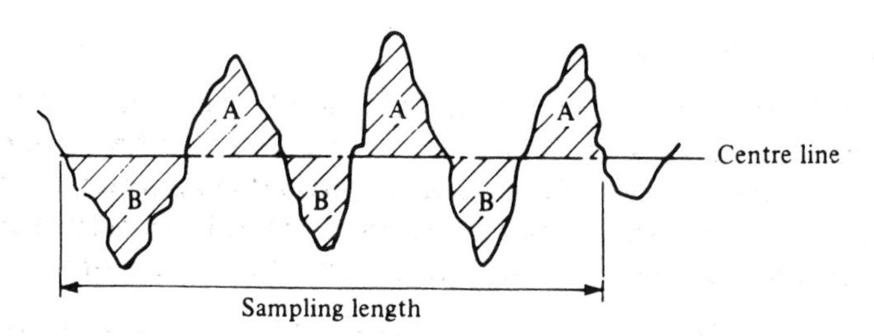

Figure 19.1 R_a – the sum of the areas marked A equals the sum of those marked B

Table 19.1 *Typical* R_a *values*

Surface texture	R_a *(μm)*
Very rough	50
Rough	25
Semi-rough	12.5
Medium	6.3
Semi-fine	3.2
Fine	1.6
Coarse-ground	0.8
Medium-ground	0.4
Fine-ground	0.2
Super-fine	0.1

Table 19.2 *Roughness values for different processes*

Process	R_a (μm)
Sand casting	12.5 to 25
Die casting	0.8 to 1.6
Investment casting	1.6 to 3.2
Hot rolling	12.5 to 25
Forging	3.2 to 12.5
Extrusion	0.8 to 3.2
Cold rolling	0.4 to 3.2
Turning	0.4 to 6.
Planing and shaping	0.8 to 15
Milling	0.8 to 6.5
Grinding	0.1 to 1.6

19.2 Casting of metals

Casting can be used for components from about 10^{-3} kg in size to 10^4 kg, with wall thicknesses from about 0.5 mm to 1 m. Castings need to have rounded corners, no abrupt changes in section and gradual sloping surfaces. Casting is likely to be the optimum method in the circumstances listed below but not for components that are simple enough to be extruded or deep drawn.

1. *The part has a large internal cavity* – there would be a considerable amount of metal to be removed if machining were used. Casting removes this need.
2. *The part has a complex internal cavity* – machining might be impossible; by casting, however, very complex internal cavities can be produced.
3. *The part is made of a material which is difficult to machine.*
4. *The metal used is expensive and so there is to be little waste.*
5. *The directional properties of a material are to be minimized* – metals subject to a manipulative process often have properties which differ in different directions.
6. *The component has a complex shape* – casting may be more economical than assembling a number of individual parts.

An important consideration in deciding whether to use casting or which casting method to use is the tooling cost for making the moulds. When many identical castings are required, a method employing a mould which may be used many times will enable the mould cost to be spread over many items and may make the process economic. Where just a one-off product is required, the mould used must be as cheap as possible since the entire cost will be defrayed against the single product.

Each casting method has important characteristics which determine its

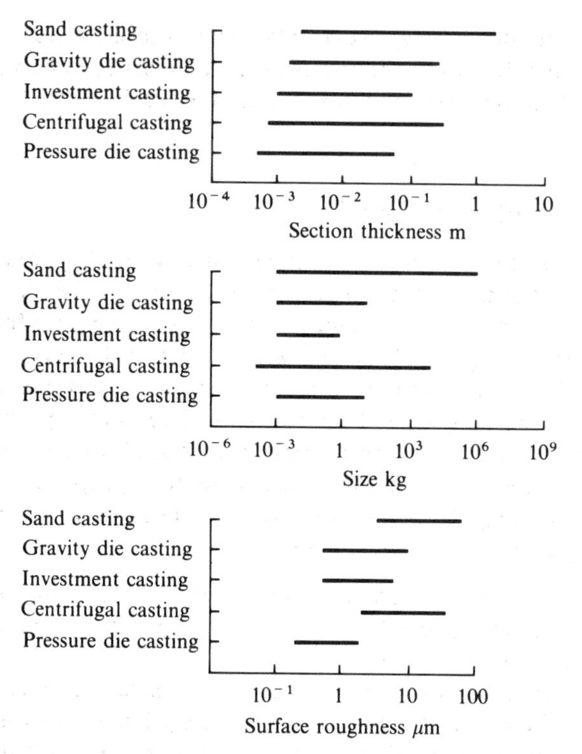

Figure 19.2 Casting processes

appropriateness in a particular situation. Figure 19.2 illustrates some of the key differences between the different casting methods. The following factors largely determine the type of casting process used.

1. *Large, heavy casting* – sand casting can be used for very large castings.
2. *Complex design* – sand casting is the most flexible method and can be used for very complex castings.
3. *Thin walls* – investment casting or pressure die casting can cope with walls as thin as 1 mm. Sand casting cannot cope with such thin walls.
4. *Small castings* – investment casting or die casting. Sand casting is not suitable for very small castings.
5. *Good reproduction of detail* – pressure die casting or investment casting, sand casting being the worst.
6. *Good surface finish* – pressure die casting or investment casting, sand casting being the worst.
7. *High melting point alloys* – sand casting or investment casting.
8. *Tooling cost* – this is highest with pressure die casting. Sand casting is cheapest. However, with large number production the tooling costs for the

metal mould can be defrayed over a large number of castings, whereas the cost of the mould for sand casting is the same no matter how many castings are made because a new mould is required for each casting.

19.3 Manipulation of metals

Manipulative processes involve the shaping of a material by means of plastic deformation methods. Such methods include forging, extruding, rolling, and drawing. Depending on the method, components can be produced from as small as about 10^{-5} kg to 10^{2} kg, with wall thicknesses from about 0.1 mm to 1 m. Figure 19.3 shows some of the characteristics of the different processes. Compared with casting, wrought products tend to have a greater degree of uniformity and reliability of mechanical properties. The manipulative processing does however tend to give a directionality of properties which is not the case with casting. Manipulative processes are likely to be the optimum method for product production when:

1 The part is to be formed from sheet metal. Depending on the form required, shearing, bending or drawing may be appropriate if the components are not too large.
2 Long lengths of constant cross-section are required. Extrusion or rolling would be the optimum methods in that long lengths of quite complex cross-section can be produced without any need for machining.
3 The part has no internal cavities. Forging might be used, particularly if better toughness and impact strength is required than is obtainable with

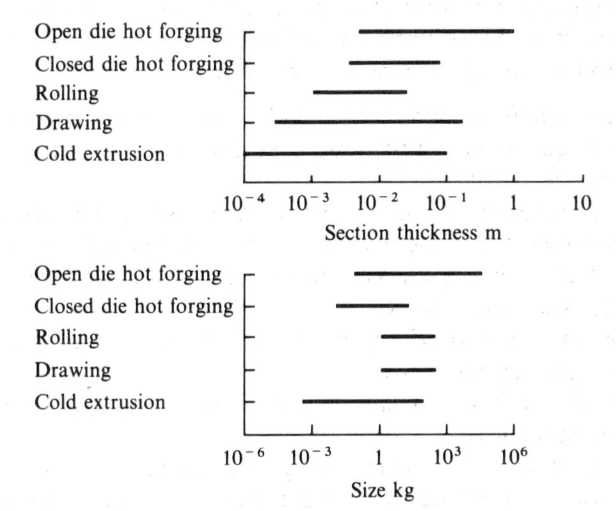

Figure 19.3 Manipulative processes

casting. Also directional properties can be imparted to the material to improve its performance in service.

4 Directional properties have to be imparted to the part.
5 Seamless cup-shaped objects or cans have to be produced. Deep drawing or impact extrusion would be the optimum methods.
6 The component is to be made from material in wire or bar form. Bending or upsetting can be used.

19.4 Powder processes

Powder processes (see section 18.4) enable large numbers of small components to be made at high rates of production and with little, if any, finishing machining required. It enables components to be made with materials which cannot be easily produced otherwise, e.g. the high melting point metals of molybdenum, tantalum and tungsten, and where there is a need for some specific degree of porosity, e.g. porous bearings to be oil-filled. The mechanical compaction of powders only however permits two-dimensional shapes to be produced, unlike casting or forging. In addition the shapes are restricted to those that are capable of being ejected from the die. Thus, for example, reverse tapers, undercuts and holes at right angles to the pressing direction have to be avoided. On a weight-for-weight basis powdered metals are more expensive than metals for use in manipulative or casting processes, however this higher cost may be offset by the absence of scrap, the elimination of finishing machining and the high rates of production.

19.5 Machining metals

When selecting a cutting process, the following factors are relevant in determining the optimum process or processes:

1 Operations should be devised so that the minimum amount of material is removed. This reduces material costs, energy costs involved in the machining and costs due to tool wear.
2 The time spent on the operation should be a minimum to keep labour costs low.
3 The skills required also effect labour costs.
4 The properties of the material being machined should be considered; in particular the hardness.
5 The process, or processes, chosen should take into account the quantity of products involved and the required rate of production.
6 The geometric form of the product should be considered in choosing the most appropriate process or processes.
7 The required surface finish and dimensional accuracy also affect the choice of process or processes.

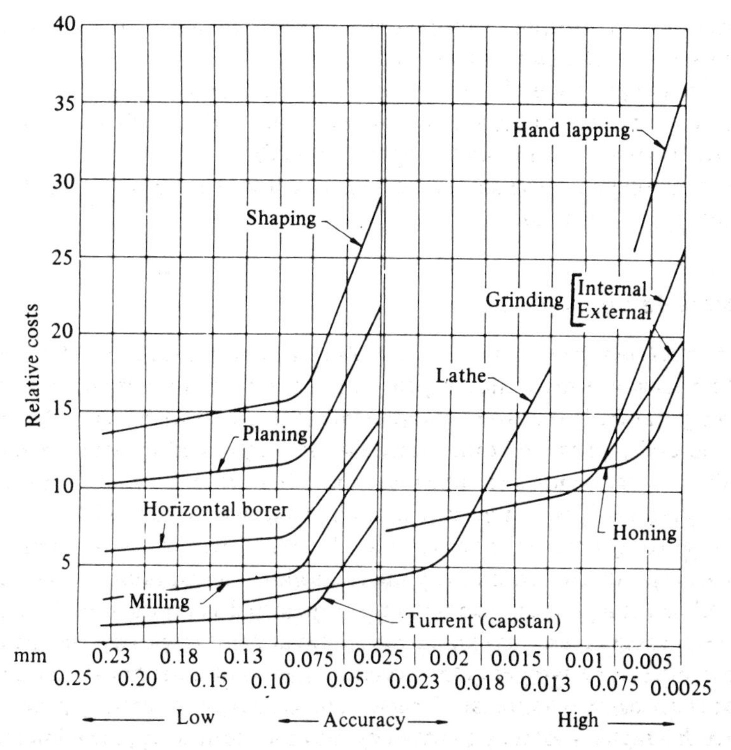

Figure 19.4 Cost of various machine and hand processes for achieving set tolerances. (Reproduced with the permission of the British Standards Institution, being taken from *Manual of British Standards in Engineering Drawing and Design*, Hutchinson, 1984)

In general, an important factor determining the time taken to cut a material, and hence the cost, is the material hardness; the harder a material, the longer it is likely to take to cut. The hardness also, however, affects the choice of tool material that can be used, and also, in the case of very hard materials, the process. Thus, for instance, grinding is a process that can be used with very hard materials, because the tool material, the abrasive particles, can be very hard.

Where a considerable amount of machining occurs, the use of free-machining grades of materials should always be considered in order to keep costs down by keeping the cutting time to as low a value as possible.

Machining operations vary quite significantly in cost, particularly if the operation is considered in terms of the cost necessary for a particular machine to achieve particular tolerances. Figure 19.4 shows how the relative costs vary for achieving set tolerances. Thus to achieve a tolerance of 0.10 mm, the rank order of the processes is:

Shaping	most expensive
Planing	
Horizontal borer	
Milling	
Turret (capstan)	least expensive

The cost of all the processes increases as the required tolerance is increased. At high tolerances, grinding is one of the cheapest processes. The different machining operations also produce different surface finishes:

	R_a (μm)	
Planing and shaping	15 –0.8	Least smooth
Drilling	8 –1.6	
Milling	6.3–0.8	
Turning	6.3–0.4	
Grinding	1.6–0.1	Most smooth

The choice of process will depend on the gemometric form of the product being produced (Table 19.3).

Machining, in general, is a relatively expensive process when compared with many other methods of forming materials. The machining process is, however, a very flexible process which allows the generation of a wide variety of forms. A significant part of the total machining cost of a product is due to setting-up times when there is a change from one machining step to another. By reducing the number of machining steps and hence the number of setting-up times, a significant saving becomes possible. Thus the careful sequencing of machining operations and the careful choice of machine to be used is important.

Table 19.3 *Choice of machining process*

Type of surface	*Suitable processes*
Plane surface	Shaping Planing Face milling Surface grinding
Externally cylindrical surface	Turning Grinding
Internally cylindrical surface	Drilling Boring Grinding
Flat and contoured surfaces and slots	Milling Grinding

19.6 Joining processes with metals

Fabrication involving joining materials enables very large structures to be assembled, much larger than can be obtained by other methods such as casting or forging. The main joining processes are essentially adhesive bonding, soldering and brazing, welding and fastening systems (see Chapter 17). The factors that determine the joining process are the materials involved, the shape of the components being joined, whether a permanent or temporary joint is required, limitations imposed by the environment and cost. Welded, brazed and adhesive joints, and some fastening joints, e.g. riveted, are generally meant to be permanent joints, while soldered joints and bolted joints are readily taken apart and rejoined.

19.7 Polymer-forming processes

Injection moulding and extrusion are the most widely used processes. Injection moulding is generally used for the mass production of small items, often with intricate shapes. Extrusion is used for products which are required in continuous lengths or which are fabricated from material of constant cross-section. The following are some of the factors that are involved in choosing a process.

1. *Rate of production* – injection moulding has the highest rate, followed by blow moulding, then rotational moulding, compression moulding, transfer moulding and thermoforming, with casting being the slowest.
2. *Capital investment required* – injection moulding requires the highest capital investment, with extrusion and blow moulding requiring less capital. Rotational moulding, compression, moulding, transfer moulding, thermoforming and casting require the least capital investment.
3. *Most economic production run* – injection moulding, extrusion and blow moulding are economic only with large production runs. Thermoforming, rotational moulding, casting and machining are used for the small production runs.
4. *Surface finish* – injection moulding, blow moulding, rotational moulding, thermoforming, transfer and compression moulding, and casting all give very good surface finishes. Extrusion gives only a fairly good finish.
5. *Metal inserts, during the process* – these are possible with injection moulding, rotational moulding, transfer moulding and casting.
6. *Dimensional accuracy* – injection moulding and transfer moulding are very good. Compression moulding good with casting; extrusion is fairly poor.
7. *Very small items* – injection moulding and machining are the best.
8. *Enclosed hollow shapes* – blow moulding and rotational moulding can be used.

9 *Intricate, complex shapes* – injection moulding, blow moulding, transfer moulding and extrusion.
10 *Threads* – injection moulding, blow moulding, casting and machining.
11 *Large formed sheets* – thermoforming.

Figure 19.5 shows some of the characteristics of the different types of polymer processes.

The assembly processes that can be used with plastics are welding, adhesive bonding, riveting, press and snap-fits, and thread systems.

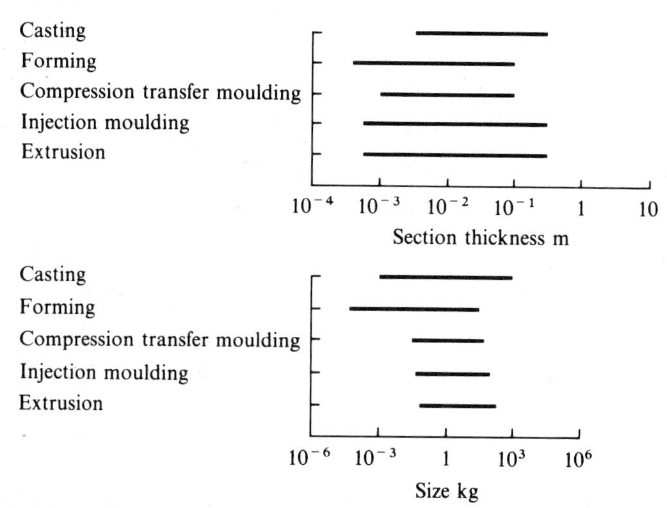

Figure 19.5 Polymer processes

19.8 The cost aspects of process selection

With sand casting a new mould has to be made for each product manufactured. With die casting the same mould can be used for a large number of components but the initial die cost is high. Which process would be the cheapest if, say, 10 products were required, or perhaps 1000 products?

The manufacturing cost, for any process, can be considered to be made up of two elements – fixed costs and variable costs. Figure 19.6 shows the graphs of costs against quantity for the two processes and how their total costs compare. For the die casting there is a higher fixed cost than for the sand casting, due to the cost of the die. The sand casting has, however, a greater cost per item produced. The comparison graph shows that, up to a quantity N, sand casting is the cheaper but that for quantities greater than N die casting is the cheaper process.

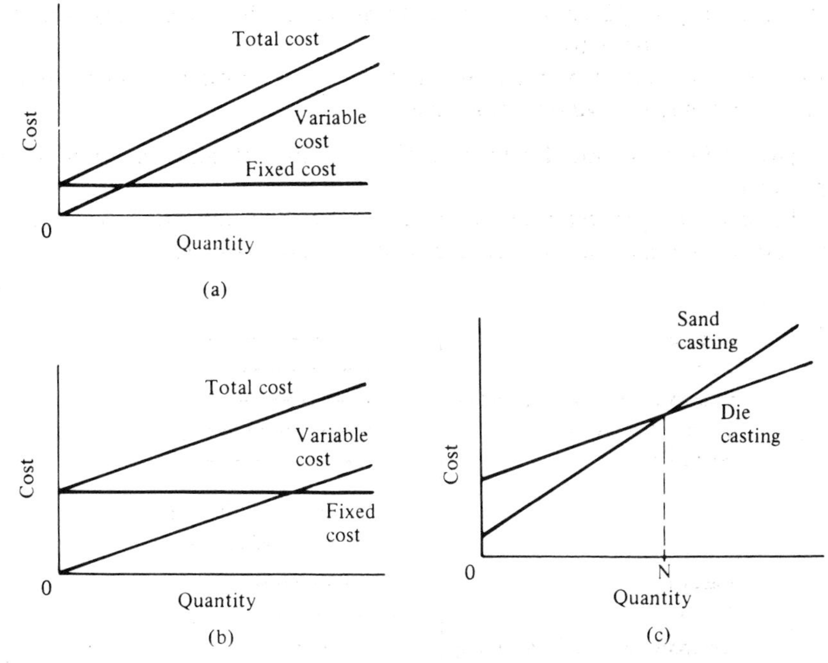

Figure 19.6 (a) Sand casting costs (b) Die casting costs (c) Total costs compared

Capital costs for installations, e.g. the cost of a machine or even a foundry, are usually defrayed over an expected lifetime of the installation. However, in any one year there will be depreciation of the asset and this is the capital cost that is defrayed against the output of the product in that year. Thus if the capital expenditure needed to purchase a machine and instal it was, say, £50 000, then in one year a depreciation of 10 per cent might be used and thus £5000 defrayed as capital cost against the quantity of product produced in that year.

Consider, as an example, a product to be made from a thermoplastic. It is to be hollow and of fairly simple form. It can be made by injection moulding or rotational moulding (other methods are also possible but for the purpose of this example only these two alternatives are considered). The cost of the installation for injection moulding is £90 000, the cost for the rotational moulding, using a fairly simple form of the equipment, is £10 000. Thus if these installations depreciate at 10 per cent per year then the costs per year are – injection moulding £9000 and rotational moulding £1000. If these installations are to be used in the year to make only the product we are concerned with, then these charges in their entirety constitute an element of the fixed costs.

Another element of the fixed costs is the cost of the dies or tools needed specifically for the product concerned. In the case of the injection moulding the cost is £2000 and for the rotational moulding £500.

Other factors we could include in the fixed costs are plant maintenance and tool or die overhaul. However, considering just the above two charges, for the installation when we assume that it only produces the above product per year and for the specific tooling if only used for the one production run, we have:

	Injection moulding	*Rotational moulding*
Installation cost	£9000	£1000
Die costs	£2000	£500
Total fixed cost	£11 000	£1500

The variable costs are the material costs, the labour costs, the power costs, and any finishing costs required. These might appear, per unit produced, as follows:

	Injection moulding	*Rotational moulding*
Direct labour cost	£0.30	£1.40
Other labour costs e.g. supervision	£0.15	£0.30
Power costs	£0.06	£0.15
Finishing costs	£1.00	zero
Material costs	£0.40	£0.40
Total variable cost	£1.91	£2.25

If 1000 units are required then the costs will be:

	Injection moulding	*Rotational moulding*
Fixed costs	£11 000	£1500
Variable costs	£1900	£2250
Total cost	£12 910	£3750

On the basis of the above costings, the rotational moulding is considerably cheaper than the injection moulding. The injection moulding installation can, however, produce 50 items per hour and the use in one year of this to produce just 1000 items is a considerable underuse of the asset. The rotational moulding installation can only produce two items per hour.

Thus if the rotational moulding installation were perhaps only used for this product but the injection moulding installation were used for a number of products, then the injection moulding installation cost would be defrayed over a greater number of products and thus that element of the fixed cost defrayed against the product would be less.

The questions to be asked when costing a product are:

1 *Is the installation to be used solely for the product concerned?* The purpose of this question is to determine whether the entire capital cost has to be

written off against the product or whether it can be spread over a number of products.

2 *Is the tooling to be used solely for the product concerned?* If specific tooling has to be developed then the entire cost will have to be put against the product.
3 *What are the direct labour costs per item?*
4 *What are the other labour costs involved?* In this category consider supervision costs, inspection costs, labouring costs, setting costs, etc.
5 *What is the power cost?*
6 *Are there any finishing processes required – if so, what are their costs?*
7 *What are the materials costs?*
8 *Are any overhead costs to be included?*

Problems

1 Suggest a casting process for the following situation. A small one-off casting is required using aluminium, there is a lot of fine detail which has to be reproduced and a good surface finish is required.
2 Compare forging and casting as methods of production.
3 What are the advantages and disadvantages of powder techniques for the production of products?
4 Suggest processes that might be used to make the following metal products:
   (a) a toothpaste tube from a very soft alloy,
   (b) the reflector concave dish for a satellite TV receiving aerial, using aluminium or mild steel,
   (c) rivets,
   (d) an aluminium can for drink storage,
   (e) a hollow hexagonal length of brass rod,
   (f) a spanner,
   (g) railway lines,
   (h) a kitchen pan,
   (i) the undercarriage for an aircraft.
5 Suggest processes that might be used to make the following polymer products.
   (a) a small plastic toy at high production rates from a thermoplastic material,
   (b) a 1 litre bottle for a soft drink, using a thermoplastic material with high production rates,
   (c) a switch cover in a thermosettting material, with high production rates,
   (d) milk churns about 340 mm diameter and 760 mm high from polyethylene,
   (e) a thermoplastic strip for use as a draught excluder with windows, long lengths being required,

(f) polythene bags with high production rates,
(g) the body for a camera, reasonably high production rates being required,
(h) the bodywork of an electric drill, threaded holes and high production rates being required,
(i) a nylon gear wheel, high production rates being required.

6 The following are the costs that would be incurred for two alternative processes for the production of a hollow thermoplastic container:

Costs	*Injection moulding* £	*Rotational moulding* £
Fixed: installation	9000	1000
: die	2000	500
Variable (per unit)		
: indirect labour	0.15	0.30
: power	0.06	0.15
Direct labour (per unit)	0.30	1.40
Direct materials (per unit)	0.40	0.40
Finishing	1.00	0

(a) Which process would be the more cost-effective if (i) 1000 (ii) 100 000 units were required?
(b) The injection-moulding process can produce fifty units per hour, the rotational moulding only two per hour. What implication might this have for using the equipment to produce a wider range of goods and so affect the choice of process?

7 The following data refer to three alternative processes by which a gearwheel can be produced (based on data given in *Engineering Materials: An Introduction* (The Open University T252 Unit 2). On the basis of the data, which process would you propose as the most cost effective if the output required was (a) 100, (b) 10 000 units?

Gravity die casting method, costs per unit
Materials: 10 g aluminium alloy plus 5 g scrap, at £1.00 per kg
Special tooling: a die at £500 which has a life of 100 000 gears
Labour: casting 0.010 hours at £4.00 per hour
finishing 0.001 hours at £4.00 per hour
Overheads to be charged at 200 per cent of labour cost

Pressure die casting method, costs per unit
Materials: 23 g zinc alloy plus 2 g scrap, at £0.45 per kg
Special tooling: a die at £4000 which has a life of 100 000 gears
Labour: casting 0.0015 hours at £4.00 per hour
finishing 0.001 hours at £4.00 per hour
Overheads to be charged at 200 per cent of labour cost

Injection moulding method, costs per unit
Materials: 4 g nylon plus 0.4 g scrap, at £2.00 per kg
Special tooling: a die at £4000 which has a life of 100 000 gears
Labour: casting 0.006 hours at £4.00 per hour
Overheads to be charged at 200 per cent of labour cost

8 A base for a special machine tool has to be made, only one having to be produced. There are two proposals: one to make a grey iron casting, and the other to make it of steel and weld together the various parts after machining. The casting will be twice the weight of the fabricated base.
   (a) Compare the two methods proposed for the base.
   (b) For the casting, the cost of making a sand pattern is £400, the labour costs, with overheads, for the making of the casting is £840, and the materials cost £250. For the fabricated base, the cost of the machining, including overheads, is £300, the cost of the welding, including overheads, is £500 with an extra £50 being required for the cost of the electrode material consumed during the welding, and the cost of the material is £120. Which method is the more economical?
   (c) Before the base can be made, it is realized that there could be a market for a small number of these bases. Plot graphs showing how the cost varies with quantity and comment on the significance of the graphs in the determination of the optimum manufacturing method.

9 The frame of a bicycle could be made of tubular mild steel welded together or produced in one piece, perhaps by casting.
   (a) Compare these two methods of producing bicycle frames.
   (b) Which method is likely to be the more economical?

10 Forging, casting or machining from the solid have been proposed as methods that could be used to produce spanners.
   (a) With forging, what would be the optimum method if the spanners were required in large quantities?
   (b) With casting, what would be the optimum method if the spanners were required in large quantities?
   (c) Compare forging, casting and machining for the production of spanners when they are required in large quantities. What method is likely to give the cheapest product?

11 A car dashboard has to be formed to a variety of contours, have holes for instruments, various control knobs and a glove compartment.
   (a) If the dashboard were to be made from a plastic, what processes could be used?
   (b) If the dashboard were to be made from a metal, what processes could be used?
   (c) For the components that are to be attached to the dashboard, what fixing methods could be used if (i) a plastic, (ii) a metal is used?
   (d) Discuss the economics of the various processes for producing the dashboard.

20

Selection criteria

20.1 Criteria

When selecting material for a particular component the properties required of the material have to be considered. However, it is seldom that the performance of the component depends on just one property. Thus a product might require tensile strength, toughness and low cost. Often the performance depends on a parameter, or performance index, which depends on a combination of properties.

For example, the material for a structural member might need to be chosen for the load F it will support and its mass, the aim being to have the maximum load per unit mass of member. The mass is the product of the density ρ and the volume of the member and thus for a length L of cross-sectional area A is $LA\rho$. The area used depends on the yield stress σ_y, being F/σ_y. The mass is thus

$$m = LA\rho = L\rho\frac{F}{\sigma_y}$$

$$\frac{F}{m} = \frac{1}{L} \times \frac{\sigma_y}{\rho}$$

Thus, since L might be assumed to be constant, the parameter determining the material is σ_y/ρ.

As another example, the material chosen for a cantilever of length L might need to be chosen for the load F it will support at its free end and the deflection y at that end, the aim being to have as stiff a cantilever as possible and so the maximum force per unit deflection. For a cantilever

$$\frac{F}{y} = \frac{F}{FL^3/3EI}$$

For a rectangular cross-section of side b then $I = b^4/12$ and so

$$\frac{F}{y}=\frac{3Eb^4/12}{L^3}=\frac{Eb^4}{4L^3}$$

The mass m of the cantilever is $b^2L\rho$, where ρ is the density. Hence

$$m=L\rho\left(\frac{4L^3}{E}\times\frac{F}{y}\right)^{1/2}=2L^{5/2}\left(\frac{F}{y}\right)^{1/2}\frac{\rho}{E^{1/2}}$$

$$\frac{(F/y)^{1/2}}{m}=\frac{1}{2L^{5/2}}\times\frac{E^{1/2}}{\rho}$$

Thus the mass is minimized and the stiffness maximized when $E^{1/2}/\rho$ is maximized. This then becomes the parameter.

In determining a parameter the technique used is

1 Write down an equation for the attribute sought.
2 Eliminate, using other property equations, any terms in that equation that is not either a property or a fixed quantity.
3 The parameter that is required is then the resulting combination of properties.

Table 20.1 shows some of the parameters used in a number of different situations.

20.2 Selecting

In considering the functions required of the materials in a component, a list of requirements is produced. Materials then can be considered in relationship to their match with the list. A reduction in the number of materials which need to be considered for a component can be arrived at in a number of ways.

1 Identification of critical properties

The critical properties without which a material would not be suitable are determined and these are used to sieve out a subset of materials for further consideration. Thus for one property there may be a requirement that its value is greater than some limit, while for another it must be less than some limit. For example in considering materials for, say, the filament of an electric light bulb, critical properties are a high melting temperature and an electrical conductor. Thus we might have the requirement that the melting point must be above 3000 K with the electrical resistivity less than, say, 2×10^{-6} Ωm. It is pointless considering any material which does not have these properties.

2 Cost per unit property

Since low cost is often a requirement, one way of comparing the properties of materials is on the basis of cost per unit property or group of properties. This

Table 20.1 *Materials selection parameters*

Component	*Requirement*	*Parameter to be maximized*
Tie	Maximize stiffness and minimize mass	$\frac{E}{\rho}$
	Maximize strength and minimize mass	$\frac{\sigma_y}{\rho}$
	Maximize stiffness and minimize cost, C=cost/kg	$\frac{E}{C\rho}$
Beam	Maximize stiffness and minimize mass	$\frac{E^{1/2}}{\rho}$
	Maximize strength and minimize mass	$\frac{\sigma_y^{2/3}}{\rho}$
Column	Maximize stiffness and minimize mass	$\frac{E^{1/2}}{\rho}$
	Maximize strength and minimize mass	$\frac{\sigma_y}{\rho}$
Pressurized cylinder	Maximize stiffness and minimize mass	$\frac{E}{\rho}$
	Maximize strength and minimize mass	$\frac{\sigma_y}{\rho}$
Diaphragm	Maximize deflection	$\frac{\sigma_y^{3/2}}{E}$
Rotating disk	Maximize strength and minimize mass	$\frac{\sigma_y}{\rho}$

Note that the yield stress σ_y is often replaced by the tensile strength. E is the tensile modulus and ρ the density.

method can be particularly useful where there is one property or group of properties which is the main requirement. Thus, for example, in selecting the materials for a car bumper the value of the yield stress/density may be the important parameter and so the materials compared on the basis of cost per unit of this parameter.

3 Merit ratings

This method involves each material being given a relative merit value for each of the relevant properties. For example, the tensile strength of material may be given a rating out of 10 or 100. A low strength material would have a low number for its rating and a high strength number a high rating. The ratings are assigned relative to some best material which is given the top number, i.e. 10 or 100 depending on the scale used. In some cases the merit rating might be for a combination of properties, e.g. strength/density. In the case of production methods being compared, the ratings might be in terms of the cost of the processes or some factor such as machinability. In order to arrive at the overall rating for a material when a number of properties have each been given merit ratings, each property is given a weighting factor and the total of the weighted values for a material obtained. The optimum material is then the one with the highest weighted values total. For example, in considering the materials which could be used for the filament in an electric light bulb there may be a requirements for high melting temperature, high electrical conductivity, high strength and sufficient ductility to enable wires to be drawn. The main requirements might be for high melting point and high electrical conductivity with high strength being a lesser requirement and high ductility desirable but not essential. Thus the properties might be weighted as melting point $\times 4$, electrical conductivity $\times 3$, strength $\times 2$ and ductility $\times 1$.

One way of arriving at merit ratings is in terms of a property being considered in terms of having a lower limit value, an upper limit value or a target value. For example, the design of the product might require that the tensile strength must be above some particular value. This value then becomes the lower limit value for that property. The merit rating for a material is then given by the value for the material divided by the lower limit value. If this value is less than 1 then the material is clearly unsuitable since the value falls below the lower limit value. Where there is an upper limit value then the merit rating is the upper limit value divided by the value for the material. If this value is less than 1 then the material is unsuitable. For example, cost might be an upper limit property. For some properties a target value, rather than lower or upper limits, might be specified. The merit value for a material is then

$$\left|\frac{\text{value for material} - \text{target value}}{\text{target value}}\right|$$

Sometimes, instead of considering cost as an upper limit value property like any other property it is used to modify the overall sum of the weighted values of the other properties. For example, a cost-modified weighted sum might be given by

$$\frac{\text{cost for a material}}{\text{upper cost limit}} \times \text{weighted sum}$$

The following sections in this chapter are case studies designed to illustrate the above selection techniques.

20.3 Case studies: critical properties

The following case studies of electrical conductors, floor joists and aircraft skins illustrate the consideration of properties in terms of properties which are critical and enable a subset of materials to be sieved out from the entire range of engineering materials.

Electrical conductors

Domestic and industrial cables are required to have small cross-sectional areas and as low a resistance as possible, i.e. a low resistivity and so high conductivity. In addition they are required to be able to be processed in long, continuous, unbroken lengths. Such a requirement argues for a highly ductile material which can be drawn. In addition, cables need to be able to bend round corners without breaking and to be cheap.

The requirement for a high electrical conductivity reduces the choice to metals, since polymeric materials and ceramics are generally insulators. The requirement for very high conductivity reduces the choice within metals to essentially just silver, copper and aluminium and within those to the metals in the highest purity and fully annealed condition. Table 20.2 shows the electrical conductivities of metals in this state. Taking cost into account rules out silver and thus copper appears to be the optimum choice.

Table 20.3 gives the electrical conductivities of various forms of copper and copper-rich alloys. The electrical conductivities have been expressed on the IACS scale, the value of 100% on this scale corresponding to the conductivity of annealed copper at 20°C with a conductivity of $5.800 \times 10^7 \Omega^{-1} m^{-1}$. On the basis of conductivity, a copper rather than a copper alloy is suggested with C101 and C103 being possibilities. Both have essentially the same tensile properties with high ductility and percentage elongations of the order of 50 to 60. Bend tests indicate that both can be bent through 180°. Both can be easily worked and so drawing can be used to produce long wires. The main distinguishing features between the two is price with C101 being cheaper. Thus the choice for everyday cables is likely to be C101.

Table 20.2 *Electrical conductivities of pure metals at 0°C*

Metal	*Conductivity* $(10^6 \Omega^{-1} m^{-1})$
Silver	68
Copper	64
Aluminium	40
Iron	11

Table 20.3 *Conductivities of copper and alloys*

BSI ref.	*Metal*	*IACS conductivity (%)*
C101	Electrolytic tough-pitch h.c copper	101.5–100
C103	Oxygen-free h.c copper	101.5–100
C105	Phosphorus deoxidized arsenical copper	50–35
C108	Copper-cadmium	80–92
CZ102	Red brass	37
CZ106	Cartridge brass	27
NS101	Leaded nickel silver	7
PB101	3 per cent phosphor bronze	15–25

The tensile strength of copper such as C101 is about 220–400 MN/m^2 (MPa). For major power transmission cables this strength is not enough. A copper alloy with about 1% cadmium has higher strength, about 250–450 MN/m^2 (MPa) when annealed and 620–700 MN/m^2 (MPa) when hard, the electrical conductivity being of the order of 80–92% IACS. For a cable sagging under its own weight, what is required is a high value of the specific strength, i.e. strength/density. For the copper cadmium alloy this is about 70–79 MPa/Mg m^{-3}. Higher specific strengths, without too much reduction in electrical conductivity, can be obtained with aluminium alloys. The material commonly used for overhead power cables is an aluminium alloy with about 0.5% magnesium and 0.5% silicon which has been heat treated to give a strength of about 260 MN/m^2 (MPa) and a specific strength of 96 MPa/Mg m^{-3}. The electrical conductivity is about 55% IACS. This is, however, not enough for major power transmission cables which have to span large distances and so such overhead cables are generally steel cored aluminium conductors, aluminium rather than copper because of its lower density.

Floor joists

Consider the properties required of floor joists in houses. They are beams which are required to support the floor boards and the loads placed on them. They have to do this without significant bending or failing. They must also have a low mass so that they do not place unnecessary loads on the supporting walls. Thus a parameter which can be used to compare joist materials with regard to stiffness is $E^{1/2}/\rho$ and one with regard to strength is $\sigma_y^{2/3}/\rho$ (see Table 20.1).

Traditionally joists are made of timber with a rectangular cross-section, with depth twice the width or standard steel I-sections. Ignoring the shape factor for the moment, Table 20.4 clearly indicates the superiority of wood when cut parallel to the grain. However the shape means that the I-section

Table 20.4 *Floor joist materials*

	Pine parallel to grain	*Mild steel*
Density (Mg/m^3)	0.5	7.9
Tensile modulus (GPa)	16	210
Strength/yield stress (MPa)	80	310
$E^{1/2}/\rho$	8.0	1.8
$\sigma_y^{2/3}/\rho$	37.1	5.8

places more of the material in the high stress region than in the low stress region. Taking this into account improves the stiffness parameter of the typical I-section in relation to the rectangular section by a factor of about $10^{1/2}$ and that for the strength by $10^{1/3}$. Even with these factors, steel does not do as well as timber. The prices per kg for the two materials are not too different to merit one being chosen rather than the other. Thus where the joists have to be not too long then wood is preferable. However, wood is a variable quantity and large lengths are often not possible. So when large lengths are required or when it is absolutely vital that a joist bears a large load with no problems, then steel I-section beams are preferred.

Aircraft skins

Aircraft fuselages are constructed using H, T and L section members to which a metal skin is attached. One way of considering a fuselage is as a pressurized cylinder. Mass is an important consideration for materials used for aircraft, the aim being to have the minimum mass possible. Thus, as indicated by Table 20.1 for a pressurized cylinder, an important parameter in determining the materials that can be used is the specific strength, i.e. yield stress/density. There are other factors involved in determining possible materials and thus the list of properties required includes:

1. *Lightness* – the lighter an aircraft the greater the range, speed or payload.
2. *Reliability* – failure of an aircraft material puts the occupants of the aircraft at risk. However, reliability cannot be ensured by using a large safety factor in the design since this would lead to an increase in weight.
3. *High specific proof or yield stress* – the 0.2 per cent proof stress per unit weight or yield stress per unit weight should be as high as possible, at the temperatures experienced by the skin in flight.
4. *High fatigue strength* – in flight and during take off and landing, aircraft are subject to much buffeting and vibratory stresses.
5. *High creep resistance* – there must be high creep resistance at the temperatures that can occur at the skin during flight.

6 *Good corrosion resistance.*
7 *Ease of forming sheets* – the sheets used for the skin will have to be formed to the shapes required of the various parts of the surface. This means reasonable ductility, though hardening can occur after forming.
8 *Ease of joining sheets* – sheets will need to be welded or riveted to form the large surfaces required.
9 *Cost* – the lowest cost material, after processing, is required which is consistent with the required properties.

Aircraft speeds of the order of 500 m/s result in a skin temperature of about 120°C, while supersonic speeds of the order of 800 m/s give temperatures of about 300°C. Military or research aircraft can attain speeds as high as 2000 m/s and give skin temperatures of the order of 500°C. Thus the speed of the aircraft will determine the temperature range over which the material must have the required properties.

Possible materials might be aluminium alloys, carbon steels, stainless steels, nickel alloys and titanium alloys. Table 20.5 shows the types of properties possible with such materials at room temperature.

For normal passenger aircraft with subsonic speeds all the above materials are feasible. However, on the basis of the specific yield stress aluminium alloys and titanium alloys are the best. On the basis of cost of material, aluminium alloys are preferable to titanium. Processing costs for titanium are also higher than for aluminium alloys. A point worth noting is that carbon steels, e.g. mild steel, are not suitable despite their low cost because they have low specific yield stresses. This is a result of their relatively high density.

To obtain a high yield stress the aluminium alloy chosen would be a heat-treatable wrought alloy. This would allow the material to be formed in the soft condition before being hardened by heat treatment, a precipitation-hardening treatment. A possible alloy would be one having 4.0 per cent copper, 0.8 per cent magnesium, 0.5 per cent silicon and 0.7 per cent manganese. When soft this has a 0.2 per cent proof stress of 90 MN m^{-2} (MPa) and when hard 400 MN m^{-2} (MPa). Aluminium alloys present one major problem – they do not make good welds. For this reason rivets have to be used to join sheets.

Table 20.5 *Typical properties*

Material	*Density (10^3 kg/m^3)*	*Yield stress (MN m^{-2})*	*Specific yield strength (kN m kg^{-1})*	*Max. use temp. (°C)*	*Relative cost per unit sheet area*
Aluminium alloys	2.8	200–400	71–140	200	1
Carbon steels	7.8	200–500	26–64	350	0.5
Stainless steels	7.7	300–1000	39–130	700	3
Nickel alloys	8.9	300–900	34–100	1000	3
Titanium alloys	4.5	700–1100	156–244	600	5

For a higher speed aircraft aluminium alloy is not suitable since it does not retain good mechanical properties at the higher temperatures. For such aircraft titanium has to be considered, despite its higher costs. Unlike aluminium, titanium does make good welds. Table 20.6 shows possible alloys and their properties.

Table 20.6 *Properties of titanium alloys*

Alloy composition (%)	*Condition*	*Yield stress ($MN\ m^{-2}$ or MPa)* 20°C	300°C	400°C	500°C
90 Ti, 8 Al, 1 Mo, 1 V	Annealed	970	630	570	520
	Solution treated and aged	1200	780	710	650
90 Ti, 6 Al, 4 V	Annealed	940	660	580	430
	Solution treated and aged	1100	710	630	490

On the basis of the mechanical properties the first of the alloys in Table 20.6 would appear to be the best choice. However, this alloy is susceptible to stress corrosion cracking in salt water environments. Aircraft flying over the sea do encounter moisture containing salt and so the second alloy in the table which does not have this problem is preferred.

For very high speed the choice has to narrow down to nickel alloys since they possess good mechanical properties even at very high temperatures.

20.4 Case studies: cost per unit property

The following case studies illustrate the use of the cost per unit property approach to reducing the number of materials to be considered. They concern materials for car bodies and the production of small components for toys and cheap clocks.

Car bodywork

The properties required of the bodywork of cars includes the following:

1 It can be formed to the shapes required.
2 It has a smooth and shiny surface.
3 Corrosion is not too significant.
4 It is not brittle, being sufficiently tough to withstand small knocks, and stiff.
5 It is cheap, taking into account raw material, processing and finishing costs.

Table 20.7 *Ductilities of carbon steel and aluminium alloys*

Material	*Percentage elongation*
0.1 per cent carbon steel	42
0.2 per cent carbon steel	37
0.3 per cent carbon steel	32
1.25 per cent Mn, aluminium alloy	30
2.24 per cent Mn, aluminium alloy	22

Processing is probably the key factor in determining the material. Thus with metals, casting is not feasible since car bodywork is too large, too thin, and too complex. Forming from sheet is the obvious method. Hot forming does present the problem of an unacceptable surface finish and so a material has to be chosen which allows for cold forming. For cold forming a highly ductile material is required. Possibilities would be low carbon steels or aluminium alloys. Table 20.7 gives the percentage elongations of possible materials in the annealed state.

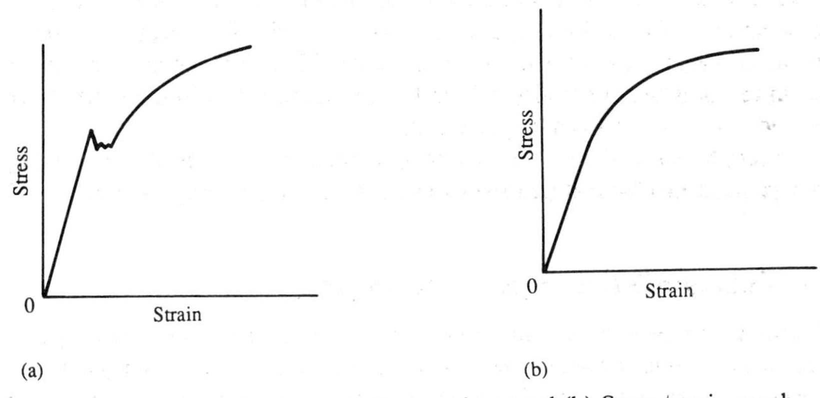

Figure 20.1 (a) Stress/strain graph for a carbon steel (b) Stress/strain graph for an aluminium alloy

From the data in Table 20.7 it can be established that carbon steels and aluminium alloys could be used, both having high enough ductility to enable sheet to be formed. The stress/strain graph of a low-carbon steel differs in form from that of an aluminium alloy in that the aluminium alloy shows a smooth transition from elastic to plastic deformation while the carbon steel shows some irregularities (Figure 20.1). The effect of this on the cold forming of carbon steel is to give some surface markings which do not occur with the aluminium alloy. The aluminium alloy thus has an advantage over the steel in that a smooth surface is required. Aluminium alloys also have another advantage – they have lower densities and so could lead to lower weight cars. The carbon steel does, however, have some advantages. It work hardens more rapidly than the aluminium alloy. A material that work hardens rapidly is less likely to form a neck. In the case of cold forming of the sheet this would show as a thin region in the formed sheet which would clearly not be desirable. The great advantage, outweighing all other considerations, is that carbon steel is much cheaper than aluminium alloy. Thus the optimum material is a low carbon steel. In practice, the material used has less than 1 per cent carbon.

Polymeric materials could be used for car bodywork. The problem with such materials is obtaining enough stiffness – polymeric materials having tensile modulus values considerably smaller than metals. One way of overcoming this is to form a composite material with glass fibre mat or cloth in a matrix of a thermoset. Unfortunately such a process of building up bodywork is essentially a manual rather than machine process and so is slow. While it can be used for one-off bodies it is not suitable for mass production. Another possibility is to produce a sandwich type of composite for panels. This could be a foamed plastic between plastic or metal sheets. While such a material could well be developed to give the required stiffness the costs are likely to be higher than if a carbon steel is used. Table 20.8 shows a comparison of the tensile modulus values of possible polymeric materials with those of carbon steel.

Table 20.8 *Comparison of polymeric materials with carbon steel*

Material	*Tensile modulus (GN m^{-2} or GPa)*
Carbon steel	220
Polypropylene (thermoplastic)	1–2
ABS (thermoplastic)	1–3
Polyester (thermoset)	2–4
Polyester with 65 per cent glass cloth	20

Table 20.9 *Cost per unit property comparisons for bodywork materials*

Material	*Relative cost/m^3 sheet*	*Tensile modulus (GPa)*	*Tensile strength (MPa)*	*Cost per unit stiffness*	*Cost per unit strength*
Low carbon steel	1.0	220	1000	0.005	0.001
Aluminium alloy (Mn)	2.2	70	200	0.03	0.01
Polypropylene	0.2	1–2	30–40	0.1–0.2	0.005–0.007
ABS	0.8	1–3	17–58	0.3–0.8	0.01–0.05
Polyester	2.0	2–4	20–70	0.5–1.0	0.03–0.1
Polyester/glass cloth	3.0	20	300	0.15	0.01

Table 20.9 shows a comparison of the materials on the basis of a cost per unit property. If cost per unit stiffness is taken as the prime factor then low carbon steel is the obvious choice. If, however, the cost per unit strength had been

taken as the prime factor then the choice of low carbon steel is not so obviously the choice.

Small components for toys and cheap clocks

The small components to be produced are such items as small shafts, gear wheels, toy car steering wheels and drive shafts, etc. The main criterion when deciding on the material is that the product must be very cheap when produced in relatively large quantities. All the components to be produced are small. In the case of metals the obvious possible process is die casting. Though the initial cost of the die is high a large number of components can be rapidly produced and so the cost per component becomes relatively low. In the case of polymeric materials the obvious process is injection moulding. This also has a high die cost but large numbers of components can be rapidly produced and hence the cost per component can be low. Both processes give a good surface finish and good dimensional accuracy.

Die casting limits the choice of metal to those with relatively low melting points, e.g. aluminium, magnesium, zinc, lead and tin alloys. Aluminium, magnesium, zinc and lead alloys are comparable in cost, with zinc tending to have the lowest cost per unit weight. Tin is more expensive than these alloys. Table 20.10 shows their general properties.

Table 20.10 *General properties of cheap die casting alloys*

Alloy	*Density* (10^3 *kg/m*^{3})	*Melting point, approx.* (°*C*)	*Tensile strength, approx. as cast* (*MN m*$^{-2}$ *or MPa*)
Aluminium	2.7	600	150
Lead	11.3	320	20
Magnesium	1.8	520	150
Zinc	6.7	380	280

Lead and zinc have the lowest melting points and so have advantages over the others for die casting. Lead, however, has the considerable disadvantage of a low tensile strength, showing considerable creep under very low loads, e.g. its own weight, and a very low fatigue strength. In the as-cast condition, zinc has the highest tensile strength. Thus zinc would seem to be the best metal choice for the components.

Zinc is very widely used for die casting. Its low melting point and excellent fluidity make it one of the easiest metals to cast. Small parts of complex shape and thin wall section can be produced. Zinc alloys have relatively good mechanical properties and are easily electroplated.

Polymeric materials are a possible alternative. Since the forming method is to be injection moulding the materials are restricted to thermoplastics. Table

Table 20.11 *Comparison of zinc and thermoplastics at 20°C*

Material	*Relative cost per m³*	*Tensile strength (MN/m²)*	*Endurance limit for 10^7 cycles (MN/m²)*	*Tensile modulus (GN/m²)*	*Creep modulus at 100 h (GN/m²)*	*Notched impact strength (kJ/m²)*
ABS	1	50	15	2.3	1.5	7
Polyamide 6	2	60	14	3.2	0.5	3
with 2.5 per cent water	2	60	14	1.4	0.5	25
Polycarbonate	2	65	7	2.3	1.6	30
Zinc alloy	3	280	65	103	51	55

20.11 shows how the properties of suitable polymeric materials compare with those of zinc.

The mechanical properties of the zinc alloy are superior to those of the thermoplastics. It has better fatigue and better creep resistance properties. Where light weight is required then polymers have the advantage, having densities of the order of one-sixth that of zinc. Where coloured surfaces are required, e.g. for decoration, then polymers have the advantage. However, if electroplating is required then zinc has the advantage. On cost per unit weight then zinc is cheaper, however on cost per unit volume the polymers are likely to be cheaper. Calculated on a basis of the unit volume cost per unit of tensile strength then zinc is considerably cheaper. You get more strength for your money with zinc than with the polymers.

On the basis of the above comparison it is likely that zinc would be used for those situations where colour of surface is not a requirement, e.g. in a toy. When colour is required then polymeric materials would be used, despite their inferior mechanical properties.

Table 20.12 shows a comparison of the materials on the basis of cost per unit property. On the basis of cost per unit strength, cost per unit stiffness or cost

Table 20.12 *Cost per unit property comparisons*

Material	*Cost/unit strength*	*Cost/unit fatigue strength*	*Cost/unit stiffness*	*Cost/unit creep strength*	*Cost/unit impact strength*
ABS	0.020	0.07	0.43	0.67	0.14
Polyamide 6	0.033	0.14	0.63	4.0	0.67
with 2.5 per cent water	0.033	0.14	1.43	4.0	0.08
Polycarbonate	0.031	0.29	0.87	1.3	0.07
Zinc alloy	0.011	0.05	0.03	0.06	0.05

Note: The costs are on the basis of costs per unit volume.

per unit creep zinc is the clear optimum choice. However, on cost per unit fatigue strength then ABS is comparable and on cost per unit impact strength a number of polymers are comparable.

20.5 Case studies: merit ratings

The following case studies on the selection of materials for gas turbine blades, tool materials, bearing materials and plastic bottles for fizzy drinks illustrate the use of merit ratings to reduce the list of materials to be considered.

Gas turbine blades

Gas turbine engines are used for the propulsion of vehicles such as warships and aircraft, in electric generators and in pumping stations along oil or gas pipelines. The engine consists of three main parts: the compressor, the combustion chamber and the turbine. Air is drawn into the compressor by fan blades and is compressed, rising in temperature as a consequence. In the compression chamber the air is mixed with fuel which burns and increases both the temperature and pressure of the air. The hot, compressed gases then pass into the turbine section where they impinge on the turbine blades with a high velocity and cause them to rotate, rather like a windmill blades rotate as a result of the wind.

The turbine blades have to rotate at high speeds, typically about 10 000 revolutions per minute, in an environment of combustion products at a temperature of the order of 1200°C. The hot gases hit the blades with velocities of the order of 700 m/s. There can also be debris in the gases as a result of objects being drawn into the airstream. The blades will be subject to large and rapid changes in temperature when the engine is started up and turned off; also different parts of a blade may be at different temperatures. As a result of the high speeds of rotation the mass of the blades leads to large centripetal forces acting on the blades and stresses of the order of 100 to 150 MN m^{-2} (MPa) occur.

The properties thus required of material for gas turbine blades are:

1 High creep resistance

This means a high rupture strength at the operating temperature. Because of the importance of blade weight what is really required is a high rupture strength per unit weight, i.e. a high specific rupture strength.

2 High thermal-fatigue resistance

Variations of temperature and mechanical vibrations can result in cyclical stresses which cause failure as a result of fatigue.

3 High corrosion resistance
The environment includes air and the products of combustion at a high temperature and thus corrosion, including oxidation, is a serious problem.

4 High strength
The turbine blades must possess a high tensile strength at the temperatures concerned. Because of the importance of blade weight what is really required is a high strength per unit weight, i.e. a high specific tensile strength.

5 Low mass
The mass of the blades should be as low as possible – the centripetal forces resulting from the mass being rotated at high speeds leads to the stresses in the blade. In addition, when the gas turbine is used in an aircraft the aim is to have as light an engine as possible.

6 A tough material
The blade is subject to impact by high speed gases and debris. In addition it is subject to differential expansion and contraction in different parts of the blade. Cracking must not occur.

7 Cost
The cost of the material will probably not be as significant as the cost of processing.

While ceramics have good high-temperature properties, they suffer from brittleness and so on this count rule themselves out. The requirement for the material to have good strength and creep resistance at high temperatures rules out most metals, leaving essentially just the nickel–chromium based alloys. Table 20.13 shows properties of a range of these alloys.

The alloys in Table 20.13 are just a selection of the nickel–chromium based alloys available. On the basis of Table 20.13 MAR-M246 and Udimet 700 would appear to be the best choices. MAR-M246 is a casting alloy while Udimet 700 is a wrought alloy. Thus Mar-M246 appears to be the choice if a casting process is to be used and Udimet 700 if forging is to be used. Because of the complex shape, casting is likely to be the most cost-effective method.

Now consider the application of merit ratings to the determination of the material for the blades. For the requirement for high specific strength we can construct a merit table, as in Table 20.14. Nimonic 115 has the highest specific strength and is given the rating of 100. Hence, for example, Nimonic 80A has a rating given by $(48.5/105.4) \times 100 = 46$. The determination of the merit ratings is then carried out for each property. Then in order to determine the overall merit rating for a material, each of the properties is given a weighting factor according to the relative importance of the property for the component concerned. Thus, specific rupture strength might be considered the most

Table 20.13 *Properties of nickel–chromium based alloys*

Name	*Composition (%)*	*Rupture strength (MPa) 1000 h at 1000°C*	*Density (kg/m^3)*
Nimonic 80A	74.7 Ni, 19.5 Cr, 1.1 Co, 1.3 Al, 2.5 Ti, 0.06 C.	30	8.22
Nimonic 90	57.4 Ni, 19.5 Cr, 18 Cr, 1.4 Al, 2.4 Ti, 0.07 C.	40	8.18
Nimonic 105	53.3 Ni, 14.5 Cr, 20 Co, 5 Mo, 1.2 Al, 4.5 Ti, 0.2 C.	57	7.99
Nimonic 115	57.3 Ni, 15 Cr, 15 Co, 3.5 Mo, 5 Al, 4 Ti, 0.15 C	100	7.85
MAR-M246	61.1 Ni, 9 Cr, 10 Co, 5.5 Al, 1.5 Ti, 0.15 C, 0.05 Zr, 0.1 Mn	190	8.44
Udimet 700	53.6 Ni, 18 Cr, 18.5 Co, 4 Mo, 2.9 Al, 2.9 Ti, 0.08 C, 6.8 Fe	119	7.91

Table 20.14 *Specific strength merit rating*

Material	*Tensile strength (MPa)*	*Density (10^3 kg m^{-3})*	*Specific strength (kPa/kg m^{-3})*	*Merit rating*
Nimonic 80A	400	8.22	48.7	46
Nimonic 90	428	8.18	52.3	50
Nimonic 115	828	7.85	105.4	100
MAR-M246	862	8.44	102.1	97
Udimet 700	690	7.91	87.2	83

important property, with perhaps oxidation resistance the least important factor. Table 20.15 shows the weighting factors we might thus arrive at. The overall rating for a material is then obtained by multiplying its relative merit rating for each property by the weighting factor for that property and then obtaining the sum of all these products. We can take cost into account by dividing the overall rating by the cost of the material and its processing or some relative cost factor C. Thus

$$\text{rating} = \frac{\Sigma(W_1M_1 + W_2M_2 + \ldots)}{C}$$

where W_1 is the weighting for property 1 where the merit rating is M_1, W_2 the weighting for property 2 where the merit rating is M_2, etc. Table 20.16 shows the results.

Thus, on the basis of the chosen weighting factors, the optimum choice is Udimet 700. The important point when using this method is determining

Table 20.15 *Weighting factors for gas turbine blade material*

Property	*Weighting factor*
Specific rupture strength	6
Thermal fatigue resistance	5
Specific strength	4
Oxidation resistance	3

Table 20.16 *Weighted rating for gas turbine blades*

Material	*Relative cost*	*Specific rupture* *M*	*WM*	*Fatigue resistance* *M*	*WM*	*Specific strength* *M*	*WM*	*Oxidation resistance* *M*	*WM*	*Overall rating* ΣWM	$\Sigma WM/C$
Nimonic 80A	64	16	96	40	200	46	184	100	300	780	12
Nimonic 90	86	21	126	20	100	50	200	100	300	726	8
Nimonic 115	100	53	318	94	470	100	400	75	225	1413	14
Mar-M246	64	100	600	27	115	97	388	35	105	1208	19
Udimet 700	99	63	378	100	500	83	332	75	225	1435	23

what weighting factors to use for the different properties. Thus, in this case, if oxidation resistance had been given the greatest weighting then Nimonic 80A or 90 might have been the optimum choice.

Tool materials

The term 'tool' is taken to include cutting tools, press tools, moulds and dies, drawing and extrusion dies, electrodes for electrical discharge machining, etc. The choice of material for a tool is dictated by the properties required of the tool concerned and cost. In considering cost, account has to be taken not only of the material and production costs for the tool but also tool life. It may be, for example, that a more expensive material might result in a longer life tool and so be more economic in the long run.

The properties required of a tool might include:

1 Room temperature hardness
2 Resistance to thermal softening, i.e. remains hard at operating temperature
3 Toughness
4 Ductility, i.e. not brittle
5 Stiffness, i.e. high tensile modulus
6 Wear resistance

7 Resistance to chemical reaction
8 High thermal conductivity
9 Low thermal expansivity
10 High electrical conductivity
11 Dimensional stability
12 Machinability and grindability
13 Availability of material at required size
14 Cost, both material and processing

A tool material might have the required properties intrinsically, e.g. diamond is a very hard material, or conferred on it by virtue of heat treatment or some other processing, e.g. coating with carbides.

The range of tool materials available include steels; non-ferrous alloys containing chromium, cobalt and tungsten; carbides; diamond and a ceramic such as alumina. Tool steels are grouped into a number of categories (the BS 4659 classification and coding being based on that of the American Iron and Steel Institute but with the AISI code just prefixed by the letter B).

1 Water-hardening tool steels (code BW or W)

These are generally plain-carbon steels with a carbon content of about 0.6 to 1.2 per cent. The steels are hardened by quenching in water. Unfortunately they are rather brittle when very hard, lacking toughness. Where medium hardness with reasonable toughness is required the steels will have about 0.7 per cent carbon. Such a steel might be used for a cold chisel where the requirement to withstand shocks is essential. Where hardness is the primary consideration and toughness not so important, e.g. taps and dies, a carbon steel with about 1.2 per cent carbon may be used.

2 Shock-resistant tool steels (code BS or S)

The most important property required of this category of steel is toughness since the materials have to withstand shock loading, e.g. as hand and pneumatic chisels, shear blades, punches, blanking dies, etc. Such steels thus have a relatively low carbon content – about 0.5 per cent. A general purpose shock-resistant tool steel is BS5 with a composition of 0.55 per cent carbon, 0.40 per cent molybdenum, 0.80 per cent manganese and 2.0 per cent silicon.

3 Cold-work tool steels

Cold-work tool steels are used where the tool is operating cold and where toughness and resistance to wear are important. This category is subdivided into three categories: oil-hardening (code BO or O), air-hardening (code BA or A) and high carbon-high chromium types (code BD or D).

Oil-hardening cold-work steels are widely used tool steels. They have a high cold hardness and a high hardenability from relatively low quenching temperatures, but cannot be used for hot working or high speed cutting (where high temperatures result) since they do not remain hard at high temperatures. An example of such a steel is BO1 which has 0.90 per cent carbon, 0.50 per cent tungsten, 0.50 per cent chromium and 1.0 per cent manganese. Such a material is used for blanking dies, bending and forming dies, taps, etc. Because oil quenching is used there is less distortion, cracking and dimensional change than with water quenching. However, these changes are even further reduced with air-hardening tool steels.

Air-hardening tool steels are used where toughness is the main requirement. They are used for blanking dies, forming rolls, stamping dies, drawing dies, etc. and since they give smaller dimensional changes after hardening and tempering than the other cold-work steels they are better for intricate die detail. An example of such a steel is BA2 which has 1.0 per cent carbon, 1.0 per cent molybdenum and 5 per cent chromium.

Cold-work, high carbon, high chromium, tool steels have excellent wear resistance. They do, however, thermally soften and so cannot be used for high speed cutting. Also, they are brittle. The high wear resistance is a result of the high chromium and high carbon content. Thus, for example, BD2 has 1.5 per cent carbon, 1.0 per cent molybdenum, 12.0 per cent chromium and 1.0 per cent vanadium. The molybdenum increases the hardenability of the steel, the vanadium refines the grain size.

4 Hot-work tool steels (code BH or H)

These are tool steels for use in applications, such as hot extrusion and hot forging, where resistance to thermal softening is required and the material is to remain hard and strong at the hot-working temperature. The material must also have good resistance to wear at temperature and be able to withstand thermal and often mechanical shock. There are three main types of hot-work steels, the distinction being in terms of the main alloying element used. The types are those based mainly on chromium (codes BH1 to BH19), those based on tungsten (codes BH20 to BH39) and those based on molybdenum (codes BH40 to BH59). Chromium, tungsten and molybdenum form carbides which are both stable and hard.

An example of a hot-work chromium tool steel is BH10 which has 0.40 per cent carbon, 2.5 per cent molybdenum, 3.25 per cent chromium and 0.40 per cent vanadium. The steel is used for extrusion and forging dies, hot shears, aluminium die-casting dies, etc. An example of a hot-work tungsten tool steel is BH21 which has 0.35 per cent carbon, 9.0 per cent tungsten and 3.5 per cent chromium. It is used for hot blanking dies, extrusion dies for brass, hot headers, etc. An example of a hot-work molybdenum tool steel is BH42 which has 0.60 per cent carbon, 6.0 per cent tungsten, 5.0 per cent molybdenum, 4.0 per cent chromium and 2.0 per cent vanadium. It is used for hot extrusion dies, hot-upsetting dies, etc.

5 High-speed tool steels
Steels that are able to work effectively at high machining speeds are called high-speed steels. The high speed results in the tool becoming hot – high-speed steels have to retain their properties at temperatures as high as 500°C. This means they must be resistant to tempering at such temperatures. The ability of a steel to resist thermal softening in the red-hot temperature range is called *red hardness*. In addition to red hardness, high-speed tool steels must also have good wear resistance and high hardness.

There are two basic types of high-speed tool steels, one where the main alloying element is tungsten (code BT or T) and the other where molybdenum is the main element (code BM or M). An example of a tungsten high-speed steel is BT4 with 0.75 per cent carbon, 18.0 per cent tungsten, 4.0 per cent chromium, 1.0 per cent vanadium and 5.0 per cent cobalt. The cobalt increases the red hardness of the alloy. Such a steel is used for lathe tools, drills, boring tools, milling cutters, etc. Molybdenum is cheaper than tungsten and so in many parts of the world molybdenum tool steels are used in preference to tungsten ones. Molybdenum tool steels are, however, susceptible to decarburization and better temperature control is required during the heat treatment. An example of a molybdenum high-speed steel is BM1 with 0.85 per cent carbon, 1.5 per cent tungsten, 8.5 per cent molybdenum, 4.0 per cent chromium and 1.0 per cent vanadium. Such a steel is used for drills, reamers, milling cutters, lathe tools, woodworking tools, saws, etc.

6 Special-purpose tool steels
In addition to the steels mentioned above there are some special-purpose low alloy steels (code BL or L). These are intended for specific tasks. An example of such a steel is BL3 which has 1.0 per cent carbon, 1.5 per cent chromium and 0.2 per cent vanadium.

7 Stellites
Stellites are cobalt alloys and have very high resistance to oxidation and chemicals with very high hardness at both normal and high temperatures. As well as use in gas turbines, internal combustion engines, etc. they are also used for tools. A stellite grade 4 used for extrusion dies for copper and brass has 53 per cent cobalt, 31 per cent chromium, 1.0 per cent carbon and 14 per cent tungsten.

8 Carbides
Tungsten, titanium and tantalum carbides are very hard with high melting points. By themselves they are brittle but when fine particles of the carbide are bonded together with cobalt metal a useful tool material is produced. The properties can be varied by varying the proportions of carbide and cobalt, the grain size of the carbide and the carbides concerned. The materials produced using essentially just tungsten carbide are used for tools for cutting cast iron

and non-ferrous alloys. By including titanium and tantalum carbides the wear resistance is improved and tools using these materials are used for machining carbon and alloy steels at high speeds. The high carbide content makes them harder than either the high-speed tool steels (these contain up to 35 per cent carbides) or stellites. Many carbide tools are made in the form of 'throwaway' inserts so that when the cutting edge becomes blunt the insert is thrown away.

9 Diamond

Diamond is the hardest known material. Its use is largely limited to the production of highly accurate, smooth, bearing surfaces. Very high cutting speeds are possible.

10 Ceramics

These are generally alumina with fine particles sintered to give cutting tips. They are usually used as 'throwaway' tips. High cutting speeds are possible.

Examples of tool material selection

In making a choice among tool materials for a particular tool, consideration has to be given to the property requirements of that tool. For instance, will the tool be operating at room temperature or high temperature? Will there be shock loads? Must abrasive wear be considered? What hardness level is required? Will the tool be required to have a long life?

Consider the requirements for a cutting tool to be used for high speed use with a homogeneous material. Because of the high speed the material will need to have a high resistance to thermal softening. Because it is a cutting operation there will need to be high hardness. Some degree of toughness is also likely to be necessary. Table 20.17 gives a comparison of cutting tool materials. The

Table 20.17 *Comparison of cutting tool materials*

Material	*Working hardness Rockwell C*	*Resistance to*		*Toughness, relative*	*Cost, relative*
		thermal softening	*abrasive wear*		
BW1	68	Low	Fair	High	1
BS1	45	Medium	Fair	Very good	3
BO1	60	Low	Medium	Medium	3
BA2	62	High	Good	Medium	4
BD2	62	High	Very good	Low	5
BH13	50	High	Fair	Very good	4
BT1	52	Very high	Very good	Low	8
BM2	60	Very high	Very good	Low	7
Stellite 4	50	Very high	Very good	Low	10
Carbides	70	Very high	Very good	Low	10
Alumina	80	Very high	Very good	Low	12

Table 20.18 *Merit rating analysis for a high speed tool*

Material	*Relative cost*	*Hardness*		*Thermal softness*		*Abrasive wear*		*Toughness*		*Rating*
		M	*WM*	*M*	*WM*	*M*	*WM*	*M*	*WM*	*ΣWM*
BW1	1	85	425	20	160	30	60	70	210	855
BS1	3	56	280	50	400	30	60	100	300	1040
BO1	3	75	375	20	160	50	100	50	150	785
BA2	4	78	390	70	560	70	140	50	150	1240
BD2	5	78	390	70	560	100	200	20	60	1210
BH13	4	63	315	70	560	30	60	100	300	1235
BT1	8	65	325	100	800	100	200	20	60	1385
BM2	7	75	375	100	800	100	200	20	60	1435
Stellite 4	10	63	315	100	800	100	200	20	60	1375
Carbides	10	88	440	100	800	100	200	20	60	1500
Alumina	12	100	500	100	800	100	200	20	60	1560
Weighting		W = 5		W = 8		W = 2		W = 3		

requirement for high resistance to thermal softening coupled with high hardness would indicate that a high-speed tool steel, e.g. BT1 or BM2, carbides or alumina are possibilities. On the basis of cost the high-speed tool steels have an advantage, however the greater hardness of the carbides and alumina would lead to a longer tool life.

Table 20.18 gives a merit rating analysis of the above data for a tool for high speed use with a homogeneous material. On this basis there is not a great deal to choose between the high-speed tool steels, stellite, carbides and alumina. If, for these materials, the cost is taken into account then the lower cost of the high-speed tool steels would make them the choice.

Consider the requirements for a punch-press die. The main requirement is probably good resistance to abrasive wear. This would suggest a high carbon–chromium tool steel, e.g. BD2. This material also has the advantage of air-quenching and thus distortion is minimized. Stellite, carbides and alumina could also be considered but the much higher cost is likely to rule them out. Table 20.19 gives a merit rating analysis for the punch-press die. The main properties required are resistance to abrasive wear and minimum distortion in hardening. Among tool steels, the optimum choice would appear to be BD2. Stellite, carbides and alumina have comparable properties but are more expensive.

Bearing materials

The main requirements of a bearing materials are:

1 *Wear resistance* – the rate of wear can be an important factor in determining the life of a bearing.

Table 20.19 *Merit rating analysis for a punch-press die*

Material	*Relative cost*	*Hardness*		*Hardening distortion*		*Abrasive wear*		*Toughness*		*Rating*
		M	*WM*	*M*	*WM*	*M*	*WM*	*M*	*WM*	*ΣWM*
BW1	1	85	340	10	50	30	240	70	210	840
BS1	3	56	224	50	250	30	240	100	300	1014
BO1	3	75	300	80	400	50	400	50	150	1250
BA2	4	78	312	100	500	70	560	50	150	1522
BD2	5	78	312	100	500	100	800	20	60	1672
BH13	4	63	252	90	450	30	240	100	300	1242
BT1	8	65	260	80	400	100	800	20	60	1520
BM2	7	75	300	80	400	100	800	20	60	1560
Stellite 4	10	63	252	100	500	100	800	20	60	1612
Carbides	10	88	352	100	500	100	800	20	60	1712
Alumina	12	100	400	100	500	100	800	20	60	1760
Weighting		W = 4		W = 5		W = 8		W = 3		

2 *Strength and stiffness* – the material must have a high elastic limit and strength to support the load, together with adequate modulus to resist deformation.

3 *Good shock resistance* – the material must be tough and ductile to be able to withstand shocks.

4 *Good fatigue resistance* – under dynamic loading a high fatigue resistance is required.

5 *Embeddability* – the bearing material must absorb dirt so that abrasive damage to the mating surface is minimized. This means a soft rather than hard material.

6 *High thermal conductivity* – the ability of the material to dissipate heat resulting from friction can be important in preventing a rise in temperature to a level which could result in melting of the bearing material and seizure of the bearing or degradation of a lubricant.

7 *Good corrosion resistance* – corrosion of the bearing material can occur for a variety of reasons, e.g. acidic oils, and result in a thinning down or deterioration of the bearing surface.

Bearing materials can be classified, in the main, into four categories: whitemetals, copper-base alloys, aluminium-base alloys, and non-metallic materials.

Whitemetals

Whitemetals are tin-base or lead-base alloys with the addition of mainly antimony or copper. They have a microstructure of hard intermetallic compounds of tin and antimony embedded in a soft matrix. (See Chapter 1 for a discussion of the frictional characteristics of such materials). Whitemetals have relatively low fatigue strength and this can limit their use to low-load conditions. Reducing the thickness of the bearing material lining can improve the fatigue properties but does require care because of the size of the hard intermetallic particle. Tin-base alloys when compared with lead-base alloys resist corrosion better, have higher thermal conductivity, have higher modulus of elasticity and have higher yield stress but are significantly more expensive. Both forms of alloy are relatively soft. Bearings are usually manufactured by casting onto prepared steel strip and then forming the resulting strip.

As an illustration, a tin-base alloy might have 89 per cent tin, 7.5 per cent antimony, 3.5 copper and very small amounts of other elements. A lead-base alloy might have 75 per cent lead, 15 per cent antimony, 10 per cent tin, 0.5 per cent copper and small amounts of other elements. Tin-base alloys are used for the main bearings for car and aeroengines. Table 20.20 shows typical properties.

Copper-base alloys

Copper-base alloys offer a wider range of strength and hardness than whitemetals. They include tin bronzes with between 10 and 18 per cent tin, leaded tin-bronze containing 1 or 2 per cent lead, phosphor bronzes and copper–lead alloys containing about 25 to 30 per cent lead. The properties of

Table 20.20 *Composition and properties of bearing metals*

Alloy and composition (%)	*Manufacturing process*	*0.1 per cent proof stress (MPa)*	*Strength (MPa)*	*Hardness HV*
Whitemetals: tin base				
89.2 Sn, 7.5 Sb, 3.3 Cu	Cast on steel strip	65	76	27
	Centrifugal casting on steel	39	70	31
Whitemetals: lead base				
84 Pb, 10 Sb, 6 Sn	Cast on steel strip	30	42	16
79.5 Pb, 10 Sb, 10 Sn, 0.5 Cu	Centrifugal casting on steel	60	73	25
Copper-base alloys				
90 Cu, 10 Sn, 0.5 P	Cast on steel	233	420	120
75 Cu, 20 Pb, 5 Sn	Cast on steel	124	233	70
80 Cu, 10 Pb, 10 Sn	Sintered on steel	249	303	120
73.5 Cu, 22 Pb, 4.5 Sn	Sintered on steel	81	121	46
Aluminium-base alloys				
92 Al, 6 Sn, 1 Cu, 1 Ni	Cast on steel	50	140	45
89.7 Al, 6 Sn, 1.5 Cu, 1.4 Ni, 0.9 Mg, 0.5 Si	Cast on steel	83	207	78

the copper–lead alloys depend on the lead content – the higher the amount of lead the lower the fatigue strength but the better the sliding properties. They have poorer corrosion resistance than whitemetals but better wear resistance, a higher modulus of elasticity and better fatigue resistance. Bearings are manufactured by casting onto steel strip or sintering copper and lead on the strip. The bronzes have higher strengths, hardness, modulus of elasticity and better fatigue resistance than the copper–lead alloys and the whitemetals. They tend to be used for high-bearing loads. Table 20.20 shows typical properties.

Aluminium-base alloys

Aluminium–tin alloys, with about 5 to 7 per cent tin, about 1 per cent copper, 1 per cent nickel and small amounts of other elements are bearing materials with a high fatigue strength. Their hardness and strength make them suitable for high-load bearings. They have the disadvantage of a high thermal expansivity which can lead to loose bushes or even seizure against other surfaces. Steelbacked aluminium bearings can be manufactured by hot rolling the aluminium alloy on the steel to permit solid-state welding. Bearings can also be sand or die cast. Table 20.20 shows typical properties.

Non-metallic bearing materials

Polymers suitable for bearing materials include phenolics, nylon, acetal, and PTFE. For some applications the polymers have fillers, e.g. graphite-filled nylon, PTFE with a silicon lubricant, acetal with PTFE filler. In addition to the fillers used to decrease the coefficient of friction, other fillers such as glass fibres are added to increase strength and dimensional stability. Polymers have the advantage of a very low coefficient of friction but present the problem of a low thermal conductivity. Polymer rubbing against polymer can lead to high rates of wear, but polymer against steel gives a very low wear rate. Polymers have thermal expansivities much greater than metals and so can present problems e.g. a higher running clearance between surfaces is needed. Polymers tend to be used under low load conditions where they have the advantage of being cheap.

Table 20.21 shows the properties of some commonly-used polymeric bearing materials. Typical applications are: PTFE bearings in car steering linkages and food processing equipment; phenolics which have the highest load rating for marine propeller shafts; acetals for electrical appliances.

Metal/non-metallic bearing materials

Graphite-impregnated metals and PTFE-impregnated metals are widely used bearing materials. Such materials are able to utilize the load-bearing and temperature advantages of metals with the low coefficient of friction and soft properties of non-metals. Thus graphite-impregnated metals can be used with load pressures up to about 40 MN m^{-2} and operating temperatures up to 500°C while PTFE impregnated metals can be used with loads up to 100 MN

Table 20.21 *Properties of polymeric bearing materials sliding against steel*

Material	*Maximum load pressure* $(MN\ m^{-2})$	*Maximum temperature* °C	*Maximum speed* (*m/s*)
Phenolics	30	150	0.5
Nylon	10	100	0.1
Acetal	10	100	0.1
PTFE	6	260	0.5
Graphite-filled nylon	7	100	0.1
Acetal with PTFE filling	10	100	0.1
Silicon-filled PTFE	10	260	0.5
Phenolic with PTFE filling	30	150	0.5

Note: The temperature is that at the bearing surface.

m^{-2} and temperatures up to 250°C. The above data refers to the material rubbing against a steel mating surface.

Strength and softness

A bearing material is inevitably a compromise between the opposing requirements of softness and high strength (see Chapter 1 and the discussion of friction). One way of achieving strength with a relatively soft bearing material is to use the soft material as a lining on a steel backing. Thus, whitemetals are generally used as a thin layer on a steel backing; likewise aluminium and copper-base alloys. Plastics may also be bonded to a steel backing. This enables them to be used at higher speeds than otherwise would be possible, because a thin layer of plastic will show less dimensional change in such a situation. This is because the steel is able to dissipate heat better than the plastic alone could and also the thinner the layer of plastic the smaller the amount by which it will expand.

In some instances, copper and aluminium-lined steel is given a very thin (of the order of 0.01 mm) coating of a lead-indium alloy. This is a very soft material and improves the embeddability.

Examples of bearing materials selection

Consider the requirement for a light load bearing when the load is steady and the speed is high. The high speed means high temperatures can be produced and so a high thermal conductivity is desirable. The light load means that a high strength material is not required. Because the load is steady, fatigue strength is not a dominant factor. Tables 20.22 give simple comparisons of the main properties of the various types of bearing materials. The requirement for high thermal conductivity rules out polymeric materials and lead-base whitemetals. Because there is no requirement for high strength the more

Table 20.22 *Comparison of bearing materials*

Material	*Hardness BH*	*Yield stress (MPa)*	*Strength (MPa)*	*Fatigue strength (MPa)*	*Modulus elasticity (GPa)*	*Density (10^3 kg m^{-3})*
Tin-base white	17–25	30–65	70–120	25–35	51–53	7.3–7.7
Lead-base white	15–20	20–60	40–110	22–30	29	9.6–10
Copper-lead	20–40	40–60	50–90	40–50	75	9.3–9.5
Phosphor bronze	70–150	130–230	280–420	90–120	80–95	8.8
Leaded tin bronze	50–80	80–150	160–300	80	95	8.8
Aluminium-base	70–75	50–90	140–210	130–170	73	2.9
Polymers	5–20	–	20–80	5–40	1–10	1.0–1.3

Material	*Thermal conductivity (W m^{-1} K^{-1})*	*Corrosion resistance relative*	*Wear resistance relative*	*Cost relative*
Tin-base white	50	5	2	7
Lead-base white	24	4	3	1
Copper-lead	42	3	5	1.5
Phosphor bronze	42	2	5	2
Leaded tin bronze	42	2	3	2
Aluminium-base	160	3	2	1.5
Polymers	0.1	5	5	0.3

highly priced copper-base alloys can be eliminated as there is no point paying for strength if it is not required. The choice would thus appear to be between aluminium-base alloys and tin-base whitemetals. The aluminium-base alloys have the advantage of a higher modulus of elasticity and much lower cost. On this basis the optimum material is likely to be an aluminium-base alloy.

Consider the requirement for a cheap, light load-bearing, intermittent use bearing when the load is steady. Because the use is intermittent thermal conductivity is not as significant a factor. This, coupled with the requirement for light load and cheapness, suggests polymers. Another possibility would be a lead-based whitemetal.

Consider the requirement for a heavy, fluctuating, load bearing to operate at slow speeds with frequent stopping and starting. Because of the fluctuating load a good fatigue strength is required. The heavy load means that high strength, coupled with high modulus of elasticity, is necessary. The heavy load, even with slow speeds, will lead to significant temperature rise and so good thermal conductivity is necessary. The above requirements would suggest phosphor bronze or aluminium-base alloys.

Table 20.23 gives a merit-rating analysis of the above data for a light load bearing when the load is steady and the speed is high. The main requirements are for a high thermal conductivity, good wear resistance and stiffness. If

Table 20.23 *Merit-rating analysis for a light load, high speed, bearing*

Material	*Relative cost*	*Thermal conductivity*		*Wear resistance*		*Modulus elastic*		*Yield stress*	
		M	*WM*	*M*	*WM*	*M*	*WM*	*M*	*WM*
Tin-base white	7	30	150	50	200	50	200	25	50
Lead-base white	1	14	70	60	240	30	120	20	40
Copper-lead	1.5	25	125	100	400	70	280	20	40
Phosphor bronze	2	25	125	90	400	100	400	100	200
Leaded tin bronze	2	25	125	60	240	100	400	45	90
Aluminium-base	1.5	100	500	40	160	70	280	30	60
Polymers	0.3	0.003	0	100	400	5	20	10	20
Weighting		W = 5		W = 4		W = 4		W = 2	

Material	*Hardness*		*Strength*		*Corrosion resistance*		*Fatigue resistance*		*Rating*
	M	*WM*	*M*	*WM*	*M*	*WM*	*M*	*WM*	*ΣWM/C*
Tin-base white	20	60	30	30	100	200	20	20	130
Lead-base white	15	45	25	25	80	160	15	15	715
Copper-lead	30	90	25	25	60	120	45	45	750
Phosphor bronze	100	300	100	100	100	200	70	70	898
Leaded tin bronze	60	180	70	70	40	80	55	55	620
Aluminium base	70	210	55	55	60	120	100	100	990
Polymers	10	30	20	20	100	100	15	15	2017
Weighting	W = 3		W = 1		W = 2		W = 1		

polymers are ignored on the count of the very low thermal conductivity then the optimum material is an aluminium-base alloy.

Plastic bottles

The properties required of a bottle to be used to contain, say, a fizzy drink are:

1 High impact strength, not brittle, tough.
2 Good barrier properties, e.g. the drink must not seep out through the container wall or lose its fizz due to a loss of carbon dioxide pressure.
3 A material which will not taint the drink.
4 Relatively stiff.
5 Able to withstand the pressure due to the 'fizz' without deforming.
6 Transparent and clear.
7 Capable of being blow moulded.
8 Cheap since the bottle is designed to be thrown away after use (this is a

difference from glass milk bottles which are designed to be reused, hence glass rather than plastic).

9 Light weight.

The blow moulding requirement means that the material must be a thermoplastic. Polymers can be grouped into three categories, with the middle class sometimes subdivided into two, for their impact properties. Category 1 are those polymeric materials which are brittle, even when unnotched. Category 2 are those materials which are tough when unnotched but brittle when notched. Category 3 are materials which are tough under all conditions (see Section 5.3). Table 20.24 indicates the materials in each category.

The choice of materials thus becomes restricted to categories 2 and 3. Hence acrylics, though clear and transparent, are not suitable materials. Of those materials in categories 2 and 3 that are clear and transparent the choice becomes low-density polyethylene, PVC or polyethylene terephthalate.

The bottles are to contain fizzy drinks, i.e. liquids containing carbon dioxide under pressure. Table 20.25, gives data concerning the permeability to water and carbon dioxide of the three materials. Despite being the tougher material, low-density polyethylene would not be a suitable material for a fizzy drink container because of its high permeability to carbon dioxide. The drink would

Table 20.24 *Impact properties of polymeric materials*

1 Brittle	*2 Tough but not brittle*	*3 Tough*
Acrylics	Polypropylene	Wet nylon
Glass-filled nylon	Cellulosics	Low-density polythene
Polystyrene	PVC	ABS (some forms)
	Dry nylon	Polycarbonate (some forms)
	Acetals	PTFE
	High density polystyrene	
	ABS (some forms)	
	Polyethylene terephthalate	
	Polycarbonate (some forms)	

Table 20.25 *Permeabilities to water and carbon dioxide*

	Permeability (10^{-8} mol mN^{-1} s^{-1})	
Polymer	*to water*	*to carbon dioxide*
Low-density polyethylene	30	5700
PVC	40	98
Polyethylene terephthalate	60	30

not retain its fizz. Polyethylene terephthalate appears to be the best choice. This material is in fact widely used for carbonated drink containers. PVC is used in France for non-carbonated drinks such as wine.

The data for the materials can be considered in terms of a limits on properties analysis. Thus there could be an upper limit on permeability of carbon dioxide of 120×10^{-18} mol $mN^{-1}\,s^{-1}$ and a similar upper limit on permeability to water. With regard to toughness, using the 1, 2, 3 scale of classification of materials, there could be a lower limit of category 2. This would then rule out brittle materials. With regard to transparency there could be a lower limit of 80 per cent transmission for a 1 mm thickness. For stiffness there could be a lower limit for the modulus of elasticity of 1.5 GPa. Table 20.26 shows how the various materials compare with the above criteria.

Table 20.26 *Limits on properties analysis of fizzy drinks bottle materials*

	CO_2 perm.		*H_2O perm.*		*Toughness*		*Transparency*		*Stiffness*		*Merit*
Material	*U/V*	*WU/V*	*U/V*	*WU/V*	*V/L*	*WV/L*	*V/L*	*WV/L*	*V/L*	*WV/L*	*parameter*
LDPE	0.02	X	4.0	12	1.5	3	0.6	X	0.1	X	X
PVC	1.2	3.6	3.0	9	1.0	2	1.2	2.4	2.0	2	19.0
PET	4.0	12	2.0	6	1.0	2	1.2	2.4	2.0	2	24.4
Acrylics	4.0	12	4.0	12	0.5	X	1.2	2.4	2.0	2	X
PP	0.1	X	7.0	21	1.0	2	0.1	X	1.0	1	X
Weighting	W=3		W=3		W=2		W=2		W=1		

Note: X is where the material is ruled as not suitable in relation to the specified limits. U is upper limit, L is lower limit, V the value for the material.

On the basis of the limits on properties analysis the optimum choice is PET, i.e. polyethylene terephthalate, with PVC being the second choice. If the drinks had not been fizzy, requiring an upper limit on permeability to carbon dioxide, then neglecting the first column in Table 20.26 the merit parameter is 12.4 for PET and 15.4 for PVC and so PVC would be the optimum choice.

20.6 Data sources and use of computers

For the selection of materials for components, data on the properties of materials is required. This may come from a range of sources such as data books, specifications issued by standards bodies, specifications issued by suppliers of materials, information available from trade associations, or computerized data bases.

Examples of data books which span a wide range of materials are:

1 *Materials selector* (1991) *Materials Engineering* (Reinhold)

2 *The Elsevier Materials Selector* (1991), Ed. N.A. Waterman and M.F. Ashby (Elsevier)
3 *ASM Engineering Materials Reference Book* (1989) (ASM)

and a concise pocket book is

4 Bolton, W. *Newnes Engineering Materials Pocket Book* (1989) (Heinemann Newnes)

Examples of data books specific to particular forms of materials are, for metals,

5 *ASM Metals Handbook* (1990) (ASM) Volumes include: Vol. 1 Irons and steels and Vol. 2 Non-ferrous alloys
6 *ASM Metals Reference Book* (1983) (ASM) This is a useful basic reference work
7 Smithells, R.J. (1987) *Metals Reference Book* (Butterworths)

A concise pocket book on metals is

8 Robb, C. *Metals Databook* (1990) (The Institute of Metals)

For polymers there are:

9 *ASM Engineering Materials Handbook, vol. 2 Engineering Plastics* (1989) (ASM)
10 *Handbook of Plastics and Elastomers* (1975), Ed. Harper, C.A. (McGraw-Hill)

For composites there are:

11 *Engineers Guide to Composite Materials* (1987), Ed. Weeton, J.W., Peters, D.M. and Thomas, K.L. (ASM)
12 *ASM Engineering Materials Handbook, vol. 1: Composites* (1987) (ASM)

Computerized data bases are essentially just lists of materials and their properties, with means to rapidly access particular materials or find materials with particular properties. Examples of such commercially available databases for a wide range of materials are:

1 *CMS: Cambridge Materials Selector* (1992) (Cambridge University Engineering Department, UK)
2 *MAT.DB* (1990) (ASM, Ohio, USA)
3 *PERITUS* (Matsel Systems Ltd, Liverpool, UK)

In addition there are databases available for specific types of material. For example, for polymers

4 *PLASCAMS 220* (RAPRA Technology Ltd, Shrewsbury, UK)
5 *dataPLAS* (1990) (Modern Plastics, New York, USA)

For copper

6 *Copperselect* (Copper Development Association Inc., Greenwich, USA)

For steels

7 *STEELMASTER* (Schwing UK Ltd, Warley, UK)

20.7 Failures

In considering the failure of a material there are a number of key questions that need to be addressed.

1 Was the material not to specification?
2 Was the right material chosen for the task?
3 Was the material correctly heat treated?
4 Was the situation in which the material was to be used, and hence the properties required of the material, wrongly diagnosed?
5 Did an abnormal situation occur, perhaps as a result of human error, and was the material therefore subject to unforeseen conditions?
6 Had the assembly of a structure, e.g. any welding, been correctly carried out?

The evidence on which answers to the above questions can be produced is derived from a consideration of the type of failure, tests on the material, and a consideration of the situation occurring when the failure happened and preceding the event. The main types of failure are:

1 Failure by fracture due to static overload, the fracture being either brittle or ductile.
2 Buckling in columns due to compressive loading.
3 Yielding under static loading which then leads to misalignment or overloading on other components.
4 Failure by fatigue fracture.
5 Creep failure.
6 Failure due to the combined effects of stresses and corrosion.
7 Failure due to excessive wear.
8 Failure due to impact loading.

The following case studies are intended to illustrate the various factors involved in material failure.

The Kings Bridge, Melbourne

In July 1962, just fifteen months after it had been opened, the Kings Bridge in Melbourne suffered a partial collapse as a loaded vehicle of about 45 tonnes (45 000 kg) was crossing. One of the spans had collapsed. Examination of the bridge showed that four girders had typical brittle fractures. The fractures had started from cracks in the parent metal at fillet welds. It was established that the ultimate collapse was just the last stage in an ever-increasing pattern of fractures which had been occurring over many months. Every crack had started in the heat-affected zone of a weld.

The material used was a high tensile, fusion welding quality, structural steel to British Standard 968.1941. The composition specified for this steel was 0.23 per cent maximum for carbon, 1.8 per cent maximum manganese, 1.0 per cent maximum chromium and the sum of the manganese and chromium to be a maximum of 2.0 per cent. Samples of the steel used to produce the girders indicated a composition at the limits indicated by the maxima. However, samples taken from the failed girders indicated compositions outside the specified limits (Table 20.27).

Thus while samples taken from the material in these batches had indicated a composition near the upper limit of the specification, what had not been taken into account was that the test pieces were only samples and the composition of the actual material used for the girders was outside the specification limits. The effect of the high carbon and high manganese contents was to make it more difficult to produce crack-free welds. In addition, the welding techniques used, together with the subsequent inspection, was found to be unsatisfactory and could also lead to cracking.

The collapse of the bridge occurred after a cold night when the temperature was about 2°C. Izod impact tests on samples from the girders indicated that at 0°C the material had impact strengths in the range 16 to 50 J, compared with the specified value of 27 J. It is thus likely that when the fractures occurred the bridge was still close to 0°C and thus the low impact values coupled with the initial cracks in the welds led to a brittle fracture.

Table 20.27 *Composition of samples*

	% carbon	*% manganese*	*% chromium*	*% (Mn + Cr)*
According to specification	0.23 max	1.8 max	1.0 max	2.0 max
Material sample, batch 55	0.21	1.70	0.23	1.93
Material sample, batch 56	0.23	1.58	0.24	1.82
Girder from batch 55	0.25	1.75	0.25	2.00
Girder from batch 56	0.26	1.70	0.25	1.95

The Sea Gem oil rig

The Sea Gem was an off-shore oil drilling rig operating in the North Sea. It consisted of a rectangular pontoon supported by ten steel tubular legs on each side. Each of the legs had a pneumatic jacking system by which it could be lowered to the sea bed and the pontoon raised clear of the water. This arrangement allowed the rig to be moved from one location to another. In December 1965, during the operation's preliminary to jacking down the pontoon so that it could be moved, the rig collapsed. Nineteen people died.

Investigation showed that the disaster was initiated by brittle fracture of tie bars. The tie bars formed the suspension links transferring the weight of the pontoon to the ten legs. The tie bars had been flame cut from 7.6 cm steel plate to the shape shown in Figure 20.2. The upper ends of the tie bars had relatively small fillet radii of 4.8 mm. Charpy tests on the steel gave low impact strengths at 0°C of 10 to 30 J. On the day concerned the temperature was about 3°C. Prior to being used in the North Sea the rig had been used in the Gulf of Mexico and the Middle East where temperatures are higher. Under such conditions the impact strength of the material was much higher. Figure 20.3 shows how the impact strength of the tie bar material varied with temperature. The consequence of the rig being in the North Sea was that the tie bar material was brittle, rather than ductile.

Investigations of the upper end of the tie bars showed that the small fillet radii gave a stress concentration factor of 7. This factor, together with the brittle state of the material, was responsible for the material failing during the stresses imposed by the jacking-down operation. After one or more tie bars failed the resulting redistribution of forces led to rapid collapse of the entire structure.

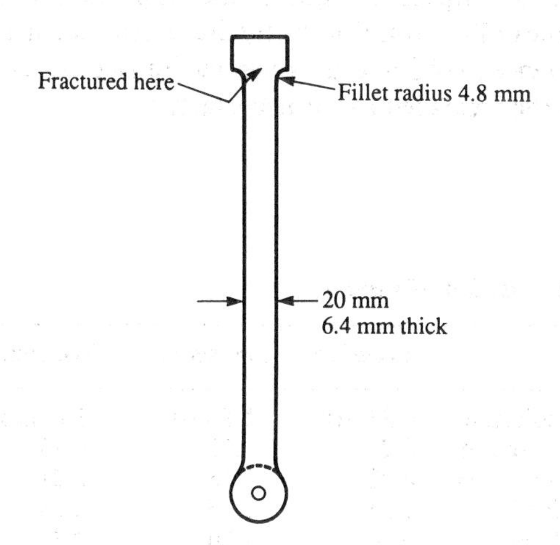

Figure 20.2 Tie bar

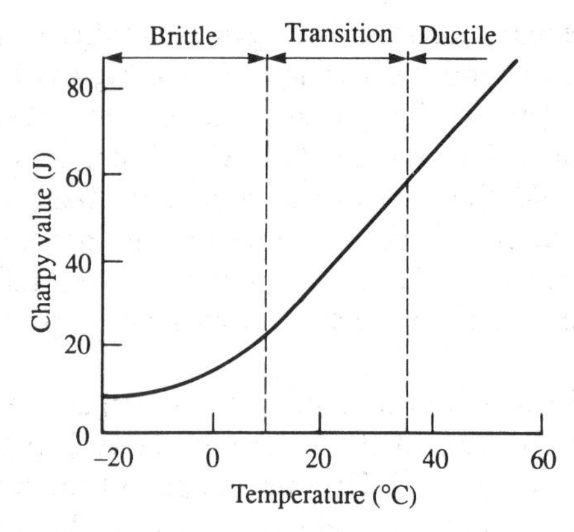

Figure 20.3 Impact strength of tie bar material

Ammonia plant pressure vessel

In December 1965 a large thick-walled cylindrical pressure vessel was being given a hydraulic test at the manufacturer's works, prior to installation in an ammonia plant at Immingham, when it fractured catastrophically with one segment weighing some 2000 kg being thrown a distance of about 46 m. The pressure vessel was about 16 m long with an internal diameter of 1.7 m and was made of 149 mm thick plates of Mn-Cr-Mo-V steel welded to a forged-end fitting. The material had been furnace stress-relieved and the joining welds locally stress-relieved. The design pressure was 35.2 MPa for the vessel with a hydrostatic test pressure of 50 MPa. The ambient temperature at the time of failure was 7°C.

Failure was found to have started in the weld region joining the forged-end fitting to the plate vessel. The origin of the failure was microcracks in the heat-affected zone. Two of three microcracks had propagated and led to the brittle fracture. Analysis of the end forging showed that considerable segregation of carbon and manganese had occurred and thus the forging was heavily banded. Table 20.28 shows the composition of a band and the cast analysis of the forging in relation to the material specification.

Table 20.28 *Composition of forging*

	% Mn	*% Cr*	*% Mo*	*% C*
As specification	1.5	0.70	0.28	0.17
Cast analysis	1.48	0.83	0.29	0.20
In the band	1.88–2.00	0.80–0.82	0.32–0.39	0.25
Outside the band	1.53–1.59	0.69–0.71	0.25–0.21	0.20

As a result of the high carbon and manganese within a band, where the heat-affected zone was in a band the hardness went as high as 400 to 470 HV. The welding technique was based on the assumption that carbon content was less than 0.20 per cent and so the pre-heat was discontinued immediately on completion of the weld. Since the carbon content in a band was higher than this the hydrogen could not disperse sufficiently to prevent cracking. Thus microcracks occurred in the banded regions of the heat-affected zone. While this accounted for the microcracks it could not explain why the cracks propagated.

The weld metal was found to have very low impact values. Charpy tests indicated values of the order of 10 J instead of the anticipated 40 J or more at the failure temperature of 7°C. Heating samples of the weld metal from the vessel to 650°C for six hours did, however, result in the higher impact values being obtained. It was considered that this could only occur if the stress-relieving treatment had not been properly carried out. Thus, the failure of the pressure vessel can be attributed to segregation in the forging and incorrectly carried out stress relief.

The Kohlbrand Bridge

The Kohlbrand Bridge is a suspension bridge of length 520 m and is part of a link road between the port of Hamburg and a major motorway. It was built in 1974. In 1976 the first signs of corrosion were detected in the suspension cables and attempts were made to halt the process by coating the strands in the cable with a plastic sealant. But within a year of this remedial action the cables near to the bridge road deck and the anchor points began to expand as the rate of corrosion accelerated. In 1978 a survey showed that rain had completely permeated the cable strands from top to bottom and that the salt applied to the road surface in winter had also worked its way into the stays. In addition it is thought that vibration from heavy vehicles crossing the bridge hastened the corrosion.

As a consequence of the rapidly deteriorating situation, all the cables were replaced, with the cable stays being heavily galvanized to give better protection against the salt. In addition, vibration dampers are being used in the cable anchor points to damp down vibration.

The Comet disasters

In 1953 and 1954 crashes occurred of the relatively new aircraft the Comet. The Comet was one of the earliest aircraft to have a pressurized fuselage so that passengers could travel in comfort and was built from aluminium alloy. It was possible to recover parts of the 1954 crash by dredging them from a depth of about 100 m in the Mediterranean Sea. The result was a vast mosaic of pieces which had to be fitted together like a jigsaw. The problem was then to identify what was the original failure and what was damage as a consequence of the aircraft crashing. At a small hole near the corner of a window evidence of fatigue damage was found and emanating from this were what were considered to be the cracks responsible for the initial failure.

As part of the investigation a pressure cabin section was subjected to pressure tests to simulate the effects of flights. The fuselage of a pressurized aircraft is rather like a cylindrical vessel which is pressurized every time the aircraft climbs and pressure-reduced every time it descends. In the simulation the fuselage was immersed in a water tank and the pressure cycles obtained by pumping water into and out of the fuselage. The fuselage used for the test had already made 1230 pressurized flights before the test. It failed after the equivalent of a further 1830 such flights. Failure occurred in a similar place to the crashed aircraft and started from a small area of fatigue damage.

The findings of the investigation were that the fundamental cause of the aircraft crash was fatigue failure occurring at a small hole near a window, the small hole acting as a stress raiser. A crack then started which slowly spread until it reached the critical length (see Section 5.3). It then self propagated.

Problems

1 What properties are required of the materials used to manufacture the following items? (a) a car tyre, (b) a car silencer, (c) the head of a hammer, (d) the blade of a screwdriver, (e) a teaspoon, (f) a domestic knife, (g) the casing of a domestic telephone, (h) the casing of a DIY electric hand drill, (i) a reusable milk bottle, (j) an aircraft undercarriage.

2 The following information is taken from a manufacturer's information sheet (Courtesy of Sterling Metals Ltd). What properties are required of the material used for a cylinder block and why is grey cast iron a feasible choice?

Cylinder blocks and transmission cases in high duty grey iron
The technology involved in the design and production of modern power units requires materials and components to be of the very highest grades. Thus the cylinder block, frequently of complicated and intricate design, must be completly sound in structure to withstand the mechanical stresses and pressures which are generated during service.

For over sixty years, the iron foundries of Sterling Metals have specialized in the production of cylinder blocks in grey iron for internal combustion engines.

Considerable attention has necessarily been devoted to the development of the correct material, which must have hard-wearing properties, evenness of grain structure through thick and thin sections, freedom from porosity and brittleness, and be capable of machining at high speeds. The following are details and properties of the normal specification:

Sterling Metals Cylinder Iron
BSS 1452 Grades 14 and 17

Chemical properties	
Total carbon	3.20–3.40%
Silicon	2.10–2.30%
Manganese	0.60–0.90%
Chromium	0.20–0.40%
Combined carbon	0.55–0.75%
Phosphorus	0.15% max.
Sulphur	0.12% max.

Mechanical properties
Brinell Hardness 180–240
Tensile strength 230–280 N/mm^2. Nominal diameter of test bar as cast 30.48 mm.

3 Figure 20.4 is taken from a manufacturer's information sheet (Courtesy of Darlington & Simpson Rolling Mills Ltd) for rolled-steel sections for use in the production of windowframes. What properties are required of the materials used for windowframes? What possible materials can be used? What is the case for rolled-steel sections?
4 Deloro Stellite in an advertisement showing bottles of Haig whisky state:

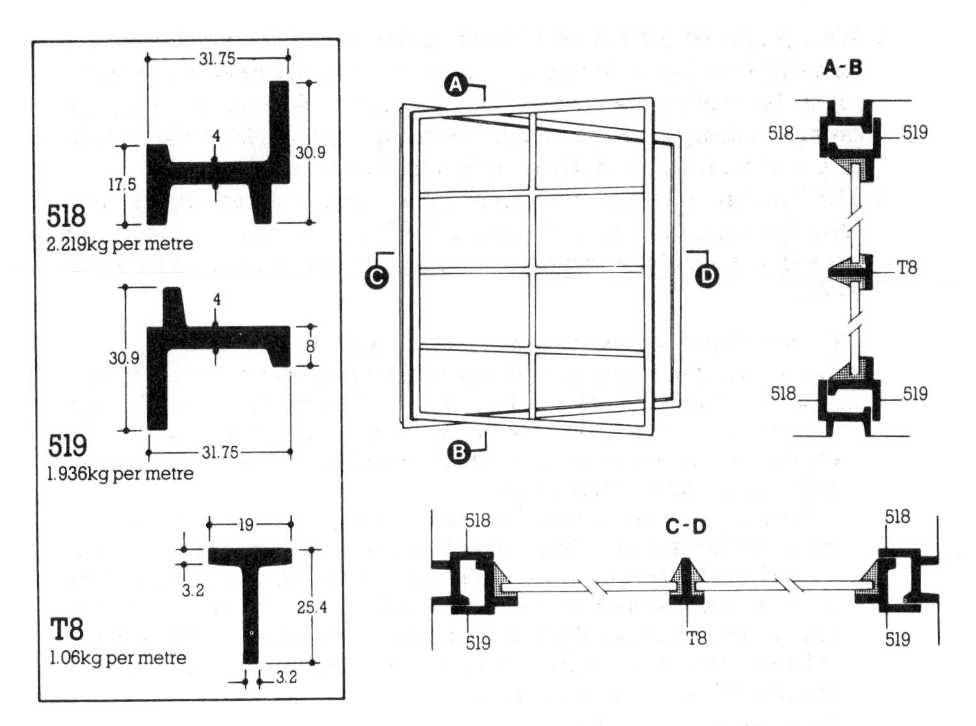

Figure 20.4 Rolled steel window sections (Courtesy of Darlington & Simpson Rolling Mills Ltd)

> We help United Glass turn out over 2500 Haig bottles an hour – a task in which high precision is even more important than high speed. (After all, the dimensional accuracy of any bottle is of paramount importance – especially when its contents are whisky.) Add to that the abrasive quality of molten glass – which creates havoc with conventional metals – and you can see why United Glass use Deloro Stellite wear resistant nickel based hardfacing alloys and castings for moulds and ancillary equipment.

What criteria determine the material used for moulds?

5 The following is some of the information supplied by a manufacturer for polycarbonate (Courtesy of Bayer UK Ltd). What other information would you require in order to make a decision regarding the suitability of the material for the production of picnic cups and saucers?

Makrolon polycarbonate
Form supplied: cylindrical granules
Colour range: in all major colours, transparent, translucent or opaque, with excellent depth of shade.
Temperature performance: limit service temperature −150°C to 135°C unfilled, −150°C to 145°C filled.
Conversion techniques: conversion from raw material: injection moulding, extrusion, casting.
Conversion from semi-fabricated form: thermoforming, e.g. vacuum deep-drawing, pressurized air forming, blow forming, embossing.
Machining: sawing, drilling, reaming, milling, turning, shaping, filing, tapping, punching, cutting.
Jointing: permanent-bonding, welding, nailing, riveting.
detachable-mechanical and snap fits, screwed joints.
Decorating: painting, printing, metallizing.
Predominant applications: electronic and electrical engineering, photographic equipment, lighting, optical equipment, mechanical and process engineering, precision engineering, safety and traffic equipment, office supplies, household and domestic ware.

6 A simple towel rail consisting of a tube or rod held by clamps at each end to the wall is proposed. The towel rail is for use in the steamy atmosphere of the bathroom. The towel rails are to be mass produced and have to be cheap. The rail needs to withstand not only the weight of wet towels but also a person partially supporting their weight against the rail without any permanent bend being produced or indeed any significant bending under such loads.
(a) What are the crucial factors that determine the type of material that can be used?
(b) Suggest some possible materials
(c) What determines whether a tube or rod is used? What determines the thickness of the tube or the diameter of the rod?
(d) What factors affect the cost of the finished component?
(e) Propose a possible specification for the material.

7 A mass production method is required for the production of spanners for

sale as cheap items for the do-it-yourself enthusiast. The material used should be relatively cheap but able to withstand the uses to which the spanner will be put.

(a) What factors determine the type of material that can be used?

(b) What processing methods are feasible, bearing in mind the need to produce a cheap product? Consider the various merits and demerits of the various processes.

(c) What calculations would be needed to ascertain the thickness of material to be used?

(d) Specify a possible material and appropriate processing.

(e) How would you test the product to find out whether it was suitable for the design purpose.

8 Examine a 13 A mains electric plug. Such plugs are mass produced and cheap. They have to be designed so that they are cheap to produce and able to cope with the various terminals, fuse and connecting screws that are needed. They have to be electrically safe.

(a) What functions are required of the plug material?

(b) What tests would you need to run on a prototype material in order to check whether it is appropriate for a plug case?

(c) What materials would you suggest for the plug case? On what basis did you decide on the materials?

(d) What processes would you suggest for the production of the plug case? How are the processes determined by the material used? How are the material and process determined by the cost?

9 Consider the design and the materials employed for the street lamp support poles. Typical materials used are reinforced concrete, carbon steel and cast iron.

(a) What functions are required of the material used as a street lamp support pole?

(b) What processes might be used for the production of street lamp support poles?

(c) What part do you think cost plays in determining the material and process used?

(d) What part do you think ease of maintenance plays in determining the material used?

(e) Consider the production of a street lamp pole as an aluminium or steel extrusion. How would such materials behave in service? What would happen to the two materials if a car collided with the pole, i.e. a sudden impact?

10 (a) Coca-Cola can be purchased in glass or plastic bottles and aluminium cans. What are the properties required of the material used for the container?

(b) What are the factors which lead to glass and aluminium being used for the containers?

(c) Consider the possibility of using a plastic container. What properties would you look for in the plastic? What factors would determine the feasibility of the plastic as the container material?
(d) What are the advantages and disadvantages of the glass and aluminium currently used for the containers? How would the plastic compare?

11 (a) What are the requirements of the materials used in a conventional toothbrush?
(b) What materials might be suitable for the handle and the brush parts?
(c) How might a toothbrush be produced?
(d) What material and process might be used to package a toothbrush so that it can be sold to the consumer in a prepackaged form?

12 (a) What mechanical properties are required of the material used for railway lines?
(b) What environmental conditions must be taken into account in considering the possible materials for railway lines?
(c) What material or materials might be suitable for railway lines?
(d) What process might be used to produce railway lines?
(e) In Britain continuous railway lines are used; how might lengths of line material be joined together to give these continuous lengths?

13 (a) Describe some typical functions of springs.
(b) What properties are required of the materials used for springs?
(c) For some specific application suggest a material that could be used. Justify the selection.

14 Car exhaust systems appear in practice to have a relatively short life before corrosion damage becomes so serious that a replacement exhaust system is required.
(a) What conditions are exhaust systems subject to in use?
(b) What properties are required of the materials used for exhaust systems?
(c) What materials might be suitable for exhaust systems?
(d) What materials are generally used?

15 Polyether foam is used widely as a packing material where goods need to be protected against shock and vibration. The material can be cut to the shape of the object being packaged and so enable the object to be completely surrounded by foam. Frequently, however, the object being packaged is not completely surrounded but held between pads or corner blocks of the foam.
(a) How is the foam able to protect the packaged object against shock?
(b) What properties are required of the foam? How rigid do you think the foam should be?
(c) What type of tests might be carried out to test the efficiency of a particular foam packaging material and the way it is used?

16 The power station of Cliff Quay, Ipswich, England, had the problem of selecting a material for chutes along which moist coal slid. The chutes were

previously made of mild steel and required frequent repair and replacement due to the development of rust and erosion by the continual passage of the coal. They were replaced by aluminium chutes and a much better performance was then obtained.

(a) Why might aluminium be less damaged by the passage of the moist coal?
(b) What other materials would you have considered worth investigating?
(c) What tests might be used to compare materials for this purpose?

17 (a) What properties are required of the material used for the hull of a small boat?
(b) Materials that have been used for small boat hulls are wood, metals and composites such as glass-reinforced polyester resin. How do the properties of these materials compare with those required?
(c) What are the processes used to produce boat hulls from these materials?
(d) What factors do you consider determine the choice of material for a small boat hull?

18 The following are materials that have been used for the components indicated. Consider what properties the materials have which make them suitable for the components and indicate other materials which might also merit consideration.
(a) A closed cell polyurethane foam with a skin for shoe soles.
(b) Low density polythene for electric wire covering.
(c) ABS for telephone handsets.
(d) A cast aluminium–copper (10%) alloy (LM12, AA 222.0) for pistons.
(e) Grey cast iron for manhole covers.
(f) A chromium steel (530M40, AISI 5140: 0.4% C–0.75% Mn–1.05% Cr) for crankshafts.
(g) A martensitic stainless steel (403, 420S29: 0.15% C–12.2% Cr–1% Mn) for steam turbine blades.
(h) A phosphor bronze (PB103: 7% Sn) for spring clips.
(i) A magnesium–aluminium–zinc casting alloy (MAG1, AZ81A: 7.6% Al–0.7% Zn–0.2% Mn) for portable electric tool cases.
(j) A precipitation hardening nickel–chromium–iron alloy (Nimonic 80A) for die casting inserts and cores.
(k) An alpha-beta titanium alloy (Ti–Al–4V) for blades and discs for aircraft turbines

19 Analyse the functional requirements and then by the use of data sources, either books or computerized, determine suitable materials and processes for the following components.
(a) Coins
(b) Rainwater guttering
(c) Car crankshaft
(d) Aircraft undercarriage
(e) A ski pole

(f) A milk container
(g) A suitcase
(h) A container for gas under pressure
(i) Rivets
(j) Pipes for domestic plumbing
(k) Domestic cooking utensils

Index